Minerals in Plants
2

by

Jan Scholten

Title	Minerals in Plants 2	
Author	Jan Scholten, M.D.	
ISBN	90-74817-13-0	
Edition	1, August 2002	
Cover design	Peter Huurdeman	
Lay-out	Jan Scholten	
Publisher	Stichting Alonnissos	
	Servaasbolwerk 13	
	3512 NK, Utrecht	
	The Netherlands	
	Telephone	31.30.2312421
	Fax	31.30.2340211
	Email	mail@alonnissos.org
	Webpage	www.alonnissos.org
Printer	Drukkerij Haasbeek BV	
	Alphen aan den Rijn, The Netherlands	

Minerals in Plants 2

Jan Scholten M.D.

I dedicate this book to humanity

Contents

Acknowledgement

I want to give special thanks to my friend John Wilmshurst from Auckland, New Zealand. His advice regarding mineral analysis was of great value.

Most thanks go to VSM, Alkmaar, Netherlands for providing the 93 dried plants. Without VSM this study would not have been done.

We give much thanks to Actalabs, 1336 Sandhill Drive, Ancaster, Ontario, Canada for analyzing the plant specimens. Special thanks go to their director Dr E. Hofman.

Literature

Beauchemin D.A., Comparison of ICP atomic spectrometry techniques, Spectroscopy, Vol 7, page 12-18, 1992.
Fogg T.R., Seawater Analysis, Applied Spectroscopy, 42, 170, 1988.
Koropchak J.A., Veber M., Conver T.W., Herries J., Applied Spectroscopy, VOl 46, 1525, 1992.
Koropchak J.A., Veber M., Herries J., Spectrochima acta, Vol 47B, 825,1992.
Montaser A., Golightly D.W., Inductively Coupled Plasmas in Analytical Atomic Spectrometry, New York, 1992.
Olesik J.W., Elemental analysis using ICP-OES and ICP/MS, an evaluation and assesment of remaining problems, Analytical Chemics, Vol 63, 2A-21A, 1991.
Scholten J, Homoeopathy and the Elements, Utrecht, 1996, ISBN 907481705X.
Scholten J., Minerals in Plants, Utrecht, 2001, ISBN 9074817068.
Smith T.R., Denton M.B., Evaluation of current nebulizers and nebulizer characterization techniques, Applied Spectroscopy, 44, page 21-44, 1990.

Introduction

Purpose

This study is a follow up of "Minerals in Plants". The purpose is to compare the medicinal properties of the plants and minerals. By analyzing the contents of minerals in plants we can see which minerals are comparatively high or low. Then we can compare the medicinal properties and the homeopathic pictures of minerals high in a plant with the properties and pictures of that plant.
We can study the picture of a plant remedy by comparing it with the minerals with a high content in the plant.

Analysis

The analysis was done by Actalabs, 1336 Sandhill Drive, Ancaster, Ontario, Canada. Their web site is www.actlabs.com. The analysis was Induced Coupled Plasma Spectroscopy on the ashes of the dried plants.

Plants

93 plants were made by available by VSM, Alkmaar, Netherlands. Their web site is www.vsm.nl . Actalabs repeated 3 analysis: Cardamine pratensis, Ocimum canum and Stachys officinalis. So 96 results were obtained. VSM delivered the dried plants, mostly in a quantity of around 30 grams. For details see the table on page 18. Osmundo regalis and Urtica urens were provided twice and thus analyzed twice. It gives a kind of comparison of plants harvested in different times. Osmundo regalis was provided as leaves and as roots. So there we can compare different parts of the plant.

Elements

59 elements were analyzed. The Kalium of all the plants were above the limits that could be analyzed. So Kalium is left out of the tables, leaving 58 elements. The amount of elements is thus greater than in "Minerals in Plants", were only 22 elements were analyzed. So comparison has a wider range. Many elements though don't have a clear picture. Comparison is more difficult there.

Layout

First come the introducing chapters.
Introduction, Layout, Discussion

Part1
Part 1 starts with a table of data about the plants.

Then follow the tables with the results of the analysis.
In each column is given:
The **element**'s name
The **value** of the element in parts per million (Indium and Rhenium in parts per billion)
The **deviation**: this is the deviation from the mean of the element in all 96 measurements of the plants, divided by the standard deviation form the distribution of those measurements.

In the **left column** the results are sorted on the name of the element.
In the **right column** the results are sorted on the deviation from the mean. The right column is the most useful for assessing which elements are relatively high in that plant.

Part 2
Part 2 starts with a table of data about the elements analyzed.

Then follow the tables with the results of the analysis.
In each column is given:
The **plant**'s name
The **value** of the element in parts per million (Indium and Rhenium in parts per billion)
The **deviation**: this is the deviation from the mean of the element in all 96 measurements of the plants, divided by the standard deviation form the distribution of those measurements.

On the **left page** the results are sorted on the name of the element.
On the **right page** the results are sorted on the deviation from the mean. The right page is the most useful for assessing which plants have a relatively high content of the element.

Discussion

The results of the analysis can depend on many factors. Some conclusions can be drawn from this study and the former one "Minerals in plants".

Consistency of the measurement: errors of analysis
Comparing the measurements of the 3 specimens that were analyzed twice can assess the consistence of the measurements. The results were generally about 10% different, with a few of them higher up to 30%. So the results look consistent.

Reliability of the measurement: errors of different analysis
Comparing the measurements of the reference coal can assess the reliability of the measurements (see the table of the elements in Part 2). The results were generally less then 5% different. So the results look reliable.

Part of plant
In general it looks as if roots and bark have a higher content of trace elements than leaves and flower (see the table of the plants in Part 1). This is very obvious in the case of Osmundo regalis, of which 2 specimens have been analyzed, one root, and one plant.

Season
The case of Osmundo regalis can be accounted for also partly by the season. The root was harvested in November, almost wintertime. Winter has connotations of contraction and concentration, which can lead to higher levels of trace elements.
The season can also be studied in the example of Urtica urens, as 2 specimens from a different season were analyzed. The differences are quite big, ranging from 30% till more than 100% difference. So the season seems to be a big influence for the mineral content of plants

Climate, stage of development
The above differences can also be accounted for by differences in climate and season, and differences in the stage of development of the plant. From this study it's difficult to differentiate those influences

Soil, air and contamination
Soil, air and contamination are equal for all the specimens as they come from the same, unpolluted garden of VSM

Results and Conclusions

This study can be used for many purposes
First is the confirmation of known remedy pictures.
Second is the study of little or unknown remedies by extrapolating the picture from the pictures of the elements that are high in the plant.
Third is the study of little or unknown elements by extrapolating the picture from the pictures of the plants that have a high content of that element.
The book can be used a s a kind of repertory: especially the right columns, which are sorted on the deviation, can give a good insight and the part which a deviation above, for instance, 1 can be used as a rubric in repertorisation.

Striking results

Hyoscyamus is high in Lithium in both studies. Belladonna, another member of the Solanaceae, also has a high content of Lithium. Tabacum of the same family is also known for it's high Lithium content. It indicates that the whole family will have a high content of lithium. This is in good accordance with the remedy pictures of the Solanaceae at one side and that of Lithium at the other side.

Syringa vulgaris showed a very high content of Indium. The plant has shown a strong feature of nostalgia (Proving and cases by Jan Scholten, not yet published). Indium has a strong nostalgia.

Verbena has the highest content of Iron of all studied plants. The Dutch name of the plant is: Iron hard.

Confusing results

The lanthanides are especially strong in a few plants. But then most of the lanthanides are strong. It looks as if differentiation of the Lanthanides is difficult.

Conclusion

From this study and the former one, one can conclude that they are worthwhile to do. Gradually some relations between plants and minerals are discovered, confirmed or extended.

Part 1

Plants

Plant	Number	Part used	Date	Time	Temp	Gram
Aconitum napellus	615001	plant, root bulb	6-15-2000	13:30	18	35
Adonis vernalis	406001	plant	4-6-2000	11:30	8	20
Agrimonia eupatoria	707001	plant	7-7-2000	13:30	16	30
Alchemilla vulgaris	529001	plant	5-28-2000	14:00	18	35
Anethum graveolens	809001	plant	8-9-2000	10:45	18	30
Angelica archangelica	228001	root	2-28-2000	12	9	25
Anthemis nobilis	816001	plant	8-16-2000	14	20	35
Aquilegia vulgaris	502001	plant	5-2-2000	8:30	14	30
Belladonna	627001	plant no twigs	6-27-2000	11	20	30
Bellis perennis	705001	plant	7-5-2000	16	17	40
Bryonia dioica	322001	plant	3-22-2000	12	9	30
Cardamine pratensis	425001	plant	4-24-2000	9:30	9	25
Cardiospermum halic.	817001	plant	8-17-2000	8	20	55
Castanea vesca	602001	leaf	6-2-2000	11	15	40
Centaurium erythraea	703001	plant + root	7-3-2000	12	23	45
Chamomilla	726001	plant + root	7-26-2000	9	16	40
Chelidonium majus	324001	roots	3-24-2000	11	8	25
Chenopodium anthelm.	822001	plant	8-22-2000	12	21	45
Cicuta virosa	530002	root	5-30-2000	9	17	30
Cimicifuga racemosa	314001	root	3-14-2000	11	8	25
Cinnamodendron cortis.	412001	bark	4-12-2000	10	12	25
Clematis erecta	622001	plant	6-22-2000	8	18	30
Collinsonia canadensis	308001	root	3-8-2000	13	10	30
Conium maculatum	605001	plant	6-5-2000	9:30	13	40
Cupressus lawsoniana	512001	twigtop	5-12-2000	9	17	40
Echinacea angustifolia	712001	plant + root	7-12-2000	9:30	16	60
Echinacea purpurea	721001	plant	7-21-2000	16	17	70
Escholtzia californica	718001	plant	7-18-2000	11	17	40
Faba vulgaris	626001	plant	7-26-2000	9:30	14	30
Fagopyrum esculentum	825001	plant	8-25-2000	8	18	30
Fraxinus excelsior	330001	bast	3-30-2000	11	8	20
Galium aparine	515001	plant	5-15-2000	14	24	35
Glechoma hederacea	426001	plant	4-26-2000	12	16	50
Gnaphalium leontopodium	606001	plant + root	6-6-2000	14	17	30
Hedera helix	628001	twigs	6-28-2000	9	14	40
Hydrophyllum virginicum	511001		6-11-2000			20
Hyoscyamus niger	622001	plant + root	6-22-2000	8	18	30
Hyssopus officinalis	815001	plant	8-15-2000	9	19	30
Lappa arctium	1031001	root	10-31-2000	8	11	70
Lapsana communis	613001	plant root	7-13-2000	11	17	40
Laurocerasus	522001	leaf	5-22-2000	14	19	35
Leonurus cardiaca	619001	plant	6-19-2000	10	25	40
Lespedeza sieboldii	905001	plant	9-5-2000	11	20	35
Linum usitatissimum	710001	plant	8-10-2000	14	17	30
Lycopus europaeus	707001	plant	7-7-2000	9	15	55
Malva sylvestris	619001	plant	6-19-2000	10	20	25
Mandragora officinalis	224001	root	2-24-2000	12	8	35

Plant	Number	Part used	Date	Time	Temp	Gram
Marrubium vulgare	626001	plant	6-26-2000	11	14	35
Melilotus officinalis	614001	plant	6-14-2000	9	15	20
Melissa officinalis	523001	leaf, twig top	5-23-2000	11	18	25
Mentha arvensis	726001	plant	7-26-2000	8	16	35
Mercurialis perennis	323001	plant + root	4-23-2000	14	5	25
Milium solis	516001	plant root	5-16-2000	9	20	20
Ocimum canum	717001	leaf	7-17-2000	14	16	20
Oenanthe crocata	519001	root	5-19-2000	10	11	35
Ononis spinosa	713001	plant	7-13-2000	8	17	40
Osmunda regalis 1	1004001	root	10-14-2000	9	13	30
Osmunda regalis 2	523001	plant	5-23-2000	9	14	30
Paeonia officinalis	426001	root	4-26-2000	10	17	40
Petroselinum crispum	61600	plant + root	6-16-2000	11	18	60
Plantago major	807001	plant	8-7-2000	11	18	40
Polygonum aviculare	627001	plant	6-27-2000	11	20	40
Primula veris	414001	plant	4-14-2000	14	12	
Prunus padus	509001	leaf	5-9-2000	14	24	20
Psoralea bituminosa	815001	plant	8-15-2000	14	23	30
Rhus toxicodendron	613001	leaf twig	6-13-2000	13	17	30
Rumex acetosa	323001	onbegr	3-23-2000	12	5	25
Ruta graveolens	605001	plant	6-5-2000	13	16	25
Salicaria purpurea	710001	plant	7-10-2000	14	17	30
Salix purpurea	328001	bark	3-28-2000	10	5	30
Salvia sclarea	622001	plant	6-22-2000	9	18	45
Sanguisorba officinalis	602001	plant	6-2-2000	9	13	35
Scrophularia nodosa	510002	plant	5-10-2000	9	19	20
Scutellaria lateriflora	727001	plant	7-27-2000	10	16	35
Solidago virgaurea	824001	flowers	8-24-2000	12	27	45
Spilanthes oleracea	822001	plant	8-22-2000	11	21	35
Stachys officinalis	621001	plant	6-21-2000	8	20	35
Stachys sylvatica	526001	plant	5-26-2000	11	16	70
Syringa vulgaris	508001	bloem	5-8-2000	11	22	30
Taraxacum officinale	418001	plant + root	4-18-2000	15	12	110
Teucrium chamaedrys	724001	plant	7-24-2000	9	15	45
Thuja occidentalis	620001	twig	6-20-2000	11	27	65
Tormentilla erecta	317001	root	3-17-2000	11	10	30
Tussilago farfara	508001	leaf	5-8-2000	10	20	30
Tussilago petasites	327001	plant	3-27-2000	12	9	30
Ulmus campestris	405001	twig	4-5-2000	11	6	30
Urtica dioica	614001	plant + root	6-14-2000	12	19	70
Urtica urens 1	607001	plant	6-7-2000	9	13	40
Urtica urens 2	810001	plant	8-10-2000	12	18	35
Verbena officinalis	703001	plant	7-3-2000	9	19	30
Vinca minor	320001	plant	3-20-2000	11	9	25
Viola tricolor	621001	plant	6-21-2000	10	24	30
Zizia aurea	512001	palnt	5-12-2000	12	20	30

Plant	English name	Ash %	Ash gr.	Mean	Mean SD
Aconitum napellus	Monkshood	9.6	.25	6752.5	0
Adonis vernalis	Pheasant's eye	15	.167	2072.4	-.1
Agrimonia eupatoria	Cockleburr	7.4	.245	4747.5	-.5
Alchemilla vulgaris	Lady's mantle	7.4	.247	6987	.2
Anethum graveolens	Dill	7.3	.249	8591.2	-.4
Angelica archangelica	Angelica	10	.249	2709.2	0
Anthemis nobilis	Chamomile	7.9	.248	5867.2	-.1
Aquilegia vulgaris		8.4	.249	3439.8	-.5
Belladonna	Devils berry	9	.246	4422.6	-.2
Bellis perennis	Daisy	29.7	.253	1238.7	.9
Bryonia dioica		5.9	.253	3222.1	.2
Cardamine pratensis	Lady's smock	7.2	.246	8076.1	.1
Cardiospermum halicacab		13	.246	5304.1	-.1
Castanea vesca	Sweet chestnut	4.9	.245	9602	.2
Centaurium erythraea	Centaury	3.8	.25	3326.9	.7
Chamomilla	Chamomile	10.2	.246	3039	-.2
Chelidonium majus	Greater celandine	28.6	.253	1689.3	1.3
Chenopodium anthelmintic	Jerusalem oak	12.6	.254	5072.4	-.5
Cicuta virosa	Water hemlock	14.9	.25	6402.8	1.3
Cimicifuga racemosa	Black Cohosh	7.5	.246	3474.5	.1
Cinnamodendron cortison	Winter's bark	5.1	.251	6888	0
Clematis erecta	Virgin's bower	5.9	.245	6029.2	-.2
Collinsonia canadensis	Stone root	4.8	.249	4308.7	.3
Conium maculatum	Poison hemlock	7.7	.249	5508.3	-.3
Cupressus lawsoniana	Cupressus laws	3.5	.247	8307.3	.8
Echinacea angustifolia	Echinacea	7.5	.255	6907.2	-.3
Echinacea purpurea		7.5	.245	9929.8	-.4
Escholtzia californica	Californian poppy	12	.25	2390.6	-.4
Faba vulgaris	Faba vulgaris	10.9	.246	4930.3	-.4
Fagopyrum esculentum	Buckwheat	9.6	.249	6721	-.5
Fraxinus excelsior	Maple leaf, Ash	7.4	.245	11315.4	-.2
Galium aparine	Cleavers	13.5	.249	5249.9	-.2
Glechoma hederacea	Ground ivy	10.1	.245	3888.1	-.2
Gnaphalium leontopodium		11	.246	4829.5	.5
Hedera helix	Ivy	6.1	.255	8458	0
Hydrophyllum virginicum		8.8	.254	4836.6	-.3
Hyoscyamus niger	Henbane	15.3	.254	3629.5	-.3
Hyssopus officinalis	Hyssop	11	.245	5807.6	-.5
Lappa arctium	Burdock	9.5	.253	3526.5	0
Lapsana communis	Nipple-wort	7.5	.249	5036.7	.2
Laurocerasus	Cherry laurel	6.5	.245	9590.6	-.3
Leonurus cardiaca	Motherwort	8.9	.247	5860.9	-.3
Lespedeza sieboldii		3.6	.251	7168.6	-.1
Linum usitatissimum	Flax	7.1	.247	7524.5	-.3
Lycopus europaeus	Bugleweed	7.4	.25	4339.4	-.4
Malva sylvestris	Mallow	11.2	.249	7998.1	-.3
Mandragora officinalis	Mandrake	15.3	.254	851.8	-.5

Plant	English name	Ash %	Ash gr.	Mean	Mean SD
Marrubium vulgare	Horehound	10.8	.251	5793.7	-.3
Melilotus officinalis		8.8	.245	12005.4	.1
Melissa officinalis	Lemon balm	9.5	.245	8497.9	.3
Mentha arvensis		11.4	.249	7588	-.1
Mercurialis perennis		11.7	.248	6765.4	.2
Milium solis	Gromwell	7.4	.251	3380	-.3
Ocimum canum	Brazilian alfavaca	16.2	.247	10177.4	0
Oenanthe crocata	Water dropwart	4.3	.25	2651.9	0
Ononis spinosa	Rest harrow	6.9	.247	5641.8	-.1
Osmunda regalis 1		5.5	.253	7747	4.1
Osmunda regalis 2		6.7	.246	1429.7	-.4
Paeonia officinalis	Peony	4.7	.255	11957.6	.3
Petroselinum crispum	Parsley	8.6	.246	5057.4	.2
Plantago major	Plantain	13.8	.252	7053.8	-.2
Polygonum aviculare		11.9	.249	6376.4	-.2
Primula veris	Cowslip	11	.246	3291	-.1
Prunus padus		6.1	.231	8063.7	0
Psoralea bituminosa		8	.248	10559	.1
Rhus toxicodendron	Poison oak	7.4	.245	10280.7	-.2
Rumex acetosa	Sheep's sorrel	7.9	.251	6539.4	1.6
Ruta graveolens	Rue	7.7	.25	7002.7	-.2
Salicaria purpurea	Salicaria purpurea	7	.246	7418.6	-.1
Salix purpurea		7.8	.254	13966.2	0
Salvia sclarea	Clary sage	15.8	.251	6670.8	-.3
Sanguisorba officinalis	Greater burnet	6.1	.249	5276.8	-.4
Scrophularia nodosa	Knotted figwort	7.9	.221	7414.5	0
Scutellaria lateriflora	Mad dog skullcap	6.5	.254	10226.5	.1
Solidago virgaurea	Goldenrod	9.4	.245	3889.5	-.5
Spilanthes oleracea		17.9	.245	5357.7	-.5
Stachys officinalis	Betony	8.5	.245	3142.8	-.5
Stachys sylvatica		8.8	.249	5340.2	-.2
Syringa vulgaris	Lilac	5.6	.253	3242.9	.3
Taraxacum officinale	Dandelion	17.8	.246	2055.5	.6
Teucrium chamaedrys	Germander	6.7	.245	6765.3	-.2
Thuja occidentalis	Western hemlock	5.5	.247	12910.2	-.1
Tormentilla erecta		6.9	.249	3243.5	.6
Tussilago farfara	Coltsfoot	12.7	.245	7415.7	-.3
Tussilago petasites	Butterbur	10.7	.25	4522	-.2
Ulmus campestris	Elm	7.4	.253	12239.5	.1
Urtica dioica	Stinging nettle	15.9	.247	7707.9	.3
Urtica urens 1		7	.25	8886.8	-.4
Urtica urens 2		19.2	.248	9864.9	-.3
Verbena officinalis	Vervain	9.8	.247	9073.6	.1
Vinca minor	Lesser periwinkle	8	.248	8969.5	.1
Viola tricolor	Violet,	16.6	.245	2337.3	-.3
Zizia aurea	Meadow parsnip	10.2	.246	6791.4	-.1

Aconitum napellus

Name	Value	Deviation	Name	Value	Deviation
Aluminium	5790	.4	Rhenium	19.34	1.5
Argentum	0	-.2	Thorium	.76	1.1
Arsenicum	2.95	-.2	Strontium	1260	.6
Barium	159.87	.1	Tantalum	.01	.6
Beryllium	.2	.3	Calcium	353296.29	.5
Bismuthum	0	-.4	Aluminium	5790	.4
Borium	266.95	-.6	Chromium	5.21	.4
Cadmium	.39	-.5	Yttrium	1.49	.4
Caesium	.4	-.1	Lanthanum	2.13	.4
Calcium	353296.29	.5	Cerium	4.63	.4
Cerium	4.63	.4	Praseodymium	.48	.4
Chromium	5.21	.4	Terbium	.06	.4
Cobaltum	1.68	-.2	Uranium	.22	.4
Cuprum	58.69	-.2	Lithium	5.47	.3
Dysprosium	.24	.3	Beryllium	.2	.3
Erbium	.12	.3	Neodymium	1.69	.3
Europium	.11	.3	Samarium	.4	.3
Ferrum	6437.88	0	Europium	.11	.3
Gadolinium	.46	.3	Gadolinium	.46	.3
Gallium	.93	-.1	Dysprosium	.24	.3
Germanium	0	-.3	Holmium	.04	.3
Hafnium	.01	-.3	Erbium	.12	.3
Holmium	.04	.3	Ytterbium	.09	.3
Indium	5.66	-.1	Vanadium	9.36	.1
Lanthanum	2.13	.4	Tellurium	.04	.1
Lithium	5.47	.3	Barium	159.87	.1
Lutetium	.01	.1	Thulium	.01	.1
Magnesium	20124.46	-.7	Lutetium	.01	.1
Manganum	343.52	-.2	Ferrum	6437.88	0
Molybdenum	8.81	-.2	Gallium	.93	-.1
Natrium	3401.9	-.4	Indium	5.66	-.1
Neodymium	1.69	.3	Caesium	.4	-.1
Niccolum	6.52	-.2	Plumbum	9.27	-.1
Niobium	.11	-.2	Manganum	343.52	-.2
Plumbum	9.27	-.1	Cobaltum	1.68	-.2
Praseodymium	.48	.4	Niccolum	6.52	-.2
Rhenium	19.34	1.5	Cuprum	58.69	-.2
Rubidium	27.42	-.4	Arsenicum	2.95	-.2
Samarium	.4	.3	Niobium	.11	-.2
Scandium	.81	-.4	Molybdenum	8.81	-.2
Selenium	0	-1.3	Argentum	0	-.2
Silicium	0	-.4	Thallium	.09	-.2
Stibium	.2	-.4	Germanium	0	-.3
Strontium	1260	.6	Hafnium	.01	-.3
Tantalum	.01	.6	Natrium	3401.9	-.4
Tellurium	.04	.1	Silicium	0	-.4
Terbium	.06	.4	Scandium	.81	-.4
Thallium	.09	-.2	Rubidium	27.42	-.4
Thorium	.76	1.1	Stibium	.2	-.4
Thulium	.01	.1	Bismuthum	0	-.4
Titanium	161.82	-.9	Zirconium	0	-.5
Tungstenium	0	-.6	Cadmium	.39	-.5
Uranium	.22	.4	Borium	266.95	-.6
Vanadium	9.36	.1	Tungstenium	0	-.6
Ytterbium	.09	.3	Magnesium	20124.46	-.7
Yttrium	1.49	.4	Zincum	228.08	-.7
Zincum	228.08	-.7	Titanium	161.82	-.9
Zirconium	0	-.5	Selenium	0	-1.3

Adonis vernalis

Name	Value	Deviation	Name	Value	Deviation
Aluminium	6110	.5	Niobium	.26	1.1
Argentum	0	-.2	Titanium	318.29	.9
Arsenicum	3.13	-.2	Chromium	7.45	.7
Barium	30.17	-.9	Thorium	.59	.7
Beryllium	.11	-.1	Tantalum	.01	.6
Bismuthum	0	-.4	Aluminium	6110	.5
Borium	107.92	-1.5	Cerium	5.06	.5
Cadmium	1.3	-.1	Praseodymium	.49	.4
Caesium	.58	.2	Niccolum	11.98	.3
Calcium	91976.54	-1.3	Lanthanum	2	.3
Cerium	5.06	.5	Samarium	.39	.3
Chromium	7.45	.7	Dysprosium	.23	.3
Cobaltum	.97	-.5	Holmium	.04	.3
Cuprum	81.4	.2	Ytterbium	.09	.3
Dysprosium	.23	.3	Cuprum	81.4	.2
Erbium	.11	.2	Yttrium	1.33	.2
Europium	.08	0	Caesium	.58	.2
Ferrum	2280.18	-.2	Neodymium	1.52	.2
Gadolinium	.43	.2	Gadolinium	.43	.2
Gallium	1.14	.1	Terbium	.05	.2
Germanium	0	-.3	Erbium	.11	.2
Hafnium	.02	0	Gallium	1.14	.1
Holmium	.04	.3	Thulium	.01	.1
Indium	4.46	-.2	Lutetium	.01	.1
Lanthanum	2	.3	Uranium	.16	.1
Lithium	3.85	-.1	Vanadium	7.05	0
Lutetium	.01	.1	Europium	.08	0
Magnesium	14556.98	-.8	Hafnium	.02	0
Manganum	191.01	-.2	Lithium	3.85	-.1
Molybdenum	1.35	-.4	Beryllium	.11	-.1
Natrium	3910.37	-.4	Cadmium	1.3	-.1
Neodymium	1.52	.2	Scandium	.98	-.2
Niccolum	11.98	.3	Manganum	191.01	-.2
Niobium	.26	1.1	Ferrum	2280.18	-.2
Plumbum	7.55	-.3	Arsenicum	3.13	-.2
Praseodymium	.49	.4	Argentum	0	-.2
Rhenium	.25	-.8	Indium	4.46	-.2
Rubidium	46.44	-.3	Thallium	.08	-.2
Samarium	.39	.3	Germanium	0	-.3
Scandium	.98	-.2	Rubidium	46.44	-.3
Selenium	0	-1.3	Plumbum	7.55	-.3
Silicium	0	-.4	Natrium	3910.37	-.4
Stibium	.1	-.7	Silicium	0	-.4
Strontium	241	-1.3	Molybdenum	1.35	-.4
Tantalum	.01	.6	Bismuthum	0	-.4
Tellurium	.02	-.8	Cobaltum	.97	-.5
Terbium	.05	.2	Zirconium	0	-.5
Thallium	.08	-.2	Zincum	284.06	-.6
Thorium	.59	.7	Tungstenium	0	-.6
Thulium	.01	.1	Stibium	.1	-.7
Titanium	318.29	.9	Magnesium	14556.98	-.8
Tungstenium	0	-.6	Tellurium	.02	-.8
Uranium	.16	.1	Rhenium	.25	-.8
Vanadium	7.05	0	Barium	30.17	-.9
Ytterbium	.09	.3	Calcium	91976.54	-1.3
Yttrium	1.33	.2	Selenium	0	-1.3
Zincum	284.06	-.6	Strontium	241	-1.3
Zirconium	0	-.5	Borium	107.92	-1.5

Agrimonia eupatoria

Name	Value	Deviation	Name	Value	Deviation
Aluminium	794	-.6	Cuprum	90.07	.4
Argentum	0	-.2	Stibium	.35	.1
Arsenicum	0	-1.1	Molybdenum	13.7	-.1
Barium	24.59	-1	Manganum	266.65	-.2
Beryllium	.03	-.6	Ferrum	1521.37	-.2
Bismuthum	0	-.4	Rubidium	91.37	-.2
Borium	313.53	-.4	Argentum	0	-.2
Cadmium	.16	-.6	Rhenium	4.79	-.2
Caesium	.06	-.6	Magnesium	36159.8	-.3
Calcium	233196.3	-.3	Calcium	233196.3	-.3
Cerium	1.46	-.6	Germanium	0	-.3
Chromium	0	-.5	Indium	4.19	-.3
Cobaltum	.73	-.6	Borium	313.53	-.4
Cuprum	90.07	.4	Silicium	0	-.4
Dysprosium	.03	-.6	Vanadium	0	-.4
Erbium	.01	-.6	Thallium	.05	-.4
Europium	.02	-.7	Plumbum	6.4	-.4
Ferrum	1521.37	-.2	Bismuthum	0	-.4
Gadolinium	.06	-.6	Natrium	1809.94	-.5
Gallium	.35	-.8	Chromium	0	-.5
Germanium	0	-.3	Yttrium	.19	-.5
Hafnium	0	-.6	Zirconium	0	-.5
Holmium	.01	-.5	Terbium	.01	-.5
Indium	4.19	-.3	Holmium	.01	-.5
Lanthanum	.32	-.6	Uranium	.03	-.5
Lithium	.71	-.8	Beryllium	.03	-.6
Lutetium	0	-.6	Aluminium	794	-.6
Magnesium	36159.8	-.3	Cobaltum	.73	-.6
Manganum	266.65	-.2	Zincum	254.48	-.6
Molybdenum	13.7	-.1	Selenium	1.18	-.6
Natrium	1809.94	-.5	Strontium	619	-.6
Neodymium	.22	-.6	Cadmium	.16	-.6
Niccolum	0	-.8	Caesium	.06	-.6
Niobium	.06	-.7	Lanthanum	.32	-.6
Plumbum	6.4	-.4	Cerium	1.46	-.6
Praseodymium	.07	-.6	Praseodymium	.07	-.6
Rhenium	4.79	-.2	Neodymium	.22	-.6
Rubidium	91.37	-.2	Samarium	.05	-.6
Samarium	.05	-.6	Gadolinium	.06	-.6
Scandium	0	-1.2	Dysprosium	.03	-.6
Selenium	1.18	-.6	Erbium	.01	-.6
Silicium	0	-.4	Thulium	0	-.6
Stibium	.35	.1	Ytterbium	.01	-.6
Strontium	619	-.6	Lutetium	0	-.6
Tantalum	0	-.7	Hafnium	0	-.6
Tellurium	.02	-.8	Tungstenium	0	-.6
Terbium	.01	-.5	Thorium	.08	-.6
Thallium	.05	-.4	Titanium	176.59	-.7
Thorium	.08	-.6	Niobium	.06	-.7
Thulium	0	-.6	Europium	.02	-.7
Titanium	176.59	-.7	Tantalum	0	-.7
Tungstenium	0	-.6	Lithium	.71	-.8
Uranium	.03	-.5	Niccolum	0	-.8
Vanadium	0	-.4	Gallium	.35	-.8
Ytterbium	.01	-.6	Tellurium	.02	-.8
Yttrium	.19	-.5	Barium	24.59	-1
Zincum	254.48	-.6	Arsenicum	0	-1.1
Zirconium	0	-.5	Scandium	0	-1.2

Alchemilla vulgaris

Name	Value	Deviation	Name	Value	Deviation
Aluminium	5830	.4	Magnesium	150000	2.1
Argentum	0	-.2	Bismuthum	.06	1.9
Arsenicum	3.34	-.1	Tungstenium	.57	1.3
Barium	85.18	-.5	Niobium	.25	1
Beryllium	.18	.2	Tellurium	.06	1
Bismuthum	.06	1.9	Titanium	306.24	.8
Borium	396.98	.1	Silicium	2365.36	.6
Cadmium	.56	-.5	Tantalum	.01	.6
Caesium	.52	.1	Hafnium	.04	.5
Calcium	231759.65	-.4	Aluminium	5830	.4
Cerium	4.56	.4	Zirconium	.85	.4
Chromium	3.5	.1	Cerium	4.56	.4
Cobaltum	1.84	-.2	Thorium	.47	.4
Cuprum	69.31	0	Scandium	1.52	.3
Dysprosium	.2	.2	Zincum	860.19	.3
Erbium	.11	.2	Gallium	1.28	.3
Europium	.09	.1	Lanthanum	2.01	.3
Ferrum	7218.18	0	Praseodymium	.47	.3
Gadolinium	.4	.2	Neodymium	1.63	.3
Gallium	1.28	.3	Holmium	.04	.3
Germanium	0	-.3	Beryllium	.18	.2
Hafnium	.04	.5	Samarium	.37	.2
Holmium	.04	.3	Gadolinium	.4	.2
Indium	7.12	.1	Terbium	.05	.2
Lanthanum	2.01	.3	Dysprosium	.2	.2
Lithium	4.05	0	Erbium	.11	.2
Lutetium	.01	.1	Ytterbium	.08	.2
Magnesium	150000	2.1	Borium	396.98	.1
Manganum	893.53	-.1	Chromium	3.5	.1
Molybdenum	28.43	.1	Yttrium	1.17	.1
Natrium	4655.21	-.3	Molybdenum	28.43	.1
Neodymium	1.63	.3	Indium	7.12	.1
Niccolum	7.91	-.1	Caesium	.52	.1
Niobium	.25	1	Europium	.09	.1
Plumbum	11.56	.1	Thulium	.01	.1
Praseodymium	.47	.3	Lutetium	.01	.1
Rhenium	2.92	-.5	Plumbum	11.56	.1
Rubidium	38.03	-.4	Uranium	.17	.1
Samarium	.37	.2	Lithium	4.05	0
Scandium	1.52	.3	Vanadium	8.13	0
Selenium	1.5	-.4	Ferrum	7218.18	0
Silicium	2365.36	.6	Cuprum	69.31	0
Stibium	.29	-.1	Manganum	893.53	-.1
Strontium	667	-.5	Niccolum	7.91	-.1
Tantalum	.01	.6	Arsenicum	3.34	-.1
Tellurium	.06	1	Stibium	.29	-.1
Terbium	.05	.2	Cobaltum	1.84	-.2
Thallium	.07	-.3	Argentum	0	-.2
Thorium	.47	.4	Natrium	4655.21	-.3
Thulium	.01	.1	Germanium	0	-.3
Titanium	306.24	.8	Thallium	.07	-.3
Tungstenium	.57	1.3	Calcium	231759.65	-.4
Uranium	.17	.1	Selenium	1.5	-.4
Vanadium	8.13	0	Rubidium	38.03	-.4
Ytterbium	.08	.2	Strontium	667	-.5
Yttrium	1.17	.1	Cadmium	.56	-.5
Zincum	860.19	.3	Barium	85.18	-.5
Zirconium	.85	.4	Rhenium	2.92	-.5

Anethum graveolens

Name	Value	Deviation	Name	Value	Deviation
Aluminium	934	-.6	Natrium	60000	3.6
Argentum	0	-.2	Calcium	402386.55	.8
Arsenicum	0	-1.1	Rhenium	13.68	.8
Barium	79.99	-.5	Strontium	1260	.6
Beryllium	.03	-.6	Cadmium	2.11	.3
Bismuthum	0	-.4	Manganum	356.94	-.2
Borium	325.7	-.3	Ferrum	2968.76	-.2
Cadmium	2.11	.3	Molybdenum	9.11	-.2
Caesium	.09	-.6	Argentum	0	-.2
Calcium	402386.55	.8	Borium	325.7	-.3
Cerium	1.51	-.6	Vanadium	1.54	-.3
Chromium	0	-.5	Germanium	0	-.3
Cobaltum	1.05	-.5	Silicium	0	-.4
Cuprum	48.32	-.4	Niccolum	5.18	-.4
Dysprosium	.04	-.5	Cuprum	48.32	-.4
Erbium	.02	-.5	Rubidium	22.05	-.4
Europium	.02	-.7	Bismuthum	0	-.4
Ferrum	2968.76	-.2	Magnesium	29332.79	-.5
Gadolinium	.08	-.6	Chromium	0	-.5
Gallium	.4	-.8	Cobaltum	1.05	-.5
Germanium	0	-.3	Zincum	372.14	-.5
Hafnium	0	-.6	Yttrium	.23	-.5
Holmium	.01	-.5	Zirconium	0	-.5
Indium	1.45	-.5	Indium	1.45	-.5
Lanthanum	.34	-.6	Barium	79.99	-.5
Lithium	1.05	-.7	Terbium	.01	-.5
Lutetium	0	-.6	Dysprosium	.04	-.5
Magnesium	29332.79	-.5	Holmium	.01	-.5
Manganum	356.94	-.2	Erbium	.02	-.5
Molybdenum	9.11	-.2	Thallium	.02	-.5
Natrium	60000	3.6	Plumbum	4.62	-.5
Neodymium	.25	-.6	Uranium	.04	-.5
Niccolum	5.18	-.4	Beryllium	.03	-.6
Niobium	.05	-.7	Aluminium	934	-.6
Plumbum	4.62	-.5	Selenium	1.21	-.6
Praseodymium	.08	-.6	Stibium	.13	-.6
Rhenium	13.68	.8	Caesium	.09	-.6
Rubidium	22.05	-.4	Lanthanum	.34	-.6
Samarium	.06	-.6	Cerium	1.51	-.6
Scandium	0	-1.2	Praseodymium	.08	-.6
Selenium	1.21	-.6	Neodymium	.25	-.6
Silicium	0	-.4	Samarium	.06	-.6
Stibium	.13	-.6	Gadolinium	.08	-.6
Strontium	1260	.6	Thulium	0	-.6
Tantalum	0	-.7	Ytterbium	.01	-.6
Tellurium	.02	-.8	Lutetium	0	-.6
Terbium	.01	-.5	Hafnium	0	-.6
Thallium	.02	-.5	Tungstenium	0	-.6
Thorium	.09	-.6	Thorium	.09	-.6
Thulium	0	-.6	Lithium	1.05	-.7
Titanium	160.64	-.9	Niobium	.05	-.7
Tungstenium	0	-.6	Europium	.02	-.7
Uranium	.04	-.5	Tantalum	0	-.7
Vanadium	1.54	-.3	Gallium	.4	-.8
Ytterbium	.01	-.6	Tellurium	.02	-.8
Yttrium	.23	-.5	Titanium	160.64	-.9
Zincum	372.14	-.5	Arsenicum	0	-1.1
Zirconium	0	-.5	Scandium	0	-1.2

Angelica archangelica

Name	Value	Deviation	Name	Value	Deviation
Aluminium	5780	.4	Titanium	441.09	2.3
Argentum	0	-.2	Niobium	.3	1.4
Arsenicum	3.08	-.2	Lutetium	.02	.8
Barium	56.12	-.7	Vanadium	22.9	.7
Beryllium	.16	.1	Chromium	6.64	.6
Bismuthum	0	-.4	Tantalum	.01	.6
Borium	279.84	-.6	Hafnium	.04	.5
Cadmium	2.28	.4	Aluminium	5780	.4
Caesium	.36	-.1	Yttrium	1.57	.4
Calcium	107612.32	-1.2	Cadmium	2.28	.4
Cerium	3.85	.1	Ytterbium	.1	.4
Chromium	6.64	.6	Zirconium	.76	.3
Cobaltum	1.18	-.4	Praseodymium	.44	.3
Cuprum	68.97	0	Samarium	.38	.3
Dysprosium	.24	.3	Dysprosium	.24	.3
Erbium	.12	.3	Holmium	.04	.3
Europium	.09	.1	Erbium	.12	.3
Ferrum	4580.4	-.1	Zincum	807.91	.2
Gadolinium	.43	.2	Gallium	1.19	.2
Gallium	1.19	.2	Lanthanum	1.76	.2
Germanium	0	-.3	Neodymium	1.51	.2
Hafnium	.04	.5	Gadolinium	.43	.2
Holmium	.04	.3	Terbium	.05	.2
Indium	2.94	-.4	Thorium	.4	.2
Lanthanum	1.76	.2	Uranium	.19	.2
Lithium	3.53	-.2	Beryllium	.16	.1
Lutetium	.02	.8	Cerium	3.85	.1
Magnesium	30031.14	-.5	Europium	.09	.1
Manganum	304.97	-.2	Thulium	.01	.1
Molybdenum	5.91	-.3	Thallium	.16	.1
Natrium	6374.07	-.2	Cuprum	68.97	0
Neodymium	1.51	.2	Scandium	1.09	-.1
Niccolum	7.09	-.2	Ferrum	4580.4	-.1
Niobium	.3	1.4	Caesium	.36	-.1
Plumbum	7.38	-.3	Lithium	3.53	-.2
Praseodymium	.44	.3	Natrium	6374.07	-.2
Rhenium	.54	-.7	Manganum	304.97	-.2
Rubidium	92.08	-.2	Niccolum	7.09	-.2
Samarium	.38	.3	Arsenicum	3.08	-.2
Scandium	1.09	-.1	Rubidium	92.08	-.2
Selenium	0	-1.3	Argentum	0	-.2
Silicium	0	-.4	Germanium	0	-.3
Stibium	.17	-.5	Molybdenum	5.91	-.3
Strontium	623	-.6	Plumbum	7.38	-.3
Tantalum	.01	.6	Silicium	0	-.4
Tellurium	.02	-.8	Cobaltum	1.18	-.4
Terbium	.05	.2	Indium	2.94	-.4
Thallium	.16	.1	Bismuthum	0	-.4
Thorium	.4	.2	Magnesium	30031.14	-.5
Thulium	.01	.1	Stibium	.17	-.5
Titanium	441.09	2.3	Borium	279.84	-.6
Tungstenium	0	-.6	Strontium	623	-.6
Uranium	.19	.2	Tungstenium	0	-.6
Vanadium	22.9	.7	Barium	56.12	-.7
Ytterbium	.1	.4	Rhenium	.54	-.7
Yttrium	1.57	.4	Tellurium	.02	-.8
Zincum	807.91	.2	Calcium	107612.32	-1.2
Zirconium	.76	.3	Selenium	0	-1.3

Anthemis nobilis

Name	Value	Deviation	Name	Value	Deviation
Aluminium	4450	.1	Natrium	60000	3.6
Argentum	0	-.2	Titanium	288.22	.6
Arsenicum	1.74	-.6	Tantalum	.01	.6
Barium	45.78	-.8	Molybdenum	42.9	.4
Beryllium	.12	-.1	Borium	435.15	.3
Bismuthum	0	-.4	Niobium	.17	.3
Borium	435.15	.3	Aluminium	4450	.1
Cadmium	1.69	.1	Scandium	1.36	.1
Caesium	.34	-.2	Selenium	2.33	.1
Calcium	248937.64	-.2	Cadmium	1.69	.1
Cerium	3.39	0	Tellurium	.04	.1
Chromium	0	-.5	Thulium	.01	.1
Cobaltum	1.19	-.4	Lutetium	.01	.1
Cuprum	64.73	-.1	Thorium	.37	.1
Dysprosium	.14	-.1	Cerium	3.39	0
Erbium	.07	-.1	Praseodymium	.33	0
Europium	.06	-.3	Samarium	.27	0
Ferrum	4953.68	-.1	Ytterbium	.06	0
Gadolinium	.29	-.1	Lithium	3.71	-.1
Gallium	.91	-.2	Beryllium	.12	-.1
Germanium	0	-.3	Ferrum	4953.68	-.1
Hafnium	.01	-.3	Cuprum	64.73	-.1
Holmium	.02	-.2	Yttrium	.9	-.1
Indium	3.32	-.3	Lanthanum	1.28	-.1
Lanthanum	1.28	-.1	Neodymium	1.09	-.1
Lithium	3.71	-.1	Gadolinium	.29	-.1
Lutetium	.01	.1	Dysprosium	.14	-.1
Magnesium	19631.53	-.7	Erbium	.07	-.1
Manganum	368.69	-.2	Rhenium	5.72	-.1
Molybdenum	42.9	.4	Thallium	.11	-.1
Natrium	60000	3.6	Plumbum	9.55	-.1
Neodymium	1.09	-.1	Calcium	248937.64	-.2
Niccolum	0	-.8	Vanadium	3.56	-.2
Niobium	.17	.3	Manganum	368.69	-.2
Plumbum	9.55	-.1	Gallium	.91	-.2
Praseodymium	.33	0	Rubidium	110.04	-.2
Rhenium	5.72	-.1	Argentum	0	-.2
Rubidium	110.04	-.2	Caesium	.34	-.2
Samarium	.27	0	Terbium	.03	-.2
Scandium	1.36	.1	Holmium	.02	-.2
Selenium	2.33	.1	Uranium	.1	-.2
Silicium	0	-.4	Germanium	0	-.3
Stibium	.24	-.3	Indium	3.32	-.3
Strontium	548	-.7	Stibium	.24	-.3
Tantalum	.01	.6	Europium	.06	-.3
Tellurium	.04	.1	Hafnium	.01	-.3
Terbium	.03	-.2	Silicium	0	-.4
Thallium	.11	-.1	Cobaltum	1.19	-.4
Thorium	.37	.1	Bismuthum	0	-.4
Thulium	.01	.1	Chromium	0	-.5
Titanium	288.22	.6	Zincum	377.62	-.5
Tungstenium	0	-.6	Zirconium	0	-.5
Uranium	.1	-.2	Arsenicum	1.74	-.6
Vanadium	3.56	-.2	Tungstenium	0	-.6
Ytterbium	.06	0	Magnesium	19631.53	-.7
Yttrium	.9	-.1	Strontium	548	-.7
Zincum	377.62	-.5	Niccolum	0	-.8
Zirconium	0	-.5	Barium	45.78	-.8

Aquilegia vulgaris

Name	Value	Deviation
Aluminium	1030	-.6
Argentum	0	-.2
Arsenicum	0	-1.1
Barium	20.05	-1
Beryllium	.03	-.6
Bismuthum	0	-.4
Borium	345.7	-.2
Cadmium	.34	-.6
Caesium	.06	-.6
Calcium	169161.37	-.8
Cerium	1.37	-.6
Chromium	0	-.5
Cobaltum	.96	-.5
Cuprum	81.88	.2
Dysprosium	.03	-.6
Erbium	.01	-.6
Europium	.02	-.7
Ferrum	1242.35	-.2
Gadolinium	.06	-.6
Gallium	.44	-.7
Germanium	0	-.3
Hafnium	0	-.6
Holmium	0	-.7
Indium	4.6	-.2
Lanthanum	.28	-.7
Lithium	1.63	-.6
Lutetium	0	-.6
Magnesium	24488.34	-.6
Manganum	340.79	-.2
Molybdenum	42.56	.4
Natrium	1519.45	-.6
Neodymium	.21	-.6
Niccolum	7.25	-.2
Niobium	.04	-.8
Plumbum	2.82	-.7
Praseodymium	.07	-.6
Rhenium	1.41	-.6
Rubidium	273.19	.3
Samarium	.05	-.6
Scandium	0	-1.2
Selenium	2.58	.3
Silicium	0	-.4
Stibium	.32	0
Strontium	366	-1
Tantalum	0	-.7
Tellurium	.04	.1
Terbium	.01	-.5
Thallium	.01	-.5
Thorium	.07	-.6
Thulium	0	-.6
Titanium	199.58	-.5
Tungstenium	.53	1.2
Uranium	.02	-.6
Vanadium	0	-.4
Ytterbium	.01	-.6
Yttrium	.13	-.6
Zincum	372.14	-.5
Zirconium	0	-.5

Name	Value	Deviation
Tungstenium	.53	1.2
Molybdenum	42.56	.4
Selenium	2.58	.3
Rubidium	273.19	.3
Cuprum	81.88	.2
Tellurium	.04	.1
Stibium	.32	0
Borium	345.7	-.2
Manganum	340.79	-.2
Ferrum	1242.35	-.2
Niccolum	7.25	-.2
Argentum	0	-.2
Indium	4.6	-.2
Germanium	0	-.3
Silicium	0	-.4
Vanadium	0	-.4
Bismuthum	0	-.4
Titanium	199.58	-.5
Chromium	0	-.5
Cobaltum	.96	-.5
Zincum	372.14	-.5
Zirconium	0	-.5
Terbium	.01	-.5
Thallium	.01	-.5
Lithium	1.63	-.6
Beryllium	.03	-.6
Natrium	1519.45	-.6
Magnesium	24488.34	-.6
Aluminium	1030	-.6
Yttrium	.13	-.6
Cadmium	.34	-.6
Caesium	.06	-.6
Cerium	1.37	-.6
Praseodymium	.07	-.6
Neodymium	.21	-.6
Samarium	.05	-.6
Gadolinium	.06	-.6
Dysprosium	.03	-.6
Erbium	.01	-.6
Thulium	0	-.6
Ytterbium	.01	-.6
Lutetium	0	-.6
Hafnium	0	-.6
Rhenium	1.41	-.6
Thorium	.07	-.6
Uranium	.02	-.6
Gallium	.44	-.7
Lanthanum	.28	-.7
Europium	.02	-.7
Holmium	0	-.7
Tantalum	0	-.7
Plumbum	2.82	-.7
Calcium	169161.37	-.8
Niobium	.04	-.8
Strontium	366	-1
Barium	20.05	-1
Arsenicum	0	-1.1
Scandium	0	-1.2

Belladonna

Name	Value	Deviation	Name	Value	Deviation
Aluminium	3050	-.2	Natrium	60000	3.6
Argentum	0	-.2	Lithium	10.71	1.5
Arsenicum	1.73	-.6	Selenium	3.37	.8
Barium	87.3	-.5	Tantalum	.01	.6
Beryllium	.07	-.3	Stibium	.37	.1
Bismuthum	0	-.4	Thulium	.01	.1
Borium	285.19	-.5	Lutetium	.01	.1
Cadmium	.86	-.3	Vanadium	4.77	-.1
Caesium	.23	-.3	Niobium	.12	-.1
Calcium	164358.34	-.8	Indium	6.05	-.1
Cerium	2.79	-.2	Praseodymium	.27	-.1
Chromium	0	-.5	Ytterbium	.05	-.1
Cobaltum	.96	-.5	Rhenium	5.58	-.1
Cuprum	49.09	-.3	Uranium	.13	-.1
Dysprosium	.12	-.2	Aluminium	3050	-.2
Erbium	.06	-.2	Scandium	1.02	-.2
Europium	.06	-.3	Manganum	285.23	-.2
Ferrum	2115.5	-.2	Ferrum	2115.5	-.2
Gadolinium	.23	-.2	Yttrium	.72	-.2
Gallium	.6	-.5	Molybdenum	11.49	-.2
Germanium	0	-.3	Argentum	0	-.2
Hafnium	0	-.6	Lanthanum	1.13	-.2
Holmium	.02	-.2	Cerium	2.79	-.2
Indium	6.05	-.1	Neodymium	.9	-.2
Lanthanum	1.13	-.2	Samarium	.2	-.2
Lithium	10.71	1.5	Gadolinium	.23	-.2
Lutetium	.01	.1	Terbium	.03	-.2
Magnesium	25268.54	-.6	Dysprosium	.12	-.2
Manganum	285.23	-.2	Holmium	.02	-.2
Molybdenum	11.49	-.2	Erbium	.06	-.2
Natrium	60000	3.6	Thorium	.26	-.2
Neodymium	.9	-.2	Beryllium	.07	-.3
Niccolum	5.55	-.3	Niccolum	5.55	-.3
Niobium	.12	-.1	Cuprum	49.09	-.3
Plumbum	7.2	-.3	Germanium	0	-.3
Praseodymium	.27	-.1	Cadmium	.86	-.3
Rhenium	5.58	-.1	Caesium	.23	-.3
Rubidium	42.77	-.4	Europium	.06	-.3
Samarium	.2	-.2	Plumbum	7.2	-.3
Scandium	1.02	-.2	Silicium	0	-.4
Selenium	3.37	.8	Rubidium	42.77	-.4
Silicium	0	-.4	Tellurium	.03	-.4
Stibium	.37	.1	Bismuthum	0	-.4
Strontium	506	-.8	Borium	285.19	-.5
Tantalum	.01	.6	Chromium	0	-.5
Tellurium	.03	-.4	Cobaltum	.96	-.5
Terbium	.03	-.2	Gallium	.6	-.5
Thallium	.02	-.5	Zirconium	0	-.5
Thorium	.26	-.2	Barium	87.3	-.5
Thulium	.01	.1	Thallium	.02	-.5
Titanium	156.29	-1	Magnesium	25268.54	-.6
Tungstenium	0	-.6	Arsenicum	1.73	-.6
Uranium	.13	-.1	Hafnium	0	-.6
Vanadium	4.77	-.1	Tungstenium	0	-.6
Ytterbium	.05	-.1	Zincum	237.02	-.7
Yttrium	.72	-.2	Calcium	164358.34	-.8
Zincum	237.02	-.7	Strontium	506	-.8
Zirconium	0	-.5	Titanium	156.29	-1

Bellis perennis

Name	Value	Deviation	Name	Value	Deviation
Aluminium	12100	1.8	Chromium	24.14	3.5
Argentum	0	-.2	Thorium	1.75	3.5
Arsenicum	6.72	.9	Praseodymium	1.76	3.4
Barium	49.29	-.8	Lanthanum	7.37	3.3
Beryllium	.46	1.7	Samarium	1.49	3.2
Bismuthum	.05	1.5	Cerium	13.34	3
Borium	66.9	-1.7	Yttrium	5.15	2.9
Cadmium	1.35	-.1	Neodymium	5.54	2.8
Caesium	1.24	1.3	Thulium	.05	2.8
Calcium	32626.53	-1.7	Lutetium	.05	2.8
Cerium	13.34	3	Gadolinium	1.53	2.7
Chromium	24.14	3.5	Holmium	.14	2.7
Cobaltum	2.96	.3	Dysprosium	.78	2.6
Cuprum	17.68	-.9	Ytterbium	.31	2.6
Dysprosium	.78	2.6	Terbium	.17	2.5
Erbium	.38	2.4	Erbium	.38	2.4
Europium	.29	2.3	Europium	.29	2.3
Ferrum	10699.85	.1	Plumbum	35.43	2.2
Gadolinium	1.53	2.7	Aluminium	12100	1.8
Gallium	2.34	1.6	Beryllium	.46	1.7
Germanium	0	-.3	Gallium	2.34	1.6
Hafnium	.03	.3	Bismuthum	.05	1.5
Holmium	.14	2.7	Uranium	.46	1.5
Indium	13.48	.7	Caesium	1.24	1.3
Lanthanum	7.37	3.3	Lithium	9.29	1.2
Lithium	9.29	1.2	Scandium	2.1	.9
Lutetium	.05	2.8	Arsenicum	6.72	.9
Magnesium	4965.97	-1	Vanadium	25.7	.8
Manganum	354.21	-.2	Indium	13.48	.7
Molybdenum	1.3	-.4	Tantalum	.01	.6
Natrium	10313.14	.1	Cobaltum	2.96	.3
Neodymium	5.54	2.8	Zirconium	.72	.3
Niccolum	9.77	.1	Hafnium	.03	.3
Niobium	.13	-.1	Natrium	10313.14	.1
Plumbum	35.43	2.2	Ferrum	10699.85	.1
Praseodymium	1.76	3.4	Niccolum	9.77	.1
Rhenium	.66	-.7	Thallium	.16	.1
Rubidium	26.31	-.4	Niobium	.13	-.1
Samarium	1.49	3.2	Cadmium	1.35	-.1
Scandium	2.1	.9	Manganum	354.21	-.2
Selenium	0	-1.3	Argentum	0	-.2
Silicium	0	-.4	Titanium	212.12	-.3
Stibium	.16	-.5	Germanium	0	-.3
Strontium	107	-1.5	Silicium	0	-.4
Tantalum	.01	.6	Rubidium	26.31	-.4
Tellurium	.02	-.8	Molybdenum	1.3	-.4
Terbium	.17	2.5	Stibium	.16	-.5
Thallium	.16	.1	Tungstenium	0	-.6
Thorium	1.75	3.5	Rhenium	.66	-.7
Thulium	.05	2.8	Zincum	127.8	-.8
Titanium	212.12	-.3	Tellurium	.02	-.8
Tungstenium	0	-.6	Barium	49.29	-.8
Uranium	.46	1.5	Cuprum	17.68	-.9
Vanadium	25.7	.8	Magnesium	4965.97	-1
Ytterbium	.31	2.6	Selenium	0	-1.3
Yttrium	5.15	2.9	Strontium	107	-1.5
Zincum	127.8	-.8	Borium	66.9	-1.7
Zirconium	.72	.3	Calcium	32626.53	-1.7

Bryonia dioica

Name	Value	Deviation	Name	Value	Deviation
Aluminium	7280	.7	Argentum	.2	4
Argentum	.2	4	Silicium	5504.86	1.9
Arsenicum	4.66	.3	Niobium	.32	1.6
Barium	89.72	-.5	Titanium	369.47	1.5
Beryllium	.23	.5	Plumbum	20.68	.9
Bismuthum	0	-.4	Scandium	2	.8
Borium	194.03	-1	Lutetium	.02	.8
Cadmium	1.01	-.2	Thorium	.65	.8
Caesium	.54	.2	Aluminium	7280	.7
Calcium	124760.08	-1.1	Tantalum	.01	.6
Cerium	5.08	.5	Beryllium	.23	.5
Chromium	4.44	.2	Lanthanum	2.3	.5
Cobaltum	1.93	-.1	Cerium	5.08	.5
Cuprum	54.33	-.3	Praseodymium	.56	.5
Dysprosium	.27	.4	Yttrium	1.59	.4
Erbium	.13	.4	Neodymium	1.86	.4
Europium	.1	.2	Samarium	.45	.4
Ferrum	5126.81	-.1	Gadolinium	.52	.4
Gadolinium	.52	.4	Terbium	.06	.4
Gallium	1.23	.2	Dysprosium	.27	.4
Germanium	.13	.3	Erbium	.13	.4
Hafnium	.03	.3	Ytterbium	.1	.4
Holmium	.04	.3	Lithium	5.34	.3
Indium	4.44	-.2	Germanium	.13	.3
Lanthanum	2.3	.5	Arsenicum	4.66	.3
Lithium	5.34	.3	Zirconium	.71	.3
Lutetium	.02	.8	Holmium	.04	.3
Magnesium	31536.29	-.4	Hafnium	.03	.3
Manganum	377.88	-.2	Uranium	.2	.3
Molybdenum	14.5	-.1	Chromium	4.44	.2
Natrium	9920.45	0	Niccolum	11.61	.2
Neodymium	1.86	.4	Gallium	1.23	.2
Niccolum	11.61	.2	Caesium	.54	.2
Niobium	.32	1.6	Europium	.1	.2
Plumbum	20.68	.9	Thulium	.01	.1
Praseodymium	.56	.5	Thallium	.16	.1
Rhenium	.53	-.7	Natrium	9920.45	0
Rubidium	152.24	-.1	Vanadium	8.86	0
Samarium	.45	.4	Zincum	685.89	0
Scandium	2	.8	Ferrum	5126.81	-.1
Selenium	0	-1.3	Cobaltum	1.93	-.1
Silicium	5504.86	1.9	Rubidium	152.24	-.1
Stibium	.1	-.7	Molybdenum	14.5	-.1
Strontium	732	-.4	Manganum	377.88	-.2
Tantalum	.01	.6	Cadmium	1.01	-.2
Tellurium	.03	-.4	Indium	4.44	-.2
Terbium	.06	.4	Cuprum	54.33	-.3
Thallium	.16	.1	Magnesium	31536.29	-.4
Thorium	.65	.8	Strontium	732	-.4
Thulium	.01	.1	Tellurium	.03	-.4
Titanium	369.47	1.5	Bismuthum	0	-.4
Tungstenium	0	-.6	Barium	89.72	-.5
Uranium	.2	.3	Tungstenium	0	-.6
Vanadium	8.86	0	Stibium	.1	-.7
Ytterbium	.1	.4	Rhenium	.53	-.7
Yttrium	1.59	.4	Borium	194.03	-1
Zincum	685.89	0	Calcium	124760.08	-1.1
Zirconium	.71	.3	Selenium	0	-1.3

Cardamine pratensis

Name	Value	Deviation	Name	Value	Deviation
Aluminium	2540	-.3	Rubidium	2590	6.4
Argentum	0	-.2	Zincum	3190	3.6
Arsenicum	1.06	-.8	Tellurium	.1	2.7
Barium	147.24	0	Tungstenium	.89	2.4
Beryllium	.08	-.3	Magnesium	150000	2.1
Bismuthum	0	-.4	Cadmium	5.86	2.1
Borium	367.68	-.1	Caesium	1.52	1.7
Cadmium	5.86	2.1	Titanium	343.14	1.2
Caesium	1.52	1.7	Stibium	.55	.7
Calcium	336765.42	.3	Selenium	2.75	.4
Cerium	3.98	.2	Calcium	336765.42	.3
Chromium	0	-.5	Molybdenum	31.19	.2
Cobaltum	1.21	-.4	Lanthanum	1.88	.2
Cuprum	63.79	-.1	Cerium	3.98	.2
Dysprosium	.07	-.4	Praseodymium	.41	.2
Erbium	.03	-.4	Plumbum	11.16	.1
Europium	.05	-.4	Barium	147.24	0
Ferrum	2249.1	-.2	Lithium	3.72	-.1
Gadolinium	.22	-.2	Borium	367.68	-.1
Gallium	.81	-.3	Manganum	700.77	-.1
Germanium	0	-.3	Cuprum	63.79	-.1
Hafnium	.01	-.3	Indium	6.03	-.1
Holmium	.01	-.5	Neodymium	1.03	-.1
Indium	6.03	-.1	Natrium	5902.97	-.2
Lanthanum	1.88	.2	Ferrum	2249.1	-.2
Lithium	3.72	-.1	Strontium	833	-.2
Lutetium	0	-.6	Argentum	0	-.2
Magnesium	150000	2.1	Gadolinium	.22	-.2
Manganum	700.77	-.1	Beryllium	.08	-.3
Molybdenum	31.19	.2	Aluminium	2540	-.3
Natrium	5902.97	-.2	Vanadium	1.84	-.3
Neodymium	1.03	-.1	Niccolum	6.23	-.3
Niccolum	6.23	-.3	Gallium	.81	-.3
Niobium	.1	-.3	Germanium	0	-.3
Plumbum	11.16	.1	Niobium	.1	-.3
Praseodymium	.41	.2	Hafnium	.01	-.3
Rhenium	1.62	-.6	Thorium	.18	-.3
Rubidium	2590	6.4	Silicium	0	-.4
Samarium	.12	-.4	Cobaltum	1.21	-.4
Scandium	.51	-.7	Yttrium	.4	-.4
Selenium	2.75	.4	Samarium	.12	-.4
Silicium	0	-.4	Europium	.05	-.4
Stibium	.55	.7	Terbium	.02	-.4
Strontium	833	-.2	Dysprosium	.07	-.4
Tantalum	0	-.7	Erbium	.03	-.4
Tellurium	.1	2.7	Bismuthum	0	-.4
Terbium	.02	-.4	Uranium	.06	-.4
Thallium	.02	-.5	Chromium	0	-.5
Thorium	.18	-.3	Zirconium	0	-.5
Thulium	0	-.6	Holmium	.01	-.5
Titanium	343.14	1.2	Ytterbium	.02	-.5
Tungstenium	.89	2.4	Thallium	.02	-.5
Uranium	.06	-.4	Thulium	0	-.6
Vanadium	1.84	-.3	Lutetium	0	-.6
Ytterbium	.02	-.5	Rhenium	1.62	-.6
Yttrium	.4	-.4	Scandium	.51	-.7
Zincum	3190	3.6	Tantalum	0	-.7
Zirconium	0	-.5	Arsenicum	1.06	-.8

Cardamine pratensis R

Name	Value	Deviation	Name	Value	Deviation
Aluminium	1590	-.5	Rubidium	2340	5.8
Argentum	0	-.2	Zincum	3110	3.5
Arsenicum	1.47	-.7	Tungstenium	.87	2.4
Barium	140.98	0	Magnesium	150000	2.1
Beryllium	.08	-.3	Cadmium	5.79	2.1
Bismuthum	0	-.4	Tellurium	.08	1.8
Borium	352.03	-.2	Caesium	1.44	1.6
Cadmium	5.79	2.1	Titanium	314.65	.9
Caesium	1.44	1.6	Stibium	.53	.6
Calcium	263806.27	-.1	Selenium	2.6	.3
Cerium	3.48	0	Plumbum	13.29	.2
Chromium	0	-.5	Molybdenum	29.81	.1
Cobaltum	1.12	-.5	Lanthanum	1.66	.1
Cuprum	61.9	-.1	Praseodymium	.36	.1
Dysprosium	.06	-.4	Indium	6.65	0
Erbium	.03	-.4	Barium	140.98	0
Europium	.04	-.5	Cerium	3.48	0
Ferrum	2081.8	-.2	Hafnium	.02	0
Gadolinium	.18	-.3	Calcium	263806.27	-.1
Gallium	.67	-.5	Manganum	696.11	-.1
Germanium	0	-.3	Cuprum	61.9	-.1
Hafnium	.02	0	Lithium	3.49	-.2
Holmium	.01	-.5	Borium	352.03	-.2
Indium	6.65	0	Ferrum	2081.8	-.2
Lanthanum	1.66	.1	Argentum	0	-.2
Lithium	3.49	-.2	Neodymium	.87	-.2
Lutetium	0	-.6	Beryllium	.08	-.3
Magnesium	150000	2.1	Natrium	5698.69	-.3
Manganum	696.11	-.1	Vanadium	1.5	-.3
Molybdenum	29.81	.1	Niccolum	6.01	-.3
Natrium	5698.69	-.3	Germanium	0	-.3
Neodymium	.87	-.2	Strontium	770	-.3
Niccolum	6.01	-.3	Gadolinium	.18	-.3
Niobium	.08	-.5	Silicium	0	-.4
Plumbum	13.29	.2	Yttrium	.35	-.4
Praseodymium	.36	.1	Terbium	.02	-.4
Rhenium	1.73	-.6	Dysprosium	.06	-.4
Rubidium	2340	5.8	Erbium	.03	-.4
Samarium	.1	-.5	Bismuthum	0	-.4
Scandium	0	-1.2	Aluminium	1590	-.5
Selenium	2.6	.3	Chromium	0	-.5
Silicium	0	-.4	Cobaltum	1.12	-.5
Stibium	.53	.6	Gallium	.67	-.5
Strontium	770	-.3	Zirconium	0	-.5
Tantalum	0	-.7	Niobium	.08	-.5
Tellurium	.08	1.8	Samarium	.1	-.5
Terbium	.02	-.4	Europium	.04	-.5
Thallium	.02	-.5	Holmium	.01	-.5
Thorium	.13	-.5	Ytterbium	.02	-.5
Thulium	0	-.6	Thallium	.02	-.5
Titanium	314.65	.9	Thorium	.13	-.5
Tungstenium	.87	2.4	Uranium	.04	-.5
Uranium	.04	-.5	Thulium	0	-.6
Vanadium	1.5	-.3	Lutetium	0	-.6
Ytterbium	.02	-.5	Rhenium	1.73	-.6
Yttrium	.35	-.4	Arsenicum	1.47	-.7
Zincum	3110	3.5	Tantalum	0	-.7
Zirconium	0	-.5	Scandium	0	-1.2

Cardiospermum halicacabum

Name	Value	Deviation	Name	Value	Deviation
Aluminium	2650	-.2	Tungstenium	.94	2.6
Argentum	0	-.2	Bismuthum	.05	1.5
Arsenicum	6.66	.9	Rhenium	18.05	1.3
Barium	62.86	-.7	Arsenicum	6.66	.9
Beryllium	.14	0	Cadmium	2.89	.7
Bismuthum	.05	1.5	Thallium	.23	.4
Borium	170.48	-1.1	Lithium	5.23	.2
Cadmium	2.89	.7	Chromium	4.08	.2
Caesium	.19	-.4	Zirconium	.69	.2
Calcium	257371.19	-.2	Niccolum	10.45	.1
Cerium	2.9	-.1	Thulium	.01	.1
Chromium	4.08	.2	Lutetium	.01	.1
Cobaltum	1.43	-.3	Beryllium	.14	0
Cuprum	36.53	-.6	Zincum	690.82	0
Dysprosium	.14	-.1	Holmium	.03	0
Erbium	.07	-.1	Ytterbium	.06	0
Europium	.06	-.3	Hafnium	.02	0
Ferrum	5084.95	-.1	Manganum	696.33	-.1
Gadolinium	.27	-.1	Ferrum	5084.95	-.1
Gallium	.67	-.5	Yttrium	.88	-.1
Germanium	0	-.3	Molybdenum	14.46	-.1
Hafnium	.02	0	Lanthanum	1.26	-.1
Holmium	.03	0	Cerium	2.9	-.1
Indium	2.5	-.4	Praseodymium	.3	-.1
Lanthanum	1.26	-.1	Neodymium	1	-.1
Lithium	5.23	.2	Samarium	.24	-.1
Lutetium	.01	.1	Gadolinium	.27	-.1
Magnesium	38657.76	-.3	Dysprosium	.14	-.1
Manganum	696.33	-.1	Erbium	.07	-.1
Molybdenum	14.46	-.1	Thorium	.28	-.1
Natrium	1204.88	-.6	Uranium	.12	-.1
Neodymium	1	-.1	Aluminium	2650	-.2
Niccolum	10.45	.1	Calcium	257371.19	-.2
Niobium	.11	-.2	Vanadium	4.22	-.2
Plumbum	6.93	-.3	Niobium	.11	-.2
Praseodymium	.3	-.1	Argentum	0	-.2
Rhenium	18.05	1.3	Terbium	.03	-.2
Rubidium	43.03	-.4	Magnesium	38657.76	-.3
Samarium	.24	-.1	Cobaltum	1.43	-.3
Scandium	.64	-.6	Germanium	0	-.3
Selenium	1.2	-.6	Europium	.06	-.3
Silicium	0	-.4	Plumbum	6.93	-.3
Stibium	.08	-.7	Silicium	0	-.4
Strontium	704	-.4	Rubidium	43.03	-.4
Tantalum	0	-.7	Strontium	704	-.4
Tellurium	.02	-.8	Indium	2.5	-.4
Terbium	.03	-.2	Caesium	.19	-.4
Thallium	.23	.4	Gallium	.67	-.5
Thorium	.28	-.1	Natrium	1204.88	-.6
Thulium	.01	.1	Scandium	.64	-.6
Titanium	176.73	-.7	Cuprum	36.53	-.6
Tungstenium	.94	2.6	Selenium	1.2	-.6
Uranium	.12	-.1	Titanium	176.73	-.7
Vanadium	4.22	-.2	Stibium	.08	-.7
Ytterbium	.06	0	Barium	62.86	-.7
Yttrium	.88	-.1	Tantalum	0	-.7
Zincum	690.82	0	Tellurium	.02	-.8
Zirconium	.69	.2	Borium	170.48	-1.1

Castanea vesca

Name	Value	Deviation	Name	Value	Deviation
Aluminium	4260	.1	Manganum	150000	9.5
Argentum	0	-.2	Gallium	4.92	4.7
Arsenicum	5.38	.5	Borium	870.09	2.6
Barium	320.35	1.4	Magnesium	150000	2.1
Beryllium	.25	.6	Niccolum	25.04	1.5
Bismuthum	0	-.4	Barium	320.35	1.4
Borium	870.09	2.6	Cobaltum	5.45	1.3
Cadmium	2.53	.5	Beryllium	.25	.6
Caesium	.25	-.3	Arsenicum	5.38	.5
Calcium	229370.67	-.4	Cadmium	2.53	.5
Cerium	2.12	-.4	Natrium	15185.98	.4
Chromium	0	-.5	Zincum	972.54	.4
Cobaltum	5.45	1.3	Lithium	5.06	.2
Cuprum	21.73	-.8	Germanium	.1	.2
Dysprosium	.11	-.2	Aluminium	4260	.1
Erbium	.05	-.3	Titanium	251.35	.1
Europium	.08	0	Rubidium	194.57	.1
Ferrum	4750.59	-.1	Europium	.08	0
Gadolinium	.2	-.3	Hafnium	.02	0
Gallium	4.92	4.7	Thallium	.13	0
Germanium	.1	.2	Ferrum	4750.59	-.1
Hafnium	.02	0	Stibium	.31	-.1
Holmium	.02	-.2	Vanadium	3.47	-.2
Indium	4.78	-.2	Yttrium	.66	-.2
Lanthanum	.75	-.4	Molybdenum	10.74	-.2
Lithium	5.06	.2	Argentum	0	-.2
Lutetium	0	-.6	Indium	4.78	-.2
Magnesium	150000	2.1	Dysprosium	.11	-.2
Manganum	150000	9.5	Holmium	.02	-.2
Molybdenum	10.74	-.2	Uranium	.09	-.2
Natrium	15185.98	.4	Caesium	.25	-.3
Neodymium	.63	-.3	Neodymium	.63	-.3
Niccolum	25.04	1.5	Samarium	.16	-.3
Niobium	.07	-.6	Gadolinium	.2	-.3
Plumbum	4.73	-.5	Erbium	.05	-.3
Praseodymium	.18	-.4	Silicium	0	-.4
Rhenium	3.31	-.4	Calcium	229370.67	-.4
Rubidium	194.57	.1	Tellurium	.03	-.4
Samarium	.16	-.3	Lanthanum	.75	-.4
Scandium	.53	-.7	Cerium	2.12	-.4
Selenium	0	-1.3	Praseodymium	.18	-.4
Silicium	0	-.4	Terbium	.02	-.4
Stibium	.31	-.1	Ytterbium	.03	-.4
Strontium	637	-.6	Rhenium	3.31	-.4
Tantalum	0	-.7	Bismuthum	0	-.4
Tellurium	.03	-.4	Chromium	0	-.5
Terbium	.02	-.4	Zirconium	0	-.5
Thallium	.13	0	Plumbum	4.73	-.5
Thorium	.12	-.5	Thorium	.12	-.5
Thulium	0	-.6	Strontium	637	-.6
Titanium	251.35	.1	Niobium	.07	-.6
Tungstenium	0	-.6	Thulium	0	-.6
Uranium	.09	-.2	Lutetium	0	-.6
Vanadium	3.47	-.2	Tungstenium	0	-.6
Ytterbium	.03	-.4	Scandium	.53	-.7
Yttrium	.66	-.2	Tantalum	0	-.7
Zincum	972.54	.4	Cuprum	21.73	-.8
Zirconium	0	-.5	Selenium	0	-1.3

Centaurium erythraea

Name	Value	Deviation	Name	Value	Deviation
Aluminium	9940	1.3	Tantalum	.03	3.4
Argentum	0	-.2	Silicium	8694.45	3.2
Arsenicum	3.93	.1	Niobium	.5	3.1
Barium	312.48	1.3	Gallium	3.45	2.9
Beryllium	.29	.8	Scandium	3.38	2.2
Bismuthum	0	-.4	Titanium	425.9	2.2
Borium	505.57	.6	Aluminium	9940	1.3
Cadmium	4.28	1.3	Cadmium	4.28	1.3
Caesium	.71	.4	Barium	312.48	1.3
Calcium	132941.54	-1	Thorium	.81	1.2
Cerium	6.49	1	Tungstenium	.51	1.1
Chromium	8.61	.9	Lanthanum	3.27	1
Cobaltum	2.84	.2	Cerium	6.49	1
Cuprum	54.37	-.3	Praseodymium	.75	1
Dysprosium	.31	.6	Chromium	8.61	.9
Erbium	.17	.7	Europium	.16	.9
Europium	.16	.9	Beryllium	.29	.8
Ferrum	7567.77	0	Neodymium	2.37	.8
Gadolinium	.62	.7	Samarium	.58	.8
Gallium	3.45	2.9	Holmium	.06	.8
Germanium	.14	.4	Ytterbium	.14	.8
Hafnium	.05	.8	Lutetium	.02	.8
Holmium	.06	.8	Hafnium	.05	.8
Indium	10.15	.4	Uranium	.3	.8
Lanthanum	3.27	1	Yttrium	1.99	.7
Lithium	6.66	.6	Zirconium	1.17	.7
Lutetium	.02	.8	Gadolinium	.62	.7
Magnesium	21186.01	-.6	Erbium	.17	.7
Manganum	704.16	-.1	Thulium	.02	.7
Molybdenum	12.7	-.2	Lithium	6.66	.6
Natrium	9207.62	0	Borium	505.57	.6
Neodymium	2.37	.8	Terbium	.07	.6
Niccolum	12.74	.3	Dysprosium	.31	.6
Niobium	.5	3.1	Plumbum	15.65	.5
Plumbum	15.65	.5	Germanium	.14	.4
Praseodymium	.75	1	Indium	10.15	.4
Rhenium	3.16	-.4	Caesium	.71	.4
Rubidium	38.92	-.4	Niccolum	12.74	.3
Samarium	.58	.8	Zincum	895.94	.3
Scandium	3.38	2.2	Vanadium	12.4	.2
Selenium	1.52	-.4	Cobaltum	2.84	.2
Silicium	8694.45	3.2	Arsenicum	3.93	.1
Stibium	.37	.1	Stibium	.37	.1
Strontium	364	-1	Tellurium	.04	.1
Tantalum	.03	3.4	Natrium	9207.62	0
Tellurium	.04	.1	Ferrum	7567.77	0
Terbium	.07	.6	Manganum	704.16	-.1
Thallium	.04	-.4	Molybdenum	12.7	-.2
Thorium	.81	1.2	Argentum	0	-.2
Thulium	.02	.7	Cuprum	54.37	-.3
Titanium	425.9	2.2	Selenium	1.52	-.4
Tungstenium	.51	1.1	Rubidium	38.92	-.4
Uranium	.3	.8	Rhenium	3.16	-.4
Vanadium	12.4	.2	Thallium	.04	-.4
Ytterbium	.14	.8	Bismuthum	0	-.4
Yttrium	1.99	.7	Magnesium	21186.01	-.6
Zincum	895.94	.3	Calcium	132941.54	-1
Zirconium	1.17	.7	Strontium	364	-1

Chamomilla - Matricaria chamomilla

Name	Value	Deviation	Name	Value	Deviation
Aluminium	3660	0	Tantalum	.01	.6
Argentum	0	-.2	Selenium	2.99	.5
Arsenicum	2.01	-.5	Gallium	1.37	.4
Barium	25.59	-1	Rhenium	9.39	.3
Beryllium	.1	-.2	Borium	403.98	.1
Bismuthum	0	-.4	Natrium	10148.74	.1
Borium	403.98	.1	Thulium	.01	.1
Cadmium	1.45	0	Lutetium	.01	.1
Caesium	.3	-.2	Aluminium	3660	0
Calcium	136975.57	-1	Cuprum	65.93	0
Cerium	2.92	-.1	Cadmium	1.45	0
Chromium	0	-.5	Yttrium	.77	-.1
Cobaltum	1.43	-.3	Niobium	.13	-.1
Cuprum	65.93	0	Cerium	2.92	-.1
Dysprosium	.14	-.1	Praseodymium	.27	-.1
Erbium	.07	-.1	Neodymium	.94	-.1
Europium	.06	-.3	Samarium	.23	-.1
Ferrum	2268.16	-.2	Dysprosium	.14	-.1
Gadolinium	.26	-.2	Erbium	.07	-.1
Gallium	1.37	.4	Ytterbium	.05	-.1
Germanium	0	-.3	Thorium	.3	-.1
Hafnium	0	-.6	Uranium	.12	-.1
Holmium	.02	-.2	Beryllium	.1	-.2
Indium	2.98	-.4	Scandium	1.05	-.2
Lanthanum	1.1	-.2	Manganum	664.07	-.2
Lithium	3.08	-.3	Ferrum	2268.16	-.2
Lutetium	.01	.1	Niccolum	6.83	-.2
Magnesium	20973.75	-.7	Rubidium	117.61	-.2
Manganum	664.07	-.2	Molybdenum	11.77	-.2
Molybdenum	11.77	-.2	Argentum	0	-.2
Natrium	10148.74	.1	Caesium	.3	-.2
Neodymium	.94	-.1	Lanthanum	1.1	-.2
Niccolum	6.83	-.2	Gadolinium	.26	-.2
Niobium	.13	-.1	Terbium	.03	-.2
Plumbum	4.86	-.5	Holmium	.02	-.2
Praseodymium	.27	-.1	Lithium	3.08	-.3
Rhenium	9.39	.3	Vanadium	2.22	-.3
Rubidium	117.61	-.2	Cobaltum	1.43	-.3
Samarium	.23	-.1	Germanium	0	-.3
Scandium	1.05	-.2	Europium	.06	-.3
Selenium	2.99	.5	Silicium	0	-.4
Silicium	0	-.4	Indium	2.98	-.4
Stibium	.18	-.4	Stibium	.18	-.4
Strontium	374	-1	Thallium	.05	-.4
Tantalum	.01	.6	Bismuthum	0	-.4
Tellurium	.02	-.8	Titanium	198.17	-.5
Terbium	.03	-.2	Chromium	0	-.5
Thallium	.05	-.4	Zincum	325.74	-.5
Thorium	.3	-.1	Arsenicum	2.01	-.5
Thulium	.01	.1	Zirconium	0	-.5
Titanium	198.17	-.5	Plumbum	4.86	-.5
Tungstenium	0	-.6	Hafnium	0	-.6
Uranium	.12	-.1	Tungstenium	0	-.6
Vanadium	2.22	-.3	Magnesium	20973.75	-.7
Ytterbium	.05	-.1	Tellurium	.02	-.8
Yttrium	.77	-.1	Calcium	136975.57	-1
Zincum	325.74	-.5	Strontium	374	-1
Zirconium	0	-.5	Barium	25.59	-1

Chelidonium majus

Name	Value	Deviation	Name	Value	Deviation
Aluminium	20800	3.6	Thallium	1.94	7.3
Argentum	0	-.2	Thorium	2.12	4.4
Arsenicum	7.56	1.2	Chromium	25.65	3.8
Barium	65.58	-.6	Aluminium	20800	3.6
Beryllium	.67	2.8	Cerium	14.56	3.4
Bismuthum	.07	2.3	Praseodymium	1.71	3.3
Borium	92.2	-1.6	Neodymium	6.2	3.2
Cadmium	1.5	0	Lanthanum	6.93	3
Caesium	1.24	1.3	Samarium	1.44	3
Calcium	45142.56	-1.6	Gallium	3.42	2.9
Cerium	14.56	3.4	Europium	.34	2.9
Chromium	25.65	3.8	Gadolinium	1.6	2.9
Cobaltum	3.9	.6	Terbium	.19	2.9
Cuprum	27.85	-.7	Holmium	.15	2.9
Dysprosium	.83	2.8	Ytterbium	.34	2.9
Erbium	.42	2.7	Beryllium	.67	2.8
Europium	.34	2.9	Dysprosium	.83	2.8
Ferrum	15150.76	.2	Lutetium	.05	2.8
Gadolinium	1.6	2.9	Erbium	.42	2.7
Gallium	3.42	2.9	Yttrium	4.76	2.6
Germanium	0	-.3	Uranium	.69	2.6
Hafnium	.06	1.1	Bismuthum	.07	2.3
Holmium	.15	2.9	Lithium	14	2.2
Indium	18.46	1.3	Thulium	.04	2.1
Lanthanum	6.93	3	Scandium	2.7	1.5
Lithium	14	2.2	Vanadium	39.69	1.4
Lutetium	.05	2.8	Plumbum	25.67	1.4
Magnesium	12575.63	-.8	Indium	18.46	1.3
Manganum	988.28	-.1	Caesium	1.24	1.3
Molybdenum	4.54	-.3	Arsenicum	7.56	1.2
Natrium	2079.82	-.5	Hafnium	.06	1.1
Neodymium	6.2	3.2	Cobaltum	3.9	.6
Niccolum	14.29	.5	Tantalum	.01	.6
Niobium	.14	0	Niccolum	14.29	.5
Plumbum	25.67	1.4	Zirconium	.78	.3
Praseodymium	1.71	3.3	Ferrum	15150.76	.2
Rhenium	.55	-.7	Tellurium	.04	.1
Rubidium	22.74	-.4	Niobium	.14	0
Samarium	1.44	3	Cadmium	1.5	0
Scandium	2.7	1.5	Manganum	988.28	-.1
Selenium	1.17	-.6	Argentum	0	-.2
Silicium	0	-.4	Germanium	0	-.3
Stibium	.07	-.8	Molybdenum	4.54	-.3
Strontium	253	-1.2	Silicium	0	-.4
Tantalum	.01	.6	Titanium	201.69	-.4
Tellurium	.04	.1	Rubidium	22.74	-.4
Terbium	.19	2.9	Natrium	2079.82	-.5
Thallium	1.94	7.3	Zincum	369.09	-.5
Thorium	2.12	4.4	Selenium	1.17	-.6
Thulium	.04	2.1	Barium	65.58	-.6
Titanium	201.69	-.4	Tungstenium	0	-.6
Tungstenium	0	-.6	Cuprum	27.85	-.7
Uranium	.69	2.6	Rhenium	.55	-.7
Vanadium	39.69	1.4	Magnesium	12575.63	-.8
Ytterbium	.34	2.9	Stibium	.07	-.8
Yttrium	4.76	2.6	Strontium	253	-1.2
Zincum	369.09	-.5	Borium	92.2	-1.6
Zirconium	.78	.3	Calcium	45142.56	-1.6

Chenopodium anthelminticum

Name	Value	Deviation
Aluminium	488	-.7
Argentum	0	-.2
Arsenicum	0	-1.1
Barium	23.08	-1
Beryllium	.03	-.6
Bismuthum	0	-.4
Borium	365.77	-.1
Cadmium	.19	-.6
Caesium	.06	-.6
Calcium	265763.31	-.1
Cerium	1.34	-.6
Chromium	0	-.5
Cobaltum	.9	-.6
Cuprum	60	-.2
Dysprosium	.02	-.6
Erbium	.01	-.6
Europium	.01	-.8
Ferrum	1722.05	-.2
Gadolinium	.06	-.6
Gallium	.56	-.6
Germanium	0	-.3
Hafnium	0	-.6
Holmium	0	-.7
Indium	2.65	-.4
Lanthanum	.34	-.6
Lithium	.52	-.8
Lutetium	0	-.6
Magnesium	21621.05	-.6
Manganum	133.77	-.2
Molybdenum	10.05	-.2
Natrium	2047.22	-.5
Neodymium	.21	-.6
Niccolum	0	-.8
Niobium	.05	-.7
Plumbum	2.54	-.7
Praseodymium	.07	-.6
Rhenium	6.45	0
Rubidium	43.26	-.3
Samarium	.04	-.6
Scandium	0	-1.2
Selenium	1.65	-.3
Silicium	0	-.4
Stibium	.24	-.3
Strontium	838	-.2
Tantalum	0	-.7
Tellurium	.04	.1
Terbium	.01	-.5
Thallium	.03	-.4
Thorium	.1	-.5
Thulium	0	-.6
Titanium	160.34	-.9
Tungstenium	0	-.6
Uranium	.02	-.6
Vanadium	0	-.4
Ytterbium	.01	-.6
Yttrium	.11	-.6
Zincum	902.18	.3
Zirconium	0	-.5

Name	Value	Deviation
Zincum	902.18	.3
Tellurium	.04	.1
Rhenium	6.45	0
Borium	365.77	-.1
Calcium	265763.31	-.1
Manganum	133.77	-.2
Ferrum	1722.05	-.2
Cuprum	60	-.2
Strontium	838	-.2
Molybdenum	10.05	-.2
Argentum	0	-.2
Germanium	0	-.3
Selenium	1.65	-.3
Rubidium	43.26	-.3
Stibium	.24	-.3
Silicium	0	-.4
Vanadium	0	-.4
Indium	2.65	-.4
Thallium	.03	-.4
Bismuthum	0	-.4
Natrium	2047.22	-.5
Chromium	0	-.5
Zirconium	0	-.5
Terbium	.01	-.5
Thorium	.1	-.5
Beryllium	.03	-.6
Magnesium	21621.05	-.6
Cobaltum	.9	-.6
Gallium	.56	-.6
Yttrium	.11	-.6
Cadmium	.19	-.6
Caesium	.06	-.6
Lanthanum	.34	-.6
Cerium	1.34	-.6
Praseodymium	.07	-.6
Neodymium	.21	-.6
Samarium	.04	-.6
Gadolinium	.06	-.6
Dysprosium	.02	-.6
Erbium	.01	-.6
Thulium	0	-.6
Ytterbium	.01	-.6
Lutetium	0	-.6
Hafnium	0	-.6
Tungstenium	0	-.6
Uranium	.02	-.6
Aluminium	488	-.7
Niobium	.05	-.7
Holmium	0	-.7
Tantalum	0	-.7
Plumbum	2.54	-.7
Lithium	.52	-.8
Niccolum	0	-.8
Europium	.01	-.8
Titanium	160.34	-.9
Barium	23.08	-1
Arsenicum	0	-1.1
Scandium	0	-1.2

Cicuta virosa

Name	Value	Deviation	Name	Value	Deviation
Aluminium	7050	.7	Tellurium	.13	4.1
Argentum	0	-.2	Cuprum	278.16	3.8
Arsenicum	11.81	2.5	Bismuthum	.11	3.8
Barium	133.9	-.1	Natrium	60000	3.6
Beryllium	.37	1.2	Cobaltum	9.91	3.1
Bismuthum	.11	3.8	Zirconium	3.43	3.1
Borium	199.11	-1	Gallium	3.37	2.8
Cadmium	.87	-.3	Uranium	.73	2.8
Caesium	.7	.4	Niobium	.44	2.6
Calcium	112275.19	-1.1	Hafnium	.11	2.6
Cerium	10.8	2.3	Arsenicum	11.81	2.5
Chromium	4.8	.3	Cerium	10.8	2.3
Cobaltum	9.91	3.1	Magnesium	150000	2.1
Cuprum	278.16	3.8	Lutetium	.04	2.1
Dysprosium	.53	1.6	Tungstenium	.76	2
Erbium	.28	1.6	Lanthanum	4.82	1.9
Europium	.23	1.6	Ytterbium	.24	1.9
Ferrum	17096.27	.3	Yttrium	3.49	1.8
Gadolinium	1.04	1.6	Praseodymium	1.09	1.8
Gallium	3.37	2.8	Terbium	.13	1.8
Germanium	0	-.3	Neodymium	3.91	1.7
Hafnium	.11	2.6	Samarium	.92	1.7
Holmium	.1	1.7	Holmium	.1	1.7
Indium	9.04	.3	Europium	.23	1.6
Lanthanum	4.82	1.9	Gadolinium	1.04	1.6
Lithium	10.49	1.4	Dysprosium	.53	1.6
Lutetium	.04	2.1	Erbium	.28	1.6
Magnesium	150000	2.1	Plumbum	28.39	1.6
Manganum	22600	1.3	Lithium	10.49	1.4
Molybdenum	1.63	-.4	Thulium	.03	1.4
Natrium	60000	3.6	Thorium	.89	1.4
Neodymium	3.91	1.7	Manganum	22600	1.3
Niccolum	18.17	.8	Beryllium	.37	1.2
Niobium	.44	2.6	Vanadium	35.37	1.2
Plumbum	28.39	1.6	Titanium	337.62	1.1
Praseodymium	1.09	1.8	Niccolum	18.17	.8
Rhenium	.6	-.7	Aluminium	7050	.7
Rubidium	263.4	.2	Scandium	1.87	.7
Samarium	.92	1.7	Tantalum	.01	.6
Scandium	1.87	.7	Caesium	.7	.4
Selenium	2.28	.1	Chromium	4.8	.3
Silicium	0	-.4	Ferrum	17096.27	.3
Stibium	.2	-.4	Indium	9.04	.3
Strontium	664	-.5	Rubidium	263.4	.2
Tantalum	.01	.6	Selenium	2.28	.1
Tellurium	.13	4.1	Barium	133.9	-.1
Terbium	.13	1.8	Thallium	.12	-.1
Thallium	.12	-.1	Argentum	0	-.2
Thorium	.89	1.4	Germanium	0	-.3
Thulium	.03	1.4	Cadmium	.87	-.3
Titanium	337.62	1.1	Silicium	0	-.4
Tungstenium	.76	2	Molybdenum	1.63	-.4
Uranium	.73	2.8	Stibium	.2	-.4
Vanadium	35.37	1.2	Strontium	664	-.5
Ytterbium	.24	1.9	Zincum	293.39	-.6
Yttrium	3.49	1.8	Rhenium	.6	-.7
Zincum	293.39	-.6	Borium	199.11	-1
Zirconium	3.43	3.1	Calcium	112275.19	-1.1

Cimicifuga racemosa

Name	Value	Deviation	Name	Value	Deviation
Aluminium	2660	-.2	Arsenicum	13.97	3.1
Argentum	0	-.2	Niccolum	23.51	1.3
Arsenicum	13.97	3.1	Tungstenium	.53	1.2
Barium	122.55	-.2	Cobaltum	4.83	1
Beryllium	.23	.5	Cuprum	124.5	1
Bismuthum	0	-.4	Vanadium	28.9	.9
Borium	301.07	-.4	Lutetium	.02	.8
Cadmium	2.25	.4	Gallium	1.64	.7
Caesium	.17	-.4	Yttrium	1.84	.6
Calcium	166268.35	-.8	Thallium	.29	.6
Cerium	4.34	.3	Beryllium	.23	.5
Chromium	2.57	-.1	Cadmium	2.25	.4
Cobaltum	4.83	1	Terbium	.06	.4
Cuprum	124.5	1	Erbium	.13	.4
Dysprosium	.24	.3	Ytterbium	.1	.4
Erbium	.13	.4	Uranium	.22	.4
Europium	.1	.2	Lanthanum	2.04	.3
Ferrum	2374.43	-.2	Cerium	4.34	.3
Gadolinium	.45	.3	Praseodymium	.45	.3
Gallium	1.64	.7	Samarium	.38	.3
Germanium	0	-.3	Gadolinium	.45	.3
Hafnium	.02	0	Dysprosium	.24	.3
Holmium	.04	.3	Holmium	.04	.3
Indium	2.21	-.5	Strontium	1040	.2
Lanthanum	2.04	.3	Neodymium	1.56	.2
Lithium	1.77	-.6	Europium	.1	.2
Lutetium	.02	.8	Thulium	.01	.1
Magnesium	23185.79	-.6	Plumbum	11.6	.1
Manganum	850.64	-.1	Hafnium	.02	0
Molybdenum	6.33	-.3	Titanium	234.03	-.1
Natrium	3694.19	-.4	Chromium	2.57	-.1
Neodymium	1.56	.2	Manganum	850.64	-.1
Niccolum	23.51	1.3	Rubidium	118.86	-.1
Niobium	.08	-.5	Aluminium	2660	-.2
Plumbum	11.6	.1	Ferrum	2374.43	-.2
Praseodymium	.45	.3	Argentum	0	-.2
Rhenium	.32	-.8	Barium	122.55	-.2
Rubidium	118.86	-.1	Germanium	0	-.3
Samarium	.38	.3	Molybdenum	6.33	-.3
Scandium	.65	-.6	Thorium	.2	-.3
Selenium	0	-1.3	Borium	301.07	-.4
Silicium	0	-.4	Natrium	3694.19	-.4
Stibium	.12	-.6	Silicium	0	-.4
Strontium	1040	.2	Zincum	430	-.4
Tantalum	0	-.7	Caesium	.17	-.4
Tellurium	.02	-.8	Bismuthum	0	-.4
Terbium	.06	.4	Zirconium	0	-.5
Thallium	.29	.6	Niobium	.08	-.5
Thorium	.2	-.3	Indium	2.21	-.5
Thulium	.01	.1	Lithium	1.77	-.6
Titanium	234.03	-.1	Magnesium	23185.79	-.6
Tungstenium	.53	1.2	Scandium	.65	-.6
Uranium	.22	.4	Stibium	.12	-.6
Vanadium	28.9	.9	Tantalum	0	-.7
Ytterbium	.1	.4	Calcium	166268.35	-.8
Yttrium	1.84	.6	Tellurium	.02	-.8
Zincum	430	-.4	Rhenium	.32	-.8
Zirconium	0	-.5	Selenium	0	-1.3

Cinnamodendron cortisonum-Drymis winteri

Name	Value	Deviation	Name	Value	Deviation
Aluminium	1610	-.5	Thallium	.74	2.4
Argentum	0	-.2	Magnesium	150000	2.1
Arsenicum	0	-1.1	Zincum	2090	2
Barium	228.36	.7	Cuprum	159.67	1.7
Beryllium	.05	-.5	Selenium	4.19	1.3
Bismuthum	0	-.4	Rhenium	18.03	1.3
Borium	457.92	.4	Strontium	1490	1
Cadmium	2.43	.4	Stibium	.57	.7
Caesium	.16	-.4	Barium	228.36	.7
Calcium	227850.39	-.4	Plumbum	17.04	.6
Cerium	2.05	-.4	Tellurium	.05	.5
Chromium	0	-.5	Lithium	5.98	.4
Cobaltum	1.23	-.4	Borium	457.92	.4
Cuprum	159.67	1.7	Cadmium	2.43	.4
Dysprosium	.08	-.4	Natrium	11119.78	.1
Erbium	.04	-.3	Rubidium	206.17	.1
Europium	.06	-.3	Thulium	.01	.1
Ferrum	1614.75	-.2	Lutetium	.01	.1
Gadolinium	.14	-.4	Manganum	2460	0
Gallium	.97	-.1	Niccolum	8.33	-.1
Germanium	0	-.3	Gallium	.97	-.1
Hafnium	0	-.6	Molybdenum	15.13	-.1
Holmium	.01	-.5	Indium	5.85	-.1
Indium	5.85	-.1	Ferrum	1614.75	-.2
Lanthanum	.69	-.4	Argentum	0	-.2
Lithium	5.98	.4	Germanium	0	-.3
Lutetium	.01	.1	Yttrium	.55	-.3
Magnesium	150000	2.1	Europium	.06	-.3
Manganum	2460	0	Erbium	.04	-.3
Molybdenum	15.13	-.1	Silicium	0	-.4
Natrium	11119.78	.1	Calcium	227850.39	-.4
Neodymium	.56	-.4	Vanadium	0	-.4
Niccolum	8.33	-.1	Cobaltum	1.23	-.4
Niobium	.08	-.5	Caesium	.16	-.4
Plumbum	17.04	.6	Lanthanum	.69	-.4
Praseodymium	.17	-.4	Cerium	2.05	-.4
Rhenium	18.03	1.3	Praseodymium	.17	-.4
Rubidium	206.17	.1	Neodymium	.56	-.4
Samarium	.12	-.4	Samarium	.12	-.4
Scandium	.57	-.6	Gadolinium	.14	-.4
Selenium	4.19	1.3	Terbium	.02	-.4
Silicium	0	-.4	Dysprosium	.08	-.4
Stibium	.57	.7	Ytterbium	.03	-.4
Strontium	1490	1	Bismuthum	0	-.4
Tantalum	0	-.7	Uranium	.05	-.4
Tellurium	.05	.5	Beryllium	.05	-.5
Terbium	.02	-.4	Aluminium	1610	-.5
Thallium	.74	2.4	Chromium	0	-.5
Thorium	.09	-.6	Zirconium	0	-.5
Thulium	.01	.1	Niobium	.08	-.5
Titanium	131.22	-1.3	Holmium	.01	-.5
Tungstenium	0	-.6	Scandium	.57	-.6
Uranium	.05	-.4	Hafnium	0	-.6
Vanadium	0	-.4	Tungstenium	0	-.6
Ytterbium	.03	-.4	Thorium	.09	-.6
Yttrium	.55	-.3	Tantalum	0	-.7
Zincum	2090	2	Arsenicum	0	-1.1
Zirconium	0	-.5	Titanium	131.22	-1.3

Clematis erecta

Name	Value	Deviation	Name	Value	Deviation
Aluminium	2410	-.3	Rhenium	65.39	6.9
Argentum	0	-.2	Tellurium	.07	1.4
Arsenicum	5.73	.6	Arsenicum	5.73	.6
Barium	120.56	-.2	Stibium	.48	.5
Beryllium	.05	-.5	Thallium	.26	.5
Bismuthum	0	-.4	Borium	459.35	.4
Borium	459.35	.4	Calcium	308022.94	.2
Cadmium	.72	-.4	Niccolum	10.17	.1
Caesium	.11	-.5	Selenium	2.14	0
Calcium	308022.94	.2	Indium	6.73	0
Cerium	1.76	-.5	Gallium	1	-.1
Chromium	0	-.5	Rubidium	126.74	-.1
Cobaltum	1.03	-.5	Strontium	878	-.1
Cuprum	59.01	-.2	Manganum	547.61	-.2
Dysprosium	.07	-.4	Ferrum	2315.4	-.2
Erbium	.03	-.4	Cuprum	59.01	-.2
Europium	.04	-.5	Argentum	0	-.2
Ferrum	2315.4	-.2	Barium	120.56	-.2
Gadolinium	.12	-.5	Plumbum	7.73	-.2
Gallium	1	-.1	Natrium	4750.16	-.3
Germanium	0	-.3	Aluminium	2410	-.3
Hafnium	0	-.6	Vanadium	1.89	-.3
Holmium	.01	-.5	Germanium	0	-.3
Indium	6.73	0	Molybdenum	6.14	-.3
Lanthanum	.52	-.5	Uranium	.07	-.3
Lithium	1.72	-.6	Silicium	0	-.4
Lutetium	0	-.6	Scandium	.81	-.4
Magnesium	29308.86	-.5	Zincum	407.57	-.4
Manganum	547.61	-.2	Yttrium	.36	-.4
Molybdenum	6.14	-.3	Cadmium	.72	-.4
Natrium	4750.16	-.3	Samarium	.11	-.4
Neodymium	.41	-.5	Terbium	.02	-.4
Niccolum	10.17	.1	Dysprosium	.07	-.4
Niobium	.08	-.5	Erbium	.03	-.4
Plumbum	7.73	-.2	Bismuthum	0	-.4
Praseodymium	.12	-.5	Beryllium	.05	-.5
Rhenium	65.39	6.9	Magnesium	29308.86	-.5
Rubidium	126.74	-.1	Chromium	0	-.5
Samarium	.11	-.4	Cobaltum	1.03	-.5
Scandium	.81	-.4	Zirconium	0	-.5
Selenium	2.14	0	Niobium	.08	-.5
Silicium	0	-.4	Caesium	.11	-.5
Stibium	.48	.5	Lanthanum	.52	-.5
Strontium	878	-.1	Cerium	1.76	-.5
Tantalum	0	-.7	Praseodymium	.12	-.5
Tellurium	.07	1.4	Neodymium	.41	-.5
Terbium	.02	-.4	Europium	.04	-.5
Thallium	.26	.5	Gadolinium	.12	-.5
Thorium	.13	-.5	Holmium	.01	-.5
Thulium	0	-.6	Ytterbium	.02	-.5
Titanium	172.93	-.8	Thorium	.13	-.5
Tungstenium	0	-.6	Lithium	1.72	-.6
Uranium	.07	-.3	Thulium	0	-.6
Vanadium	1.89	-.3	Lutetium	0	-.6
Ytterbium	.02	-.5	Hafnium	0	-.6
Yttrium	.36	-.4	Tungstenium	0	-.6
Zincum	407.57	-.4	Tantalum	0	-.7
Zirconium	0	-.5	Titanium	172.93	-.8

Collinsonia canadensis

Name	Value	Deviation	Name	Value	Deviation
Aluminium	5600	.4	Magnesium	150000	2.1
Argentum	0	-.2	Titanium	404.26	1.9
Arsenicum	8.85	1.6	Arsenicum	8.85	1.6
Barium	131.96	-.1	Cuprum	149.68	1.5
Beryllium	.27	.7	Vanadium	37.01	1.3
Bismuthum	0	-.4	Tungstenium	.51	1.1
Borium	195.58	-1	Chromium	8.39	.9
Cadmium	1.18	-.2	Zirconium	1.32	.9
Caesium	.39	-.1	Yttrium	2.07	.8
Calcium	84231.72	-1.3	Lutetium	.02	.8
Cerium	5.13	.5	Beryllium	.27	.7
Chromium	8.39	.9	Thulium	.02	.7
Cobaltum	1.84	-.2	Gallium	1.54	.6
Cuprum	149.68	1.5	Praseodymium	.6	.6
Dysprosium	.29	.5	Samarium	.5	.6
Erbium	.14	.5	Ytterbium	.12	.6
Europium	.12	.4	Lanthanum	2.44	.5
Ferrum	3426.24	-.1	Cerium	5.13	.5
Gadolinium	.54	.5	Neodymium	1.94	.5
Gallium	1.54	.6	Gadolinium	.54	.5
Germanium	.1	.2	Dysprosium	.29	.5
Hafnium	.04	.5	Holmium	.05	.5
Holmium	.05	.5	Erbium	.14	.5
Indium	3.22	-.4	Hafnium	.04	.5
Lanthanum	2.44	.5	Thorium	.51	.5
Lithium	4.36	0	Uranium	.25	.5
Lutetium	.02	.8	Aluminium	5600	.4
Magnesium	150000	2.1	Niobium	.18	.4
Manganum	493.39	-.2	Europium	.12	.4
Molybdenum	12.19	-.2	Terbium	.06	.4
Natrium	3828.36	-.4	Niccolum	12.79	.3
Neodymium	1.94	.5	Zincum	870.04	.3
Niccolum	12.79	.3	Scandium	1.39	.2
Niobium	.18	.4	Germanium	.1	.2
Plumbum	9.43	-.1	Stibium	.4	.2
Praseodymium	.6	.6	Thallium	.18	.2
Rhenium	1.4	-.6	Lithium	4.36	0
Rubidium	43.67	-.3	Ferrum	3426.24	-.1
Samarium	.5	.6	Caesium	.39	-.1
Scandium	1.39	.2	Barium	131.96	-.1
Selenium	1.18	-.6	Plumbum	9.43	-.1
Silicium	0	-.4	Manganum	493.39	-.2
Stibium	.4	.2	Cobaltum	1.84	-.2
Strontium	409	-1	Molybdenum	12.19	-.2
Tantalum	0	-.7	Argentum	0	-.2
Tellurium	.03	-.4	Cadmium	1.18	-.2
Terbium	.06	.4	Rubidium	43.67	-.3
Thallium	.18	.2	Natrium	3828.36	-.4
Thorium	.51	.5	Silicium	0	-.4
Thulium	.02	.7	Indium	3.22	-.4
Titanium	404.26	1.9	Tellurium	.03	-.4
Tungstenium	.51	1.1	Bismuthum	0	-.4
Uranium	.25	.5	Selenium	1.18	-.6
Vanadium	37.01	1.3	Rhenium	1.4	-.6
Ytterbium	.12	.6	Tantalum	0	-.7
Yttrium	2.07	.8	Borium	195.58	-1
Zincum	870.04	.3	Strontium	409	-1
Zirconium	1.32	.9	Calcium	84231.72	-1.3

Conium maculatum

Name	Value	Deviation	Name	Value	Deviation
Aluminium	486	-.7	Argentum	.2	4
Argentum	.2	4	Rhenium	20.88	1.7
Arsenicum	4.15	.1	Tungstenium	.6	1.4
Barium	50.22	-.8	Selenium	4.14	1.2
Beryllium	.01	-.7	Stibium	.59	.8
Bismuthum	0	-.4	Chromium	4.26	.2
Borium	374.6	-.1	Germanium	.11	.2
Cadmium	.38	-.5	Cuprum	74.94	.1
Caesium	.1	-.5	Arsenicum	4.15	.1
Calcium	270868.31	-.1	Indium	6.47	0
Cerium	1.59	-.5	Borium	374.6	-.1
Chromium	4.26	.2	Calcium	270868.31	-.1
Cobaltum	.63	-.7	Titanium	229.07	-.1
Cuprum	74.94	.1	Magnesium	40377.49	-.2
Dysprosium	.02	-.6	Vanadium	3.23	-.2
Erbium	.01	-.6	Manganum	526.85	-.2
Europium	.02	-.7	Ferrum	1429.06	-.2
Ferrum	1429.06	-.2	Rubidium	83.81	-.2
Gadolinium	.06	-.6	Molybdenum	12.14	-.2
Gallium	.66	-.5	Thallium	.06	-.3
Germanium	.11	.2	Natrium	3894.69	-.4
Hafnium	0	-.6	Silicium	0	-.4
Holmium	0	-.7	Plumbum	6.04	-.4
Indium	6.47	0	Bismuthum	0	-.4
Lanthanum	.34	-.6	Zincum	346.44	-.5
Lithium	1.13	-.7	Gallium	.66	-.5
Lutetium	0	-.6	Strontium	671	-.5
Magnesium	40377.49	-.2	Zirconium	0	-.5
Manganum	526.85	-.2	Cadmium	.38	-.5
Molybdenum	12.14	-.2	Caesium	.1	-.5
Natrium	3894.69	-.4	Cerium	1.59	-.5
Neodymium	.2	-.6	Terbium	.01	-.5
Niccolum	0	-.8	Yttrium	.16	-.6
Niobium	.05	-.7	Lanthanum	.34	-.6
Plumbum	6.04	-.4	Praseodymium	.07	-.6
Praseodymium	.07	-.6	Neodymium	.2	-.6
Rhenium	20.88	1.7	Samarium	.04	-.6
Rubidium	83.81	-.2	Gadolinium	.06	-.6
Samarium	.04	-.6	Dysprosium	.02	-.6
Scandium	0	-1.2	Erbium	.01	-.6
Selenium	4.14	1.2	Thulium	0	-.6
Silicium	0	-.4	Ytterbium	.01	-.6
Stibium	.59	.8	Lutetium	0	-.6
Strontium	671	-.5	Hafnium	0	-.6
Tantalum	0	-.7	Uranium	.02	-.6
Tellurium	0	-1.7	Lithium	1.13	-.7
Terbium	.01	-.5	Beryllium	.01	-.7
Thallium	.06	-.3	Aluminium	486	-.7
Thorium	.04	-.7	Cobaltum	.63	-.7
Thulium	0	-.6	Niobium	.05	-.7
Titanium	229.07	-.1	Europium	.02	-.7
Tungstenium	.6	1.4	Holmium	0	-.7
Uranium	.02	-.6	Tantalum	0	-.7
Vanadium	3.23	-.2	Thorium	.04	-.7
Ytterbium	.01	-.6	Niccolum	0	-.8
Yttrium	.16	-.6	Barium	50.22	-.8
Zincum	346.44	-.5	Scandium	0	-1.2
Zirconium	0	-.5	Tellurium	0	-1.7

Cupressus lawsoniana

Name	Value	Deviation	Name	Value	Deviation
Aluminium	3920	0	Stibium	2.39	6.2
Argentum	.2	4	Niccolum	52.95	4.1
Arsenicum	8.16	1.3	Argentum	.2	4
Barium	140.36	0	Lithium	21.35	3.9
Beryllium	.13	0	Indium	43.53	3.9
Bismuthum	0	-.4	Tellurium	.11	3.2
Borium	694.06	1.6	Zincum	2330	2.4
Cadmium	2.37	.4	Zirconium	2.2	1.8
Caesium	.24	-.3	Titanium	390.61	1.7
Calcium	415111.81	.9	Niobium	.33	1.7
Cerium	3.82	.1	Borium	694.06	1.6
Chromium	5.98	.5	Gallium	2.31	1.5
Cobaltum	5.83	1.4	Cobaltum	5.83	1.4
Cuprum	129.65	1.1	Tungstenium	.6	1.4
Dysprosium	.15	-.1	Plumbum	26.56	1.4
Erbium	.08	0	Arsenicum	8.16	1.3
Europium	.08	0	Cuprum	129.65	1.1
Ferrum	10816.19	.1	Hafnium	.06	1.1
Gadolinium	.34	0	Strontium	1490	1
Gallium	2.31	1.5	Calcium	415111.81	.9
Germanium	.11	.2	Manganum	15400	.8
Hafnium	.06	1.1	Tantalum	.01	.6
Holmium	.03	0	Chromium	5.98	.5
Indium	43.53	3.9	Cadmium	2.37	.4
Lanthanum	2.03	.3	Lanthanum	2.03	.3
Lithium	21.35	3.9	Scandium	1.45	.2
Lutetium	.01	.1	Germanium	.11	.2
Magnesium	24141.96	-.6	Praseodymium	.42	.2
Manganum	15400	.8	Vanadium	11.07	.1
Molybdenum	18.03	-.1	Ferrum	10816.19	.1
Natrium	6923.31	-.2	Cerium	3.82	.1
Neodymium	1.29	.1	Neodymium	1.29	.1
Niccolum	52.95	4.1	Thulium	.01	.1
Niobium	.33	1.7	Lutetium	.01	.1
Plumbum	26.56	1.4	Beryllium	.13	0
Praseodymium	.42	.2	Aluminium	3920	0
Rhenium	6.2	-.1	Selenium	2.18	0
Rubidium	109.6	-.2	Yttrium	.96	0
Samarium	.28	0	Barium	140.36	0
Scandium	1.45	.2	Samarium	.28	0
Selenium	2.18	0	Europium	.08	0
Silicium	0	-.4	Gadolinium	.34	0
Stibium	2.39	6.2	Terbium	.04	0
Strontium	1490	1	Holmium	.03	0
Tantalum	.01	.6	Erbium	.08	0
Tellurium	.11	3.2	Ytterbium	.06	0
Terbium	.04	0	Thorium	.34	0
Thallium	.02	-.5	Molybdenum	18.03	-.1
Thorium	.34	0	Dysprosium	.15	-.1
Thulium	.01	.1	Rhenium	6.2	-.1
Titanium	390.61	1.7	Uranium	.12	-.1
Tungstenium	.6	1.4	Natrium	6923.31	-.2
Uranium	.12	-.1	Rubidium	109.6	-.2
Vanadium	11.07	.1	Caesium	.24	-.3
Ytterbium	.06	0	Silicium	0	-.4
Yttrium	.96	0	Bismuthum	0	-.4
Zincum	2330	2.4	Thallium	.02	-.5
Zirconium	2.2	1.8	Magnesium	24141.96	-.6

Echinacea angustifolia

Name	Value	Deviation	Name	Value	Deviation
Aluminium	1960	-.4	Tungstenium	.69	1.8
Argentum	0	-.2	Borium	590.32	1.1
Arsenicum	1.43	-.7	Rhenium	13.01	.7
Barium	71.26	-.6	Calcium	347386.63	.4
Beryllium	.08	-.3	Strontium	1120	.3
Bismuthum	0	-.4	Scandium	1.28	.1
Borium	590.32	1.1	Chromium	3.34	0
Cadmium	.17	-.6	Lithium	3.66	-.1
Caesium	.16	-.4	Natrium	7776.43	-.1
Calcium	347386.63	.4	Manganum	301.49	-.2
Cerium	2.02	-.4	Ferrum	2023.96	-.2
Chromium	3.34	0	Selenium	1.86	-.2
Cobaltum	.99	-.5	Argentum	0	-.2
Cuprum	47.9	-.4	Beryllium	.08	-.3
Dysprosium	.08	-.4	Magnesium	38835.54	-.3
Erbium	.04	-.3	Vanadium	1.94	-.3
Europium	.04	-.5	Niccolum	5.98	-.3
Ferrum	2023.96	-.2	Germanium	0	-.3
Gadolinium	.16	-.4	Rubidium	61.78	-.3
Gallium	.62	-.5	Molybdenum	3.39	-.3
Germanium	0	-.3	Erbium	.04	-.3
Hafnium	.01	-.3	Hafnium	.01	-.3
Holmium	.01	-.5	Aluminium	1960	-.4
Indium	3.01	-.4	Silicium	0	-.4
Lanthanum	.72	-.4	Cuprum	47.9	-.4
Lithium	3.66	-.1	Yttrium	.44	-.4
Lutetium	0	-.6	Indium	3.01	-.4
Magnesium	38835.54	-.3	Tellurium	.03	-.4
Manganum	301.49	-.2	Caesium	.16	-.4
Molybdenum	3.39	-.3	Lanthanum	.72	-.4
Natrium	7776.43	-.1	Cerium	2.02	-.4
Neodymium	.55	-.4	Praseodymium	.17	-.4
Niccolum	5.98	-.3	Neodymium	.55	-.4
Niobium	.08	-.5	Samarium	.13	-.4
Plumbum	5.17	-.5	Gadolinium	.16	-.4
Praseodymium	.17	-.4	Terbium	.02	-.4
Rhenium	13.01	.7	Dysprosium	.08	-.4
Rubidium	61.78	-.3	Ytterbium	.03	-.4
Samarium	.13	-.4	Thallium	.05	-.4
Scandium	1.28	.1	Bismuthum	0	-.4
Selenium	1.86	-.2	Thorium	.16	-.4
Silicium	0	-.4	Uranium	.06	-.4
Stibium	.12	-.6	Cobaltum	.99	-.5
Strontium	1120	.3	Gallium	.62	-.5
Tantalum	0	-.7	Zirconium	0	-.5
Tellurium	.03	-.4	Niobium	.08	-.5
Terbium	.02	-.4	Europium	.04	-.5
Thallium	.05	-.4	Holmium	.01	-.5
Thorium	.16	-.4	Plumbum	5.17	-.5
Thulium	0	-.6	Cadmium	.17	-.6
Titanium	173.39	-.8	Stibium	.12	-.6
Tungstenium	.69	1.8	Barium	71.26	-.6
Uranium	.06	-.4	Thulium	0	-.6
Vanadium	1.94	-.3	Lutetium	0	-.6
Ytterbium	.03	-.4	Zincum	215.75	-.7
Yttrium	.44	-.4	Arsenicum	1.43	-.7
Zincum	215.75	-.7	Tantalum	0	-.7
Zirconium	0	-.5	Titanium	173.39	-.8

Echinacea purpurea

Name	Value	Deviation	Name	Value	Deviation
Aluminium	402	-.7	Magnesium	150000	2.1
Argentum	0	-.2	Borium	680.34	1.6
Arsenicum	0	-1.1	Calcium	418447.22	.9
Barium	62.95	-.7	Rhenium	14.71	.9
Beryllium	.05	-.5	Strontium	1280	.6
Bismuthum	0	-.4	Thallium	.26	.5
Borium	680.34	1.6	Scandium	1.39	.2
Cadmium	.16	-.6	Selenium	1.93	-.1
Caesium	.11	-.5	Manganum	565.24	-.2
Calcium	418447.22	.9	Ferrum	2135.29	-.2
Cerium	1.28	-.6	Argentum	0	-.2
Chromium	0	-.5	Niccolum	6.39	-.3
Cobaltum	1.38	-.4	Cuprum	53.61	-.3
Cuprum	53.61	-.3	Germanium	0	-.3
Dysprosium	.02	-.6	Molybdenum	6.86	-.3
Erbium	.01	-.6	Silicium	0	-.4
Europium	.02	-.7	Vanadium	0	-.4
Ferrum	2135.29	-.2	Cobaltum	1.38	-.4
Gadolinium	.06	-.6	Rubidium	23.21	-.4
Gallium	.28	-.9	Indium	2.62	-.4
Germanium	0	-.3	Tellurium	.03	-.4
Hafnium	0	-.6	Bismuthum	0	-.4
Holmium	0	-.7	Beryllium	.05	-.5
Indium	2.62	-.4	Natrium	1987.58	-.5
Lanthanum	.37	-.6	Chromium	0	-.5
Lithium	.7	-.8	Yttrium	.2	-.5
Lutetium	0	-.6	Zirconium	0	-.5
Magnesium	150000	2.1	Caesium	.11	-.5
Manganum	565.24	-.2	Terbium	.01	-.5
Molybdenum	6.86	-.3	Plumbum	4.76	-.5
Natrium	1987.58	-.5	Cadmium	.16	-.6
Neodymium	.18	-.6	Stibium	.13	-.6
Niccolum	6.39	-.3	Lanthanum	.37	-.6
Niobium	.04	-.8	Cerium	1.28	-.6
Plumbum	4.76	-.5	Praseodymium	.07	-.6
Praseodymium	.07	-.6	Neodymium	.18	-.6
Rhenium	14.71	.9	Samarium	.04	-.6
Rubidium	23.21	-.4	Gadolinium	.06	-.6
Samarium	.04	-.6	Dysprosium	.02	-.6
Scandium	1.39	.2	Erbium	.01	-.6
Selenium	1.93	-.1	Thulium	0	-.6
Silicium	0	-.4	Ytterbium	.01	-.6
Stibium	.13	-.6	Lutetium	0	-.6
Strontium	1280	.6	Hafnium	0	-.6
Tantalum	0	-.7	Tungstenium	0	-.6
Tellurium	.03	-.4	Uranium	.02	-.6
Terbium	.01	-.5	Aluminium	402	-.7
Thallium	.26	.5	Barium	62.95	-.7
Thorium	.04	-.7	Europium	.02	-.7
Thulium	0	-.6	Holmium	0	-.7
Titanium	128.61	-1.3	Tantalum	0	-.7
Tungstenium	0	-.6	Thorium	.04	-.7
Uranium	.02	-.6	Lithium	.7	-.8
Vanadium	0	-.4	Zincum	115.76	-.8
Ytterbium	.01	-.6	Niobium	.04	-.8
Yttrium	.2	-.5	Gallium	.28	-.9
Zincum	115.76	-.8	Arsenicum	0	-1.1
Zirconium	0	-.5	Titanium	128.61	-1.3

Escholtzia californica

Name	Value	Deviation	Name	Value	Deviation
Aluminium	3150	-.1	Tantalum	.01	.6
Argentum	0	-.2	Rhenium	10.87	.5
Arsenicum	0	-1.1	Rubidium	286.28	.3
Barium	109.91	-.3	Zincum	854.24	.2
Beryllium	.02	-.6	Aluminium	3150	-.1
Bismuthum	0	-.4	Caesium	.39	-.1
Borium	207.52	-.9	Manganum	487.42	-.2
Cadmium	.56	-.5	Ferrum	658.71	-.2
Caesium	.39	-.1	Selenium	1.72	-.2
Calcium	104626.29	-1.2	Molybdenum	10.34	-.2
Cerium	1.67	-.5	Argentum	0	-.2
Chromium	0	-.5	Germanium	0	-.3
Cobaltum	.55	-.7	Barium	109.91	-.3
Cuprum	41.42	-.5	Dysprosium	.09	-.3
Dysprosium	.09	-.3	Hafnium	.01	-.3
Erbium	.03	-.4	Silicium	0	-.4
Europium	.04	-.5	Scandium	.78	-.4
Ferrum	658.71	-.2	Titanium	201.28	-.4
Gadolinium	.17	-.4	Vanadium	0	-.4
Gallium	.71	-.4	Gallium	.71	-.4
Germanium	0	-.3	Neodymium	.47	-.4
Hafnium	.01	-.3	Samarium	.13	-.4
Holmium	.01	-.5	Gadolinium	.17	-.4
Indium	1.7	-.5	Terbium	.02	-.4
Lanthanum	.42	-.6	Erbium	.03	-.4
Lithium	.75	-.8	Thallium	.03	-.4
Lutetium	0	-.6	Bismuthum	0	-.4
Magnesium	25483.15	-.6	Uranium	.06	-.4
Manganum	487.42	-.2	Natrium	2157.27	-.5
Molybdenum	10.34	-.2	Chromium	0	-.5
Natrium	2157.27	-.5	Cuprum	41.42	-.5
Neodymium	.47	-.4	Yttrium	.28	-.5
Niccolum	0	-.8	Zirconium	0	-.5
Niobium	.08	-.5	Niobium	.08	-.5
Plumbum	2.07	-.8	Cadmium	.56	-.5
Praseodymium	.13	-.5	Indium	1.7	-.5
Rhenium	10.87	.5	Cerium	1.67	-.5
Rubidium	286.28	.3	Praseodymium	.13	-.5
Samarium	.13	-.4	Europium	.04	-.5
Scandium	.78	-.4	Holmium	.01	-.5
Selenium	1.72	-.2	Ytterbium	.02	-.5
Silicium	0	-.4	Thorium	.12	-.5
Stibium	.12	-.6	Beryllium	.02	-.6
Strontium	357	-1.1	Magnesium	25483.15	-.6
Tantalum	.01	.6	Stibium	.12	-.6
Tellurium	.01	-1.3	Lanthanum	.42	-.6
Terbium	.02	-.4	Thulium	0	-.6
Thallium	.03	-.4	Lutetium	0	-.6
Thorium	.12	-.5	Tungstenium	0	-.6
Thulium	0	-.6	Cobaltum	.55	-.7
Titanium	201.28	-.4	Lithium	.75	-.8
Tungstenium	0	-.6	Niccolum	0	-.8
Uranium	.06	-.4	Plumbum	2.07	-.8
Vanadium	0	-.4	Borium	207.52	-.9
Ytterbium	.02	-.5	Arsenicum	0	-1.1
Yttrium	.28	-.5	Strontium	357	-1.1
Zincum	854.24	.2	Calcium	104626.29	-1.2
Zirconium	0	-.5	Tellurium	.01	-1.3

Faba vulgaris

Name	Value	Deviation	Name	Value	Deviation
Aluminium	999	-.6	Tungstenium	.59	1.4
Argentum	0	-.2	Zincum	1130	.6
Arsenicum	4.15	.1	Natrium	16240.82	.5
Barium	26.63	-1	Tellurium	.05	.5
Beryllium	.04	-.5	Arsenicum	4.15	.1
Bismuthum	0	-.4	Molybdenum	28.01	.1
Borium	225.64	-.8	Rubidium	156.59	0
Cadmium	.3	-.6	Selenium	1.88	-.1
Caesium	.37	-.1	Caesium	.37	-.1
Calcium	236671.79	-.3	Manganum	424.03	-.2
Cerium	1.61	-.5	Ferrum	2560.69	-.2
Chromium	0	-.5	Argentum	0	-.2
Cobaltum	1.27	-.4	Rhenium	4.8	-.2
Cuprum	28.4	-.7	Thallium	.09	-.2
Dysprosium	.05	-.5	Calcium	236671.79	-.3
Erbium	.03	-.4	Germanium	0	-.3
Europium	.03	-.6	Indium	3.52	-.3
Ferrum	2560.69	-.2	Stibium	.23	-.3
Gadolinium	.11	-.5	Hafnium	.01	-.3
Gallium	.37	-.8	Plumbum	7.45	-.3
Germanium	0	-.3	Silicium	0	-.4
Hafnium	.01	-.3	Vanadium	0	-.4
Holmium	.01	-.5	Cobaltum	1.27	-.4
Indium	3.52	-.3	Yttrium	.36	-.4
Lanthanum	.51	-.5	Erbium	.03	-.4
Lithium	1.83	-.5	Bismuthum	0	-.4
Lutetium	0	-.6	Lithium	1.83	-.5
Magnesium	26642.78	-.5	Beryllium	.04	-.5
Manganum	424.03	-.2	Magnesium	26642.78	-.5
Molybdenum	28.01	.1	Scandium	.67	-.5
Natrium	16240.82	.5	Chromium	0	-.5
Neodymium	.38	-.5	Strontium	650	-.5
Niccolum	0	-.8	Zirconium	0	-.5
Niobium	.07	-.6	Lanthanum	.51	-.5
Plumbum	7.45	-.3	Cerium	1.61	-.5
Praseodymium	.11	-.5	Praseodymium	.11	-.5
Rhenium	4.8	-.2	Neodymium	.38	-.5
Rubidium	156.59	0	Samarium	.09	-.5
Samarium	.09	-.5	Gadolinium	.11	-.5
Scandium	.67	-.5	Terbium	.01	-.5
Selenium	1.88	-.1	Dysprosium	.05	-.5
Silicium	0	-.4	Holmium	.01	-.5
Stibium	.23	-.3	Ytterbium	.02	-.5
Strontium	650	-.5	Uranium	.04	-.5
Tantalum	0	-.7	Aluminium	999	-.6
Tellurium	.05	.5	Niobium	.07	-.6
Terbium	.01	-.5	Cadmium	.3	-.6
Thallium	.09	-.2	Europium	.03	-.6
Thorium	.08	-.6	Thulium	0	-.6
Thulium	0	-.6	Lutetium	0	-.6
Titanium	140.96	-1.1	Thorium	.08	-.6
Tungstenium	.59	1.4	Cuprum	28.4	-.7
Uranium	.04	-.5	Tantalum	0	-.7
Vanadium	0	-.4	Borium	225.64	-.8
Ytterbium	.02	-.5	Niccolum	0	-.8
Yttrium	.36	-.4	Gallium	.37	-.8
Zincum	1130	.6	Barium	26.63	-1
Zirconium	0	-.5	Titanium	140.96	-1.1

Fagopyrum esculentum

Name	Value	Deviation	Name	Value	Deviation
Aluminium	804	-.6	Calcium	345075.14	.4
Argentum	0	-.2	Strontium	1180	.4
Arsenicum	1.6	-.7	Rhenium	10.31	.4
Barium	67.43	-.6	Selenium	2.22	.1
Beryllium	.1	-.2	Tellurium	.04	.1
Bismuthum	0	-.4	Thallium	.14	0
Borium	173.41	-1.1	Molybdenum	13.66	-.1
Cadmium	1.34	-.1	Cadmium	1.34	-.1
Caesium	.15	-.5	Beryllium	.1	-.2
Calcium	345075.14	.4	Manganum	273.1	-.2
Cerium	1.19	-.7	Ferrum	1835.34	-.2
Chromium	0	-.5	Rubidium	96	-.2
Cobaltum	.9	-.6	Argentum	0	-.2
Cuprum	5.29	-1.1	Magnesium	37809.67	-.3
Dysprosium	.03	-.6	Germanium	0	-.3
Erbium	.02	-.5	Silicium	0	-.4
Europium	.02	-.7	Vanadium	0	-.4
Ferrum	1835.34	-.2	Bismuthum	0	-.4
Gadolinium	.07	-.6	Natrium	2084.61	-.5
Gallium	.38	-.8	Titanium	200.97	-.5
Germanium	0	-.3	Chromium	0	-.5
Hafnium	0	-.6	Yttrium	.23	-.5
Holmium	.01	-.5	Zirconium	0	-.5
Indium	0	-.7	Caesium	.15	-.5
Lanthanum	.31	-.6	Terbium	.01	-.5
Lithium	0	-1	Holmium	.01	-.5
Lutetium	0	-.6	Erbium	.02	-.5
Magnesium	37809.67	-.3	Uranium	.03	-.5
Manganum	273.1	-.2	Aluminium	804	-.6
Molybdenum	13.66	-.1	Cobaltum	.9	-.6
Natrium	2084.61	-.5	Stibium	.13	-.6
Neodymium	.2	-.6	Barium	67.43	-.6
Niccolum	0	-.8	Lanthanum	.31	-.6
Niobium	.04	-.8	Praseodymium	.06	-.6
Plumbum	3.33	-.6	Neodymium	.2	-.6
Praseodymium	.06	-.6	Samarium	.05	-.6
Rhenium	10.31	.4	Gadolinium	.07	-.6
Rubidium	96	-.2	Dysprosium	.03	-.6
Samarium	.05	-.6	Thulium	0	-.6
Scandium	0	-1.2	Ytterbium	.01	-.6
Selenium	2.22	.1	Lutetium	0	-.6
Silicium	0	-.4	Hafnium	0	-.6
Stibium	.13	-.6	Tungstenium	0	-.6
Strontium	1180	.4	Plumbum	3.33	-.6
Tantalum	0	-.7	Zincum	175.03	-.7
Tellurium	.04	.1	Arsenicum	1.6	-.7
Terbium	.01	-.5	Indium	0	-.7
Thallium	.14	0	Cerium	1.19	-.7
Thorium	.04	-.7	Europium	.02	-.7
Thulium	0	-.6	Tantalum	0	-.7
Titanium	200.97	-.5	Thorium	.04	-.7
Tungstenium	0	-.6	Niccolum	0	-.8
Uranium	.03	-.5	Gallium	.38	-.8
Vanadium	0	-.4	Niobium	.04	-.8
Ytterbium	.01	-.6	Lithium	0	-1
Yttrium	.23	-.5	Borium	173.41	-1.1
Zincum	175.03	-.7	Cuprum	5.29	-1.1
Zirconium	0	-.5	Scandium	0	-1.2

Fraxinus excelsior

Name	Value	Deviation	Name	Value	Deviation
Aluminium	2020	-.4	Calcium	629499.14	2.3
Argentum	0	-.2	Strontium	2250	2.3
Arsenicum	1.77	-.6	Plumbum	20.58	.9
Barium	194.45	.4	Zincum	1140	.7
Beryllium	.07	-.3	Stibium	.52	.6
Bismuthum	0	-.4	Barium	194.45	.4
Borium	305.37	-.4	Indium	8.96	.3
Cadmium	.72	-.4	Chromium	3.9	.1
Caesium	.08	-.6	Thulium	.01	.1
Calcium	629499.14	2.3	Lutetium	.01	.1
Cerium	2.6	-.2	Ferrum	7759.64	0
Chromium	3.9	.1	Vanadium	5.18	-.1
Cobaltum	1.96	-.1	Cobaltum	1.96	-.1
Cuprum	44.86	-.4	Lanthanum	1.2	-.1
Dysprosium	.1	-.3	Praseodymium	.28	-.1
Erbium	.04	-.3	Manganum	350.61	-.2
Europium	.06	-.3	Argentum	0	-.2
Ferrum	7759.64	0	Cerium	2.6	-.2
Gadolinium	.21	-.3	Neodymium	.78	-.2
Gallium	.45	-.7	Holmium	.02	-.2
Germanium	0	-.3	Beryllium	.07	-.3
Hafnium	0	-.6	Germanium	0	-.3
Holmium	.02	-.2	Yttrium	.59	-.3
Indium	8.96	.3	Niobium	.1	-.3
Lanthanum	1.2	-.1	Molybdenum	3.61	-.3
Lithium	1.52	-.6	Samarium	.17	-.3
Lutetium	.01	.1	Europium	.06	-.3
Magnesium	9778.88	-.9	Gadolinium	.21	-.3
Manganum	350.61	-.2	Dysprosium	.1	-.3
Molybdenum	3.61	-.3	Erbium	.04	-.3
Natrium	2779.29	-.5	Ytterbium	.04	-.3
Neodymium	.78	-.2	Borium	305.37	-.4
Niccolum	0	-.8	Aluminium	2020	-.4
Niobium	.1	-.3	Silicium	0	-.4
Plumbum	20.58	.9	Cuprum	44.86	-.4
Praseodymium	.28	-.1	Rubidium	13.53	-.4
Rhenium	.54	-.7	Cadmium	.72	-.4
Rubidium	13.53	-.4	Tellurium	.03	-.4
Samarium	.17	-.3	Terbium	.02	-.4
Scandium	.62	-.6	Thallium	.04	-.4
Selenium	0	-1.3	Bismuthum	0	-.4
Silicium	0	-.4	Thorium	.17	-.4
Stibium	.52	.6	Uranium	.06	-.4
Strontium	2250	2.3	Natrium	2779.29	-.5
Tantalum	0	-.7	Zirconium	0	-.5
Tellurium	.03	-.4	Lithium	1.52	-.6
Terbium	.02	-.4	Scandium	.62	-.6
Thallium	.04	-.4	Arsenicum	1.77	-.6
Thorium	.17	-.4	Caesium	.08	-.6
Thulium	.01	.1	Hafnium	0	-.6
Titanium	98.28	-1.6	Tungstenium	0	-.6
Tungstenium	0	-.6	Gallium	.45	-.7
Uranium	.06	-.4	Tantalum	0	-.7
Vanadium	5.18	-.1	Rhenium	.54	-.7
Ytterbium	.04	-.3	Niccolum	0	-.8
Yttrium	.59	-.3	Magnesium	9778.88	-.9
Zincum	1140	.7	Selenium	0	-1.3
Zirconium	0	-.5	Titanium	98.28	-1.6

Galium aparine

Name	Value	Deviation	Name	Value	Deviation
Aluminium	1950	-.4	Barium	317.87	1.4
Argentum	0	-.2	Zirconium	.66	.2
Arsenicum	4.22	.1	Molybdenum	31.34	.2
Barium	317.87	1.4	Scandium	1.32	.1
Beryllium	.07	-.3	Chromium	3.42	.1
Bismuthum	0	-.4	Arsenicum	4.22	.1
Borium	193.8	-1	Europium	.09	.1
Cadmium	.94	-.3	Thulium	.01	.1
Caesium	.2	-.4	Lutetium	.01	.1
Calcium	274154.23	-.1	Thallium	.17	.1
Cerium	2.85	-.2	Strontium	951	0
Chromium	3.42	.1	Hafnium	.02	0
Cobaltum	.95	-.5	Calcium	274154.23	-.1
Cuprum	47.2	-.4	Manganum	1020	-.1
Dysprosium	.1	-.3	Lanthanum	1.2	-.1
Erbium	.05	-.3	Praseodymium	.27	-.1
Europium	.09	.1	Plumbum	9.04	-.1
Ferrum	2282.85	-.2	Vanadium	4.35	-.2
Gadolinium	.22	-.2	Ferrum	2282.85	-.2
Gallium	.62	-.5	Rubidium	81.98	-.2
Germanium	0	-.3	Yttrium	.72	-.2
Hafnium	.02	0	Argentum	0	-.2
Holmium	.02	-.2	Indium	4.46	-.2
Indium	4.46	-.2	Cerium	2.85	-.2
Lanthanum	1.2	-.1	Neodymium	.84	-.2
Lithium	1.96	-.5	Samarium	.2	-.2
Lutetium	.01	.1	Gadolinium	.22	-.2
Magnesium	21163.09	-.6	Terbium	.03	-.2
Manganum	1020	-.1	Holmium	.02	-.2
Molybdenum	31.34	.2	Uranium	.1	-.2
Natrium	1693.98	-.5	Beryllium	.07	-.3
Neodymium	.84	-.2	Germanium	0	-.3
Niccolum	0	-.8	Cadmium	.94	-.3
Niobium	.09	-.4	Dysprosium	.1	-.3
Plumbum	9.04	-.1	Erbium	.05	-.3
Praseodymium	.27	-.1	Ytterbium	.04	-.3
Rhenium	2.35	-.5	Thorium	.21	-.3
Rubidium	81.98	-.2	Aluminium	1950	-.4
Samarium	.2	-.2	Silicium	0	-.4
Scandium	1.32	.1	Cuprum	47.2	-.4
Selenium	1.48	-.4	Selenium	1.48	-.4
Silicium	0	-.4	Niobium	.09	-.4
Stibium	.07	-.8	Tellurium	.03	-.4
Strontium	951	0	Caesium	.2	-.4
Tantalum	0	-.7	Bismuthum	0	-.4
Tellurium	.03	-.4	Lithium	1.96	-.5
Terbium	.03	-.2	Natrium	1693.98	-.5
Thallium	.17	.1	Cobaltum	.95	-.5
Thorium	.21	-.3	Zincum	375.68	-.5
Thulium	.01	.1	Gallium	.62	-.5
Titanium	185.08	-.6	Rhenium	2.35	-.5
Tungstenium	0	-.6	Magnesium	21163.09	-.6
Uranium	.1	-.2	Titanium	185.08	-.6
Vanadium	4.35	-.2	Tungstenium	0	-.6
Ytterbium	.04	-.3	Tantalum	0	-.7
Yttrium	.72	-.2	Niccolum	0	-.8
Zincum	375.68	-.5	Stibium	.07	-.8
Zirconium	.66	.2	Borium	193.8	-1

Glechoma hederacea

Name	Value	Deviation	Name	Value	Deviation
Aluminium	1910	-.4	Barium	601.57	3.6
Argentum	0	-.2	Selenium	4.89	1.7
Arsenicum	0	-1.1	Tellurium	.06	1
Barium	601.57	3.6	Zincum	1310	.9
Beryllium	.06	-.4	Stibium	.59	.8
Bismuthum	0	-.4	Europium	.12	.4
Borium	202.58	-1	Germanium	.11	.2
Cadmium	.33	-.6	Cuprum	72.2	.1
Caesium	.22	-.3	Indium	7.18	.1
Calcium	182997.94	-.7	Lutetium	.01	.1
Cerium	2.54	-.3	Rhenium	7.44	.1
Chromium	0	-.5	Manganum	1180	-.1
Cobaltum	1.35	-.4	Niccolum	7.82	-.1
Cuprum	72.2	.1	Niobium	.12	-.1
Dysprosium	.08	-.4	Plumbum	8.97	-.1
Erbium	.04	-.3	Titanium	219.08	-.2
Europium	.12	.4	Ferrum	2078.66	-.2
Ferrum	2078.66	-.2	Rubidium	102.01	-.2
Gadolinium	.18	-.3	Molybdenum	10.26	-.2
Gallium	.63	-.5	Argentum	0	-.2
Germanium	.11	.2	Lanthanum	1.03	-.2
Hafnium	0	-.6	Natrium	5046.87	-.3
Holmium	.01	-.5	Vanadium	1.16	-.3
Indium	7.18	.1	Yttrium	.52	-.3
Lanthanum	1.03	-.2	Caesium	.22	-.3
Lithium	2.03	-.5	Cerium	2.54	-.3
Lutetium	.01	.1	Praseodymium	.22	-.3
Magnesium	29086.17	-.5	Neodymium	.71	-.3
Manganum	1180	-.1	Samarium	.15	-.3
Molybdenum	10.26	-.2	Gadolinium	.18	-.3
Natrium	5046.87	-.3	Erbium	.04	-.3
Neodymium	.71	-.3	Thallium	.07	-.3
Niccolum	7.82	-.1	Thorium	.19	-.3
Niobium	.12	-.1	Uranium	.07	-.3
Plumbum	8.97	-.1	Beryllium	.06	-.4
Praseodymium	.22	-.3	Aluminium	1910	-.4
Rhenium	7.44	.1	Silicium	0	-.4
Rubidium	102.01	-.2	Scandium	.85	-.4
Samarium	.15	-.3	Cobaltum	1.35	-.4
Scandium	.85	-.4	Terbium	.02	-.4
Selenium	4.89	1.7	Dysprosium	.08	-.4
Silicium	0	-.4	Ytterbium	.03	-.4
Stibium	.59	.8	Bismuthum	0	-.4
Strontium	641	-.5	Lithium	2.03	-.5
Tantalum	0	-.7	Magnesium	29086.17	-.5
Tellurium	.06	1	Chromium	0	-.5
Terbium	.02	-.4	Gallium	.63	-.5
Thallium	.07	-.3	Strontium	641	-.5
Thorium	.19	-.3	Zirconium	0	-.5
Thulium	0	-.6	Holmium	.01	-.5
Titanium	219.08	-.2	Cadmium	.33	-.6
Tungstenium	0	-.6	Thulium	0	-.6
Uranium	.07	-.3	Hafnium	0	-.6
Vanadium	1.16	-.3	Tungstenium	0	-.6
Ytterbium	.03	-.4	Calcium	182997.94	-.7
Yttrium	.52	-.3	Tantalum	0	-.7
Zincum	1310	.9	Borium	202.58	-1
Zirconium	0	-.5	Arsenicum	0	-1.1

Gnaphalium leontopodium

Name	Value	Deviation	Name	Value	Deviation
Aluminium	8770	1.1	Silicium	7297.85	2.6
Argentum	0	-.2	Niobium	.42	2.5
Arsenicum	6.02	.7	Tantalum	.02	2
Barium	176.54	.2	Scandium	3.13	1.9
Beryllium	.3	.9	Thulium	.03	1.4
Bismuthum	0	-.4	Lutetium	.03	1.4
Borium	295.37	-.5	Tungstenium	.56	1.3
Cadmium	.5	-.5	Lanthanum	3.61	1.2
Caesium	.89	.7	Cerium	7.22	1.2
Calcium	229674.87	-.4	Praseodymium	.83	1.2
Cerium	7.22	1.2	Ytterbium	.18	1.2
Chromium	5.01	.3	Thorium	.83	1.2
Cobaltum	2.84	.2	Aluminium	8770	1.1
Cuprum	35.79	-.6	Yttrium	2.54	1.1
Dysprosium	.39	1	Neodymium	2.71	1
Erbium	.21	1	Samarium	.66	1
Europium	.16	.9	Gadolinium	.75	1
Ferrum	8184.59	0	Dysprosium	.39	1
Gadolinium	.75	1	Holmium	.07	1
Gallium	1.49	.5	Erbium	.21	1
Germanium	.13	.3	Beryllium	.3	.9
Hafnium	.01	-.3	Europium	.16	.9
Holmium	.07	1	Terbium	.08	.8
Indium	11.3	.5	Titanium	296.82	.7
Lanthanum	3.61	1.2	Arsenicum	6.02	.7
Lithium	5.5	.3	Selenium	3.28	.7
Lutetium	.03	1.4	Caesium	.89	.7
Magnesium	16972.49	-.7	Gallium	1.49	.5
Manganum	945.1	-.1	Indium	11.3	.5
Molybdenum	24.8	.1	Uranium	.23	.4
Natrium	6362.67	-.2	Lithium	5.5	.3
Neodymium	2.71	1	Chromium	5.01	.3
Niccolum	9.09	0	Germanium	.13	.3
Niobium	.42	2.5	Cobaltum	2.84	.2
Plumbum	12.53	.2	Barium	176.54	.2
Praseodymium	.83	1.2	Plumbum	12.53	.2
Rhenium	2.45	-.5	Molybdenum	24.8	.1
Rubidium	185.34	0	Vanadium	8.3	0
Samarium	.66	1	Ferrum	8184.59	0
Scandium	3.13	1.9	Niccolum	9.09	0
Selenium	3.28	.7	Rubidium	185.34	0
Silicium	7297.85	2.6	Manganum	945.1	-.1
Stibium	.18	-.4	Natrium	6362.67	-.2
Strontium	595	-.6	Argentum	0	-.2
Tantalum	.02	2	Hafnium	.01	-.3
Tellurium	.03	-.4	Calcium	229674.87	-.4
Terbium	.08	.8	Stibium	.18	-.4
Thallium	.05	-.4	Tellurium	.03	-.4
Thorium	.83	1.2	Thallium	.05	-.4
Thulium	.03	1.4	Bismuthum	0	-.4
Titanium	296.82	.7	Borium	295.37	-.5
Tungstenium	.56	1.3	Zirconium	0	-.5
Uranium	.23	.4	Cadmium	.5	-.5
Vanadium	8.3	0	Rhenium	2.45	-.5
Ytterbium	.18	1.2	Cuprum	35.79	-.6
Yttrium	2.54	1.1	Strontium	595	-.6
Zincum	200.16	-.7	Magnesium	16972.49	-.7
Zirconium	0	-.5	Zincum	200.16	-.7

Hedera helix

Name	Value	Deviation	Name	Value	Deviation
Aluminium	1420	-.5	Argentum	.23	4.7
Argentum	.23	4.7	Stibium	.89	1.7
Arsenicum	1.41	-.7	Barium	323.8	1.4
Barium	323.8	1.4	Calcium	426983.2	.9
Beryllium	.07	-.3	Zincum	1120	.6
Bismuthum	0	-.4	Indium	12.58	.6
Borium	373.76	-.1	Tellurium	.05	.5
Cadmium	2	.2	Strontium	1190	.4
Caesium	.52	.1	Lithium	5.42	.3
Calcium	426983.2	.9	Germanium	.1	.2
Cerium	2.18	-.4	Cadmium	2	.2
Chromium	0	-.5	Natrium	11050.82	.1
Cobaltum	1.59	-.3	Manganum	4790	.1
Cuprum	74.16	.1	Cuprum	74.16	.1
Dysprosium	.07	-.4	Rubidium	198.02	.1
Erbium	.04	-.3	Caesium	.52	.1
Europium	.08	0	Lutetium	.01	.1
Ferrum	3837.81	-.1	Europium	.08	0
Gadolinium	.17	-.4	Borium	373.76	-.1
Gallium	.78	-.3	Ferrum	3837.81	-.1
Germanium	.1	.2	Niccolum	8.3	-.1
Hafnium	.01	-.3	Plumbum	9.13	-.1
Holmium	.01	-.5	Vanadium	2.8	-.2
Indium	12.58	.6	Niobium	.11	-.2
Lanthanum	.9	-.3	Beryllium	.07	-.3
Lithium	5.42	.3	Magnesium	38933.63	-.3
Lutetium	.01	.1	Scandium	.92	-.3
Magnesium	38933.63	-.3	Titanium	210.5	-.3
Manganum	4790	.1	Cobaltum	1.59	-.3
Molybdenum	4.9	-.3	Gallium	.78	-.3
Natrium	11050.82	.1	Yttrium	.52	-.3
Neodymium	.57	-.4	Molybdenum	4.9	-.3
Niccolum	8.3	-.1	Lanthanum	.9	-.3
Niobium	.11	-.2	Praseodymium	.19	-.3
Plumbum	9.13	-.1	Erbium	.04	-.3
Praseodymium	.19	-.3	Hafnium	.01	-.3
Rhenium	2.36	-.5	Thorium	.18	-.3
Rubidium	198.02	.1	Uranium	.08	-.3
Samarium	.14	-.4	Silicium	0	-.4
Scandium	.92	-.3	Selenium	1.52	-.4
Selenium	1.52	-.4	Cerium	2.18	-.4
Silicium	0	-.4	Neodymium	.57	-.4
Stibium	.89	1.7	Samarium	.14	-.4
Strontium	1190	.4	Gadolinium	.17	-.4
Tantalum	0	-.7	Terbium	.02	-.4
Tellurium	.05	.5	Dysprosium	.07	-.4
Terbium	.02	-.4	Ytterbium	.03	-.4
Thallium	.04	-.4	Thallium	.04	-.4
Thorium	.18	-.3	Bismuthum	0	-.4
Thulium	0	-.6	Aluminium	1420	-.5
Titanium	210.5	-.3	Chromium	0	-.5
Tungstenium	0	-.6	Zirconium	0	-.5
Uranium	.08	-.3	Holmium	.01	-.5
Vanadium	2.8	-.2	Rhenium	2.36	-.5
Ytterbium	.03	-.4	Thulium	0	-.6
Yttrium	.52	-.3	Tungstenium	0	-.6
Zincum	1120	.6	Arsenicum	1.41	-.7
Zirconium	0	-.5	Tantalum	0	-.7

Hydrophyllum virginicum

Name	Value	Deviation	Name	Value	Deviation
Aluminium	779	-.6	Arsenicum	9.22	1.7
Argentum	0	-.2	Tellurium	.06	1
Arsenicum	9.22	1.7	Barium	233.79	.7
Barium	233.79	.7	Cuprum	91.01	.4
Beryllium	.05	-.5	Titanium	258.95	.2
Bismuthum	0	-.4	Niccolum	10.07	.1
Borium	348.8	-.2	Stibium	.37	.1
Cadmium	.27	-.6	Rhenium	7.58	.1
Caesium	.16	-.4	Selenium	2.17	0
Calcium	235811.97	-.3	Indium	6.73	0
Cerium	1.54	-.6	Borium	348.8	-.2
Chromium	0	-.5	Scandium	1.06	-.2
Cobaltum	.77	-.6	Manganum	464.69	-.2
Cuprum	91.01	.4	Ferrum	1748.18	-.2
Dysprosium	.05	-.5	Argentum	0	-.2
Erbium	.02	-.5	Magnesium	39268.75	-.3
Europium	.05	-.4	Calcium	235811.97	-.3
Ferrum	1748.18	-.2	Germanium	0	-.3
Gadolinium	.09	-.5	Rubidium	53.23	-.3
Gallium	.58	-.6	Strontium	768	-.3
Germanium	0	-.3	Molybdenum	6.53	-.3
Hafnium	0	-.6	Thallium	.06	-.3
Holmium	.01	-.5	Silicium	0	-.4
Indium	6.73	0	Vanadium	0	-.4
Lanthanum	.46	-.6	Yttrium	.34	-.4
Lithium	.56	-.8	Caesium	.16	-.4
Lutetium	0	-.6	Europium	.05	-.4
Magnesium	39268.75	-.3	Plumbum	6.17	-.4
Manganum	464.69	-.2	Bismuthum	0	-.4
Molybdenum	6.53	-.3	Beryllium	.05	-.5
Natrium	363.28	-.6	Chromium	0	-.5
Neodymium	.32	-.5	Zirconium	0	-.5
Niccolum	10.07	.1	Neodymium	.32	-.5
Niobium	.07	-.6	Samarium	.08	-.5
Plumbum	6.17	-.4	Gadolinium	.09	-.5
Praseodymium	.09	-.6	Terbium	.01	-.5
Rhenium	7.58	.1	Dysprosium	.05	-.5
Rubidium	53.23	-.3	Holmium	.01	-.5
Samarium	.08	-.5	Erbium	.02	-.5
Scandium	1.06	-.2	Ytterbium	.02	-.5
Selenium	2.17	0	Natrium	363.28	-.6
Silicium	0	-.4	Aluminium	779	-.6
Stibium	.37	.1	Cobaltum	.77	-.6
Strontium	768	-.3	Zincum	278.9	-.6
Tantalum	0	-.7	Gallium	.58	-.6
Tellurium	.06	1	Niobium	.07	-.6
Terbium	.01	-.5	Cadmium	.27	-.6
Thallium	.06	-.3	Lanthanum	.46	-.6
Thorium	.09	-.6	Cerium	1.54	-.6
Thulium	0	-.6	Praseodymium	.09	-.6
Titanium	258.95	.2	Thulium	0	-.6
Tungstenium	0	-.6	Lutetium	0	-.6
Uranium	.02	-.6	Hafnium	0	-.6
Vanadium	0	-.4	Tungstenium	0	-.6
Ytterbium	.02	-.5	Thorium	.09	-.6
Yttrium	.34	-.4	Uranium	.02	-.6
Zincum	278.9	-.6	Tantalum	0	-.7
Zirconium	0	-.5	Lithium	.56	-.8

Hyoscyamus niger

Name	Value	Deviation	Name	Value	Deviation
Aluminium	2430	-.3	Lithium	10.69	1.5
Argentum	0	-.2	Tellurium	.06	1
Arsenicum	1.15	-.8	Tantalum	.01	.6
Barium	89.14	-.5	Niobium	.2	.5
Beryllium	.06	-.4	Zirconium	.8	.3
Bismuthum	0	-.4	Thulium	.01	.1
Borium	156.66	-1.2	Lutetium	.01	.1
Cadmium	1.39	-.1	Hafnium	.02	0
Caesium	.22	-.3	Scandium	1.09	-.1
Calcium	173088.17	-.7	Selenium	1.95	-.1
Cerium	2.62	-.2	Rubidium	125.74	-.1
Chromium	0	-.5	Cadmium	1.39	-.1
Cobaltum	1.09	-.5	Rhenium	5.86	-.1
Cuprum	25.91	-.8	Uranium	.11	-.1
Dysprosium	.1	-.3	Vanadium	3.01	-.2
Erbium	.05	-.3	Manganum	201.53	-.2
Europium	.06	-.3	Ferrum	1818.91	-.2
Ferrum	1818.91	-.2	Argentum	0	-.2
Gadolinium	.21	-.3	Lanthanum	1.08	-.2
Gallium	.6	-.5	Cerium	2.62	-.2
Germanium	0	-.3	Praseodymium	.25	-.2
Hafnium	.02	0	Neodymium	.82	-.2
Holmium	.02	-.2	Samarium	.19	-.2
Indium	4.02	-.3	Terbium	.03	-.2
Lanthanum	1.08	-.2	Holmium	.02	-.2
Lithium	10.69	1.5	Thallium	.1	-.2
Lutetium	.01	.1	Thorium	.23	-.2
Magnesium	29484.98	-.5	Aluminium	2430	-.3
Manganum	201.53	-.2	Germanium	0	-.3
Molybdenum	4.47	-.3	Yttrium	.61	-.3
Natrium	2082.52	-.5	Molybdenum	4.47	-.3
Neodymium	.82	-.2	Indium	4.02	-.3
Niccolum	0	-.8	Caesium	.22	-.3
Niobium	.2	.5	Europium	.06	-.3
Plumbum	5.7	-.4	Gadolinium	.21	-.3
Praseodymium	.25	-.2	Dysprosium	.1	-.3
Rhenium	5.86	-.1	Erbium	.05	-.3
Rubidium	125.74	-.1	Ytterbium	.04	-.3
Samarium	.19	-.2	Beryllium	.06	-.4
Scandium	1.09	-.1	Silicium	0	-.4
Selenium	1.95	-.1	Stibium	.21	-.4
Silicium	0	-.4	Plumbum	5.7	-.4
Stibium	.21	-.4	Bismuthum	0	-.4
Strontium	531	-.7	Natrium	2082.52	-.5
Tantalum	.01	.6	Magnesium	29484.98	-.5
Tellurium	.06	1	Chromium	0	-.5
Terbium	.03	-.2	Cobaltum	1.09	-.5
Thallium	.1	-.2	Gallium	.6	-.5
Thorium	.23	-.2	Barium	89.14	-.5
Thulium	.01	.1	Zincum	266.44	-.6
Titanium	158.38	-.9	Tungstenium	0	-.6
Tungstenium	0	-.6	Calcium	173088.17	-.7
Uranium	.11	-.1	Strontium	531	-.7
Vanadium	3.01	-.2	Niccolum	0	-.8
Ytterbium	.04	-.3	Cuprum	25.91	-.8
Yttrium	.61	-.3	Arsenicum	1.15	-.8
Zincum	266.44	-.6	Titanium	158.38	-.9
Zirconium	.8	.3	Borium	156.66	-1.2

Hyssopus officinalis

Name	Value	Deviation	Name	Value	Deviation
Aluminium	1670	-.5	Calcium	301928.79	.1
Argentum	0	-.2	Thallium	.16	.1
Arsenicum	2.64	-.3	Selenium	2.15	0
Barium	31.57	-.9	Strontium	883	-.1
Beryllium	.03	-.6	Molybdenum	17.11	-.1
Bismuthum	0	-.4	Rhenium	5.83	-.1
Borium	202.97	-1	Manganum	284.63	-.2
Cadmium	.31	-.6	Ferrum	1911.16	-.2
Caesium	.05	-.6	Argentum	0	-.2
Calcium	301928.79	.1	Germanium	0	-.3
Cerium	1.54	-.6	Arsenicum	2.64	-.3
Chromium	0	-.5	Stibium	.22	-.3
Cobaltum	.9	-.6	Silicium	0	-.4
Cuprum	37.46	-.6	Vanadium	0	-.4
Dysprosium	.07	-.4	Rubidium	25.64	-.4
Erbium	.03	-.4	Yttrium	.35	-.4
Europium	.03	-.6	Indium	3.08	-.4
Ferrum	1911.16	-.2	Samarium	.11	-.4
Gadolinium	.13	-.4	Gadolinium	.13	-.4
Gallium	.48	-.7	Terbium	.02	-.4
Germanium	0	-.3	Dysprosium	.07	-.4
Hafnium	0	-.6	Erbium	.03	-.4
Holmium	.01	-.5	Plumbum	6.25	-.4
Indium	3.08	-.4	Bismuthum	0	-.4
Lanthanum	.45	-.6	Magnesium	27858.46	-.5
Lithium	.85	-.8	Aluminium	1670	-.5
Lutetium	0	-.6	Scandium	.67	-.5
Magnesium	27858.46	-.5	Chromium	0	-.5
Manganum	284.63	-.2	Zirconium	0	-.5
Molybdenum	17.11	-.1	Praseodymium	.12	-.5
Natrium	1535.51	-.6	Neodymium	.4	-.5
Neodymium	.4	-.5	Holmium	.01	-.5
Niccolum	0	-.8	Ytterbium	.02	-.5
Niobium	.07	-.6	Thorium	.11	-.5
Plumbum	6.25	-.4	Uranium	.04	-.5
Praseodymium	.12	-.5	Beryllium	.03	-.6
Rhenium	5.83	-.1	Natrium	1535.51	-.6
Rubidium	25.64	-.4	Cobaltum	.9	-.6
Samarium	.11	-.4	Cuprum	37.46	-.6
Scandium	.67	-.5	Zincum	283.15	-.6
Selenium	2.15	0	Niobium	.07	-.6
Silicium	0	-.4	Cadmium	.31	-.6
Stibium	.22	-.3	Caesium	.05	-.6
Strontium	883	-.1	Lanthanum	.45	-.6
Tantalum	0	-.7	Cerium	1.54	-.6
Tellurium	.02	-.8	Europium	.03	-.6
Terbium	.02	-.4	Thulium	0	-.6
Thallium	.16	.1	Lutetium	0	-.6
Thorium	.11	-.5	Hafnium	0	-.6
Thulium	0	-.6	Tungstenium	0	-.6
Titanium	145.6	-1.1	Gallium	.48	-.7
Tungstenium	0	-.6	Tantalum	0	-.7
Uranium	.04	-.5	Lithium	.85	-.8
Vanadium	0	-.4	Niccolum	0	-.8
Ytterbium	.02	-.5	Tellurium	.02	-.8
Yttrium	.35	-.4	Barium	31.57	-.9
Zincum	283.15	-.6	Borium	202.97	-1
Zirconium	0	-.5	Titanium	145.6	-1.1

Lappa arctium

Name	Value	Deviation
Aluminium	4410	.1
Argentum	0	-.2
Arsenicum	2.16	-.5
Barium	61.3	-.7
Beryllium	.16	.1
Bismuthum	0	-.4
Borium	256.81	-.7
Cadmium	1.41	0
Caesium	.37	-.1
Calcium	147378.66	-.9
Cerium	4.28	.3
Chromium	0	-.5
Cobaltum	2.04	-.1
Cuprum	256.69	3.4
Dysprosium	.19	.1
Erbium	.1	.1
Europium	.08	0
Ferrum	4941.56	-.1
Gadolinium	.38	.1
Gallium	1.01	0
Germanium	.11	.2
Hafnium	.03	.3
Holmium	.03	0
Indium	6.38	0
Lanthanum	1.92	.2
Lithium	4.12	0
Lutetium	.01	.1
Magnesium	41057.72	-.2
Manganum	1060	-.1
Molybdenum	14.76	-.1
Natrium	3602.84	-.4
Neodymium	1.54	.2
Niccolum	15.38	.6
Niobium	.15	.1
Plumbum	11.43	.1
Praseodymium	.46	.3
Rhenium	4.16	-.3
Rubidium	107.71	-.2
Samarium	.33	.1
Scandium	1.24	0
Selenium	3.35	.8
Silicium	0	-.4
Stibium	.55	.7
Strontium	639	-.5
Tantalum	.01	.6
Tellurium	.04	.1
Terbium	.05	.2
Thallium	.08	-.2
Thorium	.44	.3
Thulium	.01	.1
Titanium	293.41	.6
Tungstenium	0	-.6
Uranium	.16	.1
Vanadium	6.92	0
Ytterbium	.08	.2
Yttrium	1.21	.2
Zincum	382.17	-.4
Zirconium	0	-.5

Name	Value	Deviation
Cuprum	256.69	3.4
Selenium	3.35	.8
Stibium	.55	.7
Titanium	293.41	.6
Niccolum	15.38	.6
Tantalum	.01	.6
Cerium	4.28	.3
Praseodymium	.46	.3
Hafnium	.03	.3
Thorium	.44	.3
Germanium	.11	.2
Yttrium	1.21	.2
Lanthanum	1.92	.2
Neodymium	1.54	.2
Terbium	.05	.2
Ytterbium	.08	.2
Beryllium	.16	.1
Aluminium	4410	.1
Niobium	.15	.1
Tellurium	.04	.1
Samarium	.33	.1
Gadolinium	.38	.1
Dysprosium	.19	.1
Erbium	.1	.1
Thulium	.01	.1
Lutetium	.01	.1
Plumbum	11.43	.1
Uranium	.16	.1
Lithium	4.12	0
Scandium	1.24	0
Vanadium	6.92	0
Gallium	1.01	0
Cadmium	1.41	0
Indium	6.38	0
Europium	.08	0
Holmium	.03	0
Manganum	1060	-.1
Ferrum	4941.56	-.1
Cobaltum	2.04	-.1
Molybdenum	14.76	-.1
Caesium	.37	-.1
Magnesium	41057.72	-.2
Rubidium	107.71	-.2
Argentum	0	-.2
Thallium	.08	-.2
Rhenium	4.16	-.3
Natrium	3602.84	-.4
Silicium	0	-.4
Zincum	382.17	-.4
Bismuthum	0	-.4
Chromium	0	-.5
Arsenicum	2.16	-.5
Strontium	639	-.5
Zirconium	0	-.5
Tungstenium	0	-.6
Borium	256.81	-.7
Barium	61.3	-.7
Calcium	147378.66	-.9

Lapsana communis

Name	Value	Deviation	Name	Value	Deviation
Aluminium	4330	.1	Niobium	.43	2.5
Argentum	0	-.2	Tantalum	.02	2
Arsenicum	1.78	-.6	Tungstenium	.57	1.3
Barium	75.34	-.6	Cadmium	3.72	1.1
Beryllium	.15	.1	Titanium	322.72	1
Bismuthum	0	-.4	Thallium	.36	.9
Borium	290.18	-.5	Lutetium	.02	.8
Cadmium	3.72	1.1	Scandium	1.87	.7
Caesium	.37	-.1	Thulium	.02	.7
Calcium	239345.74	-.3	Yttrium	1.82	.6
Cerium	4.68	.4	Terbium	.07	.6
Chromium	2	-.2	Ytterbium	.12	.6
Cobaltum	1.69	-.2	Tellurium	.05	.5
Cuprum	64.73	-.1	Lanthanum	2.3	.5
Dysprosium	.28	.5	Praseodymium	.56	.5
Erbium	.15	.5	Neodymium	1.9	.5
Europium	.11	.3	Samarium	.47	.5
Ferrum	6719.52	0	Dysprosium	.28	.5
Gadolinium	.52	.4	Holmium	.05	.5
Gallium	.99	-.1	Erbium	.15	.5
Germanium	0	-.3	Thorium	.54	.5
Hafnium	.03	.3	Cerium	4.68	.4
Holmium	.05	.5	Gadolinium	.52	.4
Indium	5.56	-.1	Vanadium	14.68	.3
Lanthanum	2.3	.5	Zincum	874.3	.3
Lithium	3.49	-.2	Europium	.11	.3
Lutetium	.02	.8	Hafnium	.03	.3
Magnesium	33802.8	-.4	Selenium	2.42	.2
Manganum	626.33	-.2	Zirconium	.64	.2
Molybdenum	17.3	-.1	Plumbum	12.52	.2
Natrium	4703.05	-.3	Beryllium	.15	.1
Neodymium	1.9	.5	Aluminium	4330	.1
Niccolum	6.46	-.3	Uranium	.17	.1
Niobium	.43	2.5	Ferrum	6719.52	0
Plumbum	12.52	.2	Cuprum	64.73	-.1
Praseodymium	.56	.5	Gallium	.99	-.1
Rhenium	.98	-.7	Rubidium	132.17	-.1
Rubidium	132.17	-.1	Molybdenum	17.3	-.1
Samarium	.47	.5	Indium	5.56	-.1
Scandium	1.87	.7	Caesium	.37	-.1
Selenium	2.42	.2	Lithium	3.49	-.2
Silicium	0	-.4	Chromium	2	-.2
Stibium	.16	-.5	Manganum	626.33	-.2
Strontium	748	-.4	Cobaltum	1.69	-.2
Tantalum	.02	2	Argentum	0	-.2
Tellurium	.05	.5	Natrium	4703.05	-.3
Terbium	.07	.6	Calcium	239345.74	-.3
Thallium	.36	.9	Niccolum	6.46	-.3
Thorium	.54	.5	Germanium	0	-.3
Thulium	.02	.7	Magnesium	33802.8	-.4
Titanium	322.72	1	Silicium	0	-.4
Tungstenium	.57	1.3	Strontium	748	-.4
Uranium	.17	.1	Bismuthum	0	-.4
Vanadium	14.68	.3	Borium	290.18	-.5
Ytterbium	.12	.6	Stibium	.16	-.5
Yttrium	1.82	.6	Arsenicum	1.78	-.6
Zincum	874.3	.3	Barium	75.34	-.6
Zirconium	.64	.2	Rhenium	.98	-.7

Laurocerasus -Prunus laurocerasus

Name	Value	Deviation	Name	Value	Deviation
Aluminium	1650	-.5	Magnesium	150000	2.1
Argentum	0	-.2	Cobaltum	4.96	1.1
Arsenicum	1.93	-.6	Calcium	394922.74	.7
Barium	94.85	-.4	Rubidium	437.07	.7
Beryllium	.04	-.5	Borium	491.92	.6
Bismuthum	0	-.4	Niccolum	13.8	.4
Borium	491.92	.6	Strontium	1090	.3
Cadmium	1.27	-.1	Zirconium	.54	.1
Caesium	.26	-.3	Thallium	.17	.1
Calcium	394922.74	.7	Manganum	3140	0
Cerium	1.35	-.6	Stibium	.33	0
Chromium	0	-.5	Titanium	231.81	-.1
Cobaltum	4.96	1.1	Cadmium	1.27	-.1
Cuprum	53.85	-.3	Ferrum	2573.33	-.2
Dysprosium	.05	-.5	Molybdenum	9.92	-.2
Erbium	.02	-.5	Argentum	0	-.2
Europium	.03	-.6	Vanadium	1.11	-.3
Ferrum	2573.33	-.2	Cuprum	53.85	-.3
Gadolinium	.1	-.5	Gallium	.77	-.3
Gallium	.77	-.3	Germanium	0	-.3
Germanium	0	-.3	Indium	3.5	-.3
Hafnium	.01	-.3	Caesium	.26	-.3
Holmium	.01	-.5	Hafnium	.01	-.3
Indium	3.5	-.3	Lithium	2.32	-.4
Lanthanum	.36	-.6	Silicium	0	-.4
Lithium	2.32	-.4	Zincum	409.36	-.4
Lutetium	0	-.6	Tellurium	.03	-.4
Magnesium	150000	2.1	Barium	94.85	-.4
Manganum	3140	0	Bismuthum	0	-.4
Molybdenum	9.92	-.2	Beryllium	.04	-.5
Natrium	1110.86	-.6	Aluminium	1650	-.5
Neodymium	.29	-.6	Chromium	0	-.5
Niccolum	13.8	.4	Yttrium	.2	-.5
Niobium	.05	-.7	Samarium	.08	-.5
Plumbum	4.57	-.5	Gadolinium	.1	-.5
Praseodymium	.08	-.6	Terbium	.01	-.5
Rhenium	2.43	-.5	Dysprosium	.05	-.5
Rubidium	437.07	.7	Holmium	.01	-.5
Samarium	.08	-.5	Erbium	.02	-.5
Scandium	.62	-.6	Rhenium	2.43	-.5
Selenium	0	-1.3	Plumbum	4.57	-.5
Silicium	0	-.4	Thorium	.1	-.5
Stibium	.33	0	Natrium	1110.86	-.6
Strontium	1090	.3	Scandium	.62	-.6
Tantalum	0	-.7	Arsenicum	1.93	-.6
Tellurium	.03	-.4	Lanthanum	.36	-.6
Terbium	.01	-.5	Cerium	1.35	-.6
Thallium	.17	.1	Praseodymium	.08	-.6
Thorium	.1	-.5	Neodymium	.29	-.6
Thulium	0	-.6	Europium	.03	-.6
Titanium	231.81	-.1	Thulium	0	-.6
Tungstenium	0	-.6	Ytterbium	.01	-.6
Uranium	.02	-.6	Lutetium	0	-.6
Vanadium	1.11	-.3	Tungstenium	0	-.6
Ytterbium	.01	-.6	Uranium	.02	-.6
Yttrium	.2	-.5	Niobium	.05	-.7
Zincum	409.36	-.4	Tantalum	0	-.7
Zirconium	.54	.1	Selenium	0	-1.3

Leonurus cardiaca

Name	Value	Deviation	Name	Value	Deviation
Aluminium	750	-.7	Rhenium	22.72	1.9
Argentum	0	-.2	Barium	307.24	1.3
Arsenicum	3.65	0	Selenium	3.19	.7
Barium	307.24	1.3	Tantalum	.01	.6
Beryllium	.04	-.5	Tellurium	.05	.5
Bismuthum	0	-.4	Stibium	.47	.4
Borium	273.92	-.6	Cuprum	82.55	.3
Cadmium	.2	-.6	Calcium	300123.85	.1
Caesium	.11	-.5	Zirconium	.57	.1
Calcium	300123.85	.1	Arsenicum	3.65	0
Cerium	1.38	-.6	Strontium	934	0
Chromium	0	-.5	Molybdenum	20.36	0
Cobaltum	.81	-.6	Manganum	355.76	-.2
Cuprum	82.55	.3	Ferrum	2067.61	-.2
Dysprosium	.04	-.5	Argentum	0	-.2
Erbium	.02	-.5	Europium	.07	-.2
Europium	.07	-.2	Natrium	4839.01	-.3
Ferrum	2067.61	-.2	Niccolum	5.67	-.3
Gadolinium	.09	-.5	Germanium	0	-.3
Gallium	.43	-.7	Rubidium	46.63	-.3
Germanium	0	-.3	Indium	4.15	-.3
Hafnium	.01	-.3	Hafnium	.01	-.3
Holmium	.01	-.5	Thallium	.06	-.3
Indium	4.15	-.3	Silicium	0	-.4
Lanthanum	.46	-.6	Scandium	.77	-.4
Lithium	.95	-.8	Vanadium	0	-.4
Lutetium	0	-.6	Plumbum	6.43	-.4
Magnesium	29649.29	-.5	Bismuthum	0	-.4
Manganum	355.76	-.2	Uranium	.05	-.4
Molybdenum	20.36	0	Beryllium	.04	-.5
Natrium	4839.01	-.3	Magnesium	29649.29	-.5
Neodymium	.28	-.6	Chromium	0	-.5
Niccolum	5.67	-.3	Yttrium	.26	-.5
Niobium	.08	-.5	Niobium	.08	-.5
Plumbum	6.43	-.4	Caesium	.11	-.5
Praseodymium	.09	-.6	Samarium	.07	-.5
Rhenium	22.72	1.9	Gadolinium	.09	-.5
Rubidium	46.63	-.3	Terbium	.01	-.5
Samarium	.07	-.5	Dysprosium	.04	-.5
Scandium	.77	-.4	Holmium	.01	-.5
Selenium	3.19	.7	Erbium	.02	-.5
Silicium	0	-.4	Ytterbium	.02	-.5
Stibium	.47	.4	Borium	273.92	-.6
Strontium	934	0	Cobaltum	.81	-.6
Tantalum	.01	.6	Zincum	267.8	-.6
Tellurium	.05	.5	Cadmium	.2	-.6
Terbium	.01	-.5	Lanthanum	.46	-.6
Thallium	.06	-.3	Cerium	1.38	-.6
Thorium	.09	-.6	Praseodymium	.09	-.6
Thulium	0	-.6	Neodymium	.28	-.6
Titanium	163.69	-.9	Thulium	0	-.6
Tungstenium	0	-.6	Lutetium	0	-.6
Uranium	.05	-.4	Tungstenium	0	-.6
Vanadium	0	-.4	Thorium	.09	-.6
Ytterbium	.02	-.5	Aluminium	750	-.7
Yttrium	.26	-.5	Gallium	.43	-.7
Zincum	267.8	-.6	Lithium	.95	-.8
Zirconium	.57	.1	Titanium	163.69	-.9

Lespedeza sieboldii

Name	Value	Deviation	Name	Value	Deviation
Aluminium	965	-.6	Molybdenum	530.57	9.1
Argentum	0	-.2	Barium	402.84	2
Arsenicum	3.9	0	Borium	742.51	1.9
Barium	402.84	2	Tungstenium	.55	1.3
Beryllium	.03	-.6	Calcium	379866.28	.6
Bismuthum	0	-.4	Cuprum	100.84	.6
Borium	742.51	1.9	Zincum	1090	.6
Cadmium	1.15	-.2	Strontium	1300	.6
Caesium	.1	-.5	Titanium	280.53	.5
Calcium	379866.28	.6	Stibium	.48	.5
Cerium	1.43	-.6	Germanium	.11	.2
Chromium	2.58	-.1	Niccolum	10.06	.1
Cobaltum	1.15	-.5	Rhenium	7.59	.1
Cuprum	100.84	.6	Scandium	1.26	0
Dysprosium	.03	-.6	Arsenicum	3.9	0
Erbium	.02	-.5	Europium	.08	0
Europium	.08	0	Chromium	2.58	-.1
Ferrum	3347.64	-.1	Ferrum	3347.64	-.1
Gadolinium	.08	-.6	Manganum	529.03	-.2
Gallium	.46	-.7	Argentum	0	-.2
Germanium	.11	.2	Cadmium	1.15	-.2
Hafnium	0	-.6	Indium	4.39	-.2
Holmium	.01	-.5	Vanadium	1.92	-.3
Indium	4.39	-.2	Plumbum	6.78	-.3
Lanthanum	.42	-.6	Silicium	0	-.4
Lithium	.79	-.8	Rubidium	28.89	-.4
Lutetium	0	-.6	Tellurium	.03	-.4
Magnesium	24808.94	-.6	Thallium	.04	-.4
Manganum	529.03	-.2	Bismuthum	0	-.4
Molybdenum	530.57	9.1	Natrium	1740.6	-.5
Natrium	1740.6	-.5	Cobaltum	1.15	-.5
Neodymium	.27	-.6	Yttrium	.19	-.5
Niccolum	10.06	.1	Zirconium	0	-.5
Niobium	.04	-.8	Caesium	.1	-.5
Plumbum	6.78	-.3	Terbium	.01	-.5
Praseodymium	.08	-.6	Holmium	.01	-.5
Rhenium	7.59	.1	Erbium	.02	-.5
Rubidium	28.89	-.4	Uranium	.03	-.5
Samarium	.06	-.6	Beryllium	.03	-.6
Scandium	1.26	0	Magnesium	24808.94	-.6
Selenium	0	-1.3	Aluminium	965	-.6
Silicium	0	-.4	Lanthanum	.42	-.6
Stibium	.48	.5	Cerium	1.43	-.6
Strontium	1300	.6	Praseodymium	.08	-.6
Tantalum	0	-.7	Neodymium	.27	-.6
Tellurium	.03	-.4	Samarium	.06	-.6
Terbium	.01	-.5	Gadolinium	.08	-.6
Thallium	.04	-.4	Dysprosium	.03	-.6
Thorium	.06	-.6	Thulium	0	-.6
Thulium	0	-.6	Ytterbium	.01	-.6
Titanium	280.53	.5	Lutetium	0	-.6
Tungstenium	.55	1.3	Hafnium	0	-.6
Uranium	.03	-.5	Thorium	.06	-.6
Vanadium	1.92	-.3	Gallium	.46	-.7
Ytterbium	.01	-.6	Tantalum	0	-.7
Yttrium	.19	-.5	Lithium	.79	-.8
Zincum	1090	.6	Niobium	.04	-.8
Zirconium	0	-.5	Selenium	0	-1.3

Linum usitatissimum

Name	Value	Deviation	Name	Value	Deviation
Aluminium	506	-.7	Natrium	60000	3.6
Argentum	0	-.2	Bismuthum	.08	2.7
Arsenicum	5.85	.6	Magnesium	150000	2.1
Barium	19.51	-1	Borium	589.77	1.1
Beryllium	.01	-.7	Thallium	.35	.9
Bismuthum	.08	2.7	Titanium	300.8	.7
Borium	589.77	1.1	Zincum	1140	.7
Cadmium	2.75	.6	Arsenicum	5.85	.6
Caesium	.15	-.5	Cadmium	2.75	.6
Calcium	221280.44	-.4	Tellurium	.04	.1
Cerium	1.28	-.6	Rhenium	6.28	-.1
Chromium	0	-.5	Manganum	256.43	-.2
Cobaltum	.78	-.6	Ferrum	1642.83	-.2
Cuprum	37	-.6	Niccolum	7.02	-.2
Dysprosium	.02	-.6	Molybdenum	12.94	-.2
Erbium	.01	-.6	Argentum	0	-.2
Europium	.01	-.8	Germanium	0	-.3
Ferrum	1642.83	-.2	Silicium	0	-.4
Gadolinium	.05	-.6	Calcium	221280.44	-.4
Gallium	.39	-.8	Vanadium	0	-.4
Germanium	0	-.3	Rubidium	39.73	-.4
Hafnium	0	-.6	Chromium	0	-.5
Holmium	0	-.7	Selenium	1.3	-.5
Indium	1.65	-.5	Zirconium	0	-.5
Lanthanum	.23	-.7	Indium	1.65	-.5
Lithium	.69	-.8	Stibium	.16	-.5
Lutetium	0	-.6	Caesium	.15	-.5
Magnesium	150000	2.1	Uranium	.04	-.5
Manganum	256.43	-.2	Cobaltum	.78	-.6
Molybdenum	12.94	-.2	Cuprum	37	-.6
Natrium	60000	3.6	Yttrium	.09	-.6
Neodymium	.15	-.6	Cerium	1.28	-.6
Niccolum	7.02	-.2	Neodymium	.15	-.6
Niobium	.02	-1	Gadolinium	.05	-.6
Plumbum	2.54	-.7	Dysprosium	.02	-.6
Praseodymium	.05	-.7	Erbium	.01	-.6
Rhenium	6.28	-.1	Thulium	0	-.6
Rubidium	39.73	-.4	Lutetium	0	-.6
Samarium	.02	-.7	Hafnium	0	-.6
Scandium	0	-1.2	Tungstenium	0	-.6
Selenium	1.3	-.5	Beryllium	.01	-.7
Silicium	0	-.4	Aluminium	506	-.7
Stibium	.16	-.5	Strontium	564	-.7
Strontium	564	-.7	Lanthanum	.23	-.7
Tantalum	0	-.7	Praseodymium	.05	-.7
Tellurium	.04	.1	Samarium	.02	-.7
Terbium	0	-.7	Terbium	0	-.7
Thallium	.35	.9	Holmium	0	-.7
Thorium	.03	-.7	Ytterbium	0	-.7
Thulium	0	-.6	Tantalum	0	-.7
Titanium	300.8	.7	Plumbum	2.54	-.7
Tungstenium	0	-.6	Thorium	.03	-.7
Uranium	.04	-.5	Lithium	.69	-.8
Vanadium	0	-.4	Gallium	.39	-.8
Ytterbium	0	-.7	Europium	.01	-.8
Yttrium	.09	-.6	Niobium	.02	-1
Zincum	1140	.7	Barium	19.51	-1
Zirconium	0	-.5	Scandium	0	-1.2

Lycopus europaeus

Name	Value	Deviation	Name	Value	Deviation
Aluminium	1160	-.6	Tungstenium	.62	1.5
Argentum	0	-.2	Zincum	1460	1.1
Arsenicum	1.77	-.6	Borium	424.72	.2
Barium	154.84	.1	Cobaltum	2.65	.1
Beryllium	.03	-.6	Niccolum	10.02	.1
Bismuthum	0	-.4	Cuprum	76.08	.1
Borium	424.72	.2	Barium	154.84	.1
Cadmium	.17	-.6	Stibium	.34	0
Caesium	.06	-.6	Manganum	807.07	-.1
Calcium	209251.43	-.5	Molybdenum	14.31	-.1
Cerium	1.48	-.6	Ferrum	1569.71	-.2
Chromium	0	-.5	Argentum	0	-.2
Cobaltum	2.65	.1	Thallium	.08	-.2
Cuprum	76.08	.1	Germanium	0	-.3
Dysprosium	.04	-.5	Rhenium	3.91	-.3
Erbium	.02	-.5	Uranium	.07	-.3
Europium	.04	-.5	Lithium	2.37	-.4
Ferrum	1569.71	-.2	Magnesium	33333.4	-.4
Gadolinium	.09	-.5	Silicium	0	-.4
Gallium	.41	-.8	Scandium	.81	-.4
Germanium	0	-.3	Vanadium	0	-.4
Hafnium	0	-.6	Selenium	1.52	-.4
Holmium	.01	-.5	Rubidium	23.55	-.4
Indium	3.12	-.4	Indium	3.12	-.4
Lanthanum	.38	-.6	Bismuthum	0	-.4
Lithium	2.37	-.4	Natrium	2527.36	-.5
Lutetium	0	-.6	Calcium	209251.43	-.5
Magnesium	33333.4	-.4	Chromium	0	-.5
Manganum	807.07	-.1	Strontium	670	-.5
Molybdenum	14.31	-.1	Yttrium	.2	-.5
Natrium	2527.36	-.5	Zirconium	0	-.5
Neodymium	.28	-.6	Samarium	.07	-.5
Niccolum	10.02	.1	Europium	.04	-.5
Niobium	.07	-.6	Gadolinium	.09	-.5
Plumbum	4.38	-.5	Terbium	.01	-.5
Praseodymium	.08	-.6	Dysprosium	.04	-.5
Rhenium	3.91	-.3	Holmium	.01	-.5
Rubidium	23.55	-.4	Erbium	.02	-.5
Samarium	.07	-.5	Ytterbium	.02	-.5
Scandium	.81	-.4	Plumbum	4.38	-.5
Selenium	1.52	-.4	Beryllium	.03	-.6
Silicium	0	-.4	Aluminium	1160	-.6
Stibium	.34	0	Arsenicum	1.77	-.6
Strontium	670	-.5	Niobium	.07	-.6
Tantalum	0	-.7	Cadmium	.17	-.6
Tellurium	.02	-.8	Caesium	.06	-.6
Terbium	.01	-.5	Lanthanum	.38	-.6
Thallium	.08	-.2	Cerium	1.48	-.6
Thorium	.07	-.6	Praseodymium	.08	-.6
Thulium	0	-.6	Neodymium	.28	-.6
Titanium	175.8	-.7	Thulium	0	-.6
Tungstenium	.62	1.5	Lutetium	0	-.6
Uranium	.07	-.3	Hafnium	0	-.6
Vanadium	0	-.4	Thorium	.07	-.6
Ytterbium	.02	-.5	Titanium	175.8	-.7
Yttrium	.2	-.5	Tantalum	0	-.7
Zincum	1460	1.1	Gallium	.41	-.8
Zirconium	0	-.5	Tellurium	.02	-.8

Malva sylvestris

Name	Value	Deviation	Name	Value	Deviation
Aluminium	2890	-.2	Tungstenium	.54	1.2
Argentum	0	-.2	Calcium	416222.55	.9
Arsenicum	1.09	-.8	Tantalum	.01	.6
Barium	60.39	-.7	Germanium	.14	.4
Beryllium	.03	-.6	Titanium	268.61	.3
Bismuthum	0	-.4	Borium	411.07	.1
Borium	411.07	.1	Strontium	1010	.1
Cadmium	.82	-.3	Zirconium	.52	.1
Caesium	.04	-.6	Stibium	.36	.1
Calcium	416222.55	.9	Tellurium	.04	.1
Cerium	1.77	-.5	Selenium	2.07	0
Chromium	0	-.5	Hafnium	.02	0
Cobaltum	1.34	-.4	Molybdenum	16.41	-.1
Cuprum	36.04	-.6	Natrium	6061.11	-.2
Dysprosium	.08	-.4	Aluminium	2890	-.2
Erbium	.03	-.4	Manganum	384.64	-.2
Europium	.04	-.5	Ferrum	2319.82	-.2
Ferrum	2319.82	-.2	Niobium	.11	-.2
Gadolinium	.17	-.4	Argentum	0	-.2
Gallium	.7	-.4	Scandium	.92	-.3
Germanium	.14	.4	Vanadium	1.01	-.3
Hafnium	.02	0	Cadmium	.82	-.3
Holmium	.01	-.5	Magnesium	33858.33	-.4
Indium	2.85	-.4	Silicium	0	-.4
Lanthanum	.51	-.5	Cobaltum	1.34	-.4
Lithium	1.1	-.7	Gallium	.7	-.4
Lutetium	0	-.6	Rubidium	21.45	-.4
Magnesium	33858.33	-.4	Yttrium	.34	-.4
Manganum	384.64	-.2	Indium	2.85	-.4
Molybdenum	16.41	-.1	Neodymium	.48	-.4
Natrium	6061.11	-.2	Samarium	.13	-.4
Neodymium	.48	-.4	Gadolinium	.17	-.4
Niccolum	0	-.8	Terbium	.02	-.4
Niobium	.11	-.2	Dysprosium	.08	-.4
Plumbum	3.47	-.6	Erbium	.03	-.4
Praseodymium	.13	-.5	Rhenium	3	-.4
Rhenium	3	-.4	Bismuthum	0	-.4
Rubidium	21.45	-.4	Thorium	.14	-.4
Samarium	.13	-.4	Uranium	.06	-.4
Scandium	.92	-.3	Chromium	0	-.5
Selenium	2.07	0	Lanthanum	.51	-.5
Silicium	0	-.4	Cerium	1.77	-.5
Stibium	.36	.1	Praseodymium	.13	-.5
Strontium	1010	.1	Europium	.04	-.5
Tantalum	.01	.6	Holmium	.01	-.5
Tellurium	.04	.1	Ytterbium	.02	-.5
Terbium	.02	-.4	Thallium	.02	-.5
Thallium	.02	-.5	Beryllium	.03	-.6
Thorium	.14	-.4	Cuprum	36.04	-.6
Thulium	0	-.6	Zincum	305.22	-.6
Titanium	268.61	.3	Caesium	.04	-.6
Tungstenium	.54	1.2	Thulium	0	-.6
Uranium	.06	-.4	Lutetium	0	-.6
Vanadium	1.01	-.3	Plumbum	3.47	-.6
Ytterbium	.02	-.5	Lithium	1.1	-.7
Yttrium	.34	-.4	Barium	60.39	-.7
Zincum	305.22	-.6	Niccolum	0	-.8
Zirconium	.52	.1	Arsenicum	1.09	-.8

Mandragora officinalis

Name	Value	Deviation	Name	Value	Deviation
Aluminium	1040	-.6	Bismuthum	.05	1.5
Argentum	0	-.2	Thulium	.01	.1
Arsenicum	1.65	-.6	Lutetium	.01	.1
Barium	13.19	-1.1	Thallium	.13	0
Beryllium	.06	-.4	Cobaltum	2.05	-.1
Bismuthum	.05	1.5	Vanadium	3.06	-.2
Borium	93.64	-1.5	Manganum	175.72	-.2
Cadmium	1.15	-.2	Ferrum	870.46	-.2
Caesium	.05	-.6	Argentum	0	-.2
Calcium	27234.99	-1.7	Cadmium	1.15	-.2
Cerium	2	-.4	Holmium	.02	-.2
Chromium	1.45	-.3	Uranium	.09	-.2
Cobaltum	2.05	-.1	Chromium	1.45	-.3
Cuprum	1.12	-1.2	Niccolum	6.41	-.3
Dysprosium	.1	-.3	Germanium	0	-.3
Erbium	.05	-.3	Yttrium	.54	-.3
Europium	.04	-.5	Molybdenum	5.53	-.3
Ferrum	870.46	-.2	Samarium	.16	-.3
Gadolinium	.19	-.3	Gadolinium	.19	-.3
Gallium	.33	-.9	Dysprosium	.1	-.3
Germanium	0	-.3	Erbium	.05	-.3
Hafnium	.01	-.3	Ytterbium	.04	-.3
Holmium	.02	-.2	Hafnium	.01	-.3
Indium	1.73	-.5	Beryllium	.06	-.4
Lanthanum	.61	-.5	Silicium	0	-.4
Lithium	1.36	-.7	Rubidium	18.41	-.4
Lutetium	.01	.1	Cerium	2	-.4
Magnesium	16838.09	-.7	Praseodymium	.15	-.4
Manganum	175.72	-.2	Neodymium	.58	-.4
Molybdenum	5.53	-.3	Terbium	.02	-.4
Natrium	2675.51	-.5	Natrium	2675.51	-.5
Neodymium	.58	-.4	Zirconium	0	-.5
Niccolum	6.41	-.3	Indium	1.73	-.5
Niobium	.05	-.7	Lanthanum	.61	-.5
Plumbum	2.49	-.7	Europium	.04	-.5
Praseodymium	.15	-.4	Aluminium	1040	-.6
Rhenium	.56	-.7	Arsenicum	1.65	-.6
Rubidium	18.41	-.4	Caesium	.05	-.6
Samarium	.16	-.3	Tungstenium	0	-.6
Scandium	0	-1.2	Thorium	.08	-.6
Selenium	0	-1.3	Lithium	1.36	-.7
Silicium	0	-.4	Magnesium	16838.09	-.7
Stibium	.05	-.8	Niobium	.05	-.7
Strontium	162	-1.4	Tantalum	0	-.7
Tantalum	0	-.7	Rhenium	.56	-.7
Tellurium	.02	-.8	Plumbum	2.49	-.7
Terbium	.02	-.4	Stibium	.05	-.8
Thallium	.13	0	Tellurium	.02	-.8
Thorium	.08	-.6	Titanium	165.26	-.9
Thulium	.01	.1	Zincum	85.05	-.9
Titanium	165.26	-.9	Gallium	.33	-.9
Tungstenium	0	-.6	Barium	13.19	-1.1
Uranium	.09	-.2	Scandium	0	-1.2
Vanadium	3.06	-.2	Cuprum	1.12	-1.2
Ytterbium	.04	-.3	Selenium	0	-1.3
Yttrium	.54	-.3	Strontium	162	-1.4
Zincum	85.05	-.9	Borium	93.64	-1.5
Zirconium	0	-.5	Calcium	27234.99	-1.7

Marrubium vulgare

Name	Value	Deviation	Name	Value	Deviation
Aluminium	2190	-.3	Tantalum	.01	.6
Argentum	0	-.2	Arsenicum	5.44	.5
Arsenicum	5.44	.5	Borium	429.88	.2
Barium	63.16	-.7	Calcium	307417.38	.2
Beryllium	.07	-.3	Rhenium	8.75	.2
Bismuthum	0	-.4	Thulium	.01	.1
Borium	429.88	.2	Lutetium	.01	.1
Cadmium	.43	-.5	Strontium	960	0
Caesium	.1	-.5	Zirconium	.51	0
Calcium	307417.38	.2	Hafnium	.02	0
Cerium	2.3	-.3	Chromium	2.2	-.1
Chromium	2.2	-.1	Scandium	.97	-.2
Cobaltum	1	-.5	Vanadium	3.08	-.2
Cuprum	23.21	-.8	Manganum	193.06	-.2
Dysprosium	.11	-.2	Ferrum	2582.39	-.2
Erbium	.06	-.2	Niobium	.11	-.2
Europium	.05	-.4	Molybdenum	9.11	-.2
Ferrum	2582.39	-.2	Argentum	0	-.2
Gadolinium	.22	-.2	Neodymium	.79	-.2
Gallium	.44	-.7	Samarium	.19	-.2
Germanium	0	-.3	Gadolinium	.22	-.2
Hafnium	.02	0	Terbium	.03	-.2
Holmium	.02	-.2	Dysprosium	.11	-.2
Indium	3.81	-.3	Holmium	.02	-.2
Lanthanum	.87	-.3	Erbium	.06	-.2
Lithium	1.89	-.5	Uranium	.09	-.2
Lutetium	.01	.1	Beryllium	.07	-.3
Magnesium	19510.13	-.7	Aluminium	2190	-.3
Manganum	193.06	-.2	Germanium	0	-.3
Molybdenum	9.11	-.2	Yttrium	.61	-.3
Natrium	2181.71	-.5	Indium	3.81	-.3
Neodymium	.79	-.2	Lanthanum	.87	-.3
Niccolum	0	-.8	Cerium	2.3	-.3
Niobium	.11	-.2	Praseodymium	.2	-.3
Plumbum	6.13	-.4	Ytterbium	.04	-.3
Praseodymium	.2	-.3	Thorium	.18	-.3
Rhenium	8.75	.2	Silicium	0	-.4
Rubidium	31.46	-.4	Rubidium	31.46	-.4
Samarium	.19	-.2	Stibium	.21	-.4
Scandium	.97	-.2	Tellurium	.03	-.4
Selenium	1.34	-.5	Europium	.05	-.4
Silicium	0	-.4	Plumbum	6.13	-.4
Stibium	.21	-.4	Bismuthum	0	-.4
Strontium	960	0	Lithium	1.89	-.5
Tantalum	.01	.6	Natrium	2181.71	-.5
Tellurium	.03	-.4	Cobaltum	1	-.5
Terbium	.03	-.2	Selenium	1.34	-.5
Thallium	.02	-.5	Cadmium	.43	-.5
Thorium	.18	-.3	Caesium	.1	-.5
Thulium	.01	.1	Thallium	.02	-.5
Titanium	159.85	-.9	Tungstenium	0	-.6
Tungstenium	0	-.6	Magnesium	19510.13	-.7
Uranium	.09	-.2	Zincum	238.25	-.7
Vanadium	3.08	-.2	Gallium	.44	-.7
Ytterbium	.04	-.3	Barium	63.16	-.7
Yttrium	.61	-.3	Niccolum	0	-.8
Zincum	238.25	-.7	Cuprum	23.21	-.8
Zirconium	.51	0	Titanium	159.85	-.9

Melilotus officinalis

Name	Value	Deviation	Name	Value	Deviation
Aluminium	1110	-.6	Caesium	4.77	6.9
Argentum	0	-.2	Tungstenium	.81	2.2
Arsenicum	1.95	-.6	Magnesium	150000	2.1
Barium	54.58	-.7	Borium	761.29	2
Beryllium	.16	.1	Calcium	527089.08	1.6
Bismuthum	0	-.4	Niccolum	22.83	1.3
Borium	761.29	2	Strontium	1670	1.3
Cadmium	.83	-.3	Rubidium	512.29	.9
Caesium	4.77	6.9	Cobaltum	4.31	.8
Calcium	527089.08	1.6	Molybdenum	58.11	.7
Cerium	1.61	-.5	Zincum	960.74	.4
Chromium	1.45	-.3	Stibium	.45	.4
Cobaltum	4.31	.8	Beryllium	.16	.1
Cuprum	76.02	.1	Cuprum	76.02	.1
Dysprosium	.05	-.5	Selenium	2.29	.1
Erbium	.03	-.4	Zirconium	.6	.1
Europium	.03	-.6	Tellurium	.04	.1
Ferrum	6626.77	0	Ferrum	6626.77	0
Gadolinium	.13	-.4	Rhenium	6.8	0
Gallium	.43	-.7	Manganum	767.74	-.1
Germanium	0	-.3	Natrium	6365.55	-.2
Hafnium	.01	-.3	Scandium	.97	-.2
Holmium	.01	-.5	Argentum	0	-.2
Indium	2.97	-.4	Vanadium	1.45	-.3
Lanthanum	.58	-.5	Chromium	1.45	-.3
Lithium	1.59	-.6	Germanium	0	-.3
Lutetium	0	-.6	Cadmium	.83	-.3
Magnesium	150000	2.1	Hafnium	.01	-.3
Manganum	767.74	-.1	Silicium	0	-.4
Molybdenum	58.11	.7	Yttrium	.38	-.4
Natrium	6365.55	-.2	Indium	2.97	-.4
Neodymium	.38	-.5	Gadolinium	.13	-.4
Niccolum	22.83	1.3	Erbium	.03	-.4
Niobium	.05	-.7	Plumbum	6.11	-.4
Plumbum	6.11	-.4	Bismuthum	0	-.4
Praseodymium	.11	-.5	Titanium	196.83	-.5
Rhenium	6.8	0	Lanthanum	.58	-.5
Rubidium	512.29	.9	Cerium	1.61	-.5
Samarium	.09	-.5	Praseodymium	.11	-.5
Scandium	.97	-.2	Neodymium	.38	-.5
Selenium	2.29	.1	Samarium	.09	-.5
Silicium	0	-.4	Terbium	.01	-.5
Stibium	.45	.4	Dysprosium	.05	-.5
Strontium	1670	1.3	Holmium	.01	-.5
Tantalum	0	-.7	Ytterbium	.02	-.5
Tellurium	.04	.1	Thallium	.01	-.5
Terbium	.01	-.5	Uranium	.04	-.5
Thallium	.01	-.5	Lithium	1.59	-.6
Thorium	.08	-.6	Aluminium	1110	-.6
Thulium	0	-.6	Arsenicum	1.95	-.6
Titanium	196.83	-.5	Europium	.03	-.6
Tungstenium	.81	2.2	Thulium	0	-.6
Uranium	.04	-.5	Lutetium	0	-.6
Vanadium	1.45	-.3	Thorium	.08	-.6
Ytterbium	.02	-.5	Gallium	.43	-.7
Yttrium	.38	-.4	Niobium	.05	-.7
Zincum	960.74	.4	Barium	54.58	-.7
Zirconium	.6	.1	Tantalum	0	-.7

Melissa officinalis

Name	Value	Deviation	Name	Value	Deviation
Aluminium	4540	.2	Barium	520.32	3
Argentum	0	-.2	Magnesium	150000	2.1
Arsenicum	5.85	.6	Tantalum	.02	2
Barium	520.32	3	Bismuthum	.05	1.5
Beryllium	.12	-.1	Niobium	.27	1.2
Bismuthum	.05	1.5	Tungstenium	.54	1.2
Borium	379.83	0	Molybdenum	72	.9
Cadmium	.47	-.5	Titanium	311.65	.8
Caesium	.23	-.3	Rhenium	13.74	.8
Calcium	325880.37	.3	Silicium	2478.49	.6
Cerium	3.02	-.1	Scandium	1.8	.6
Chromium	3.97	.2	Zincum	1130	.6
Cobaltum	1.29	-.4	Arsenicum	5.85	.6
Cuprum	64.06	-.1	Europium	.14	.6
Dysprosium	.16	0	Zirconium	.93	.5
Erbium	.08	0	Calcium	325880.37	.3
Europium	.14	.6	Germanium	.13	.3
Ferrum	3967.03	-.1	Strontium	1140	.3
Gadolinium	.32	0	Hafnium	.03	.3
Gallium	.94	-.1	Plumbum	13.39	.3
Germanium	.13	.3	Aluminium	4540	.2
Hafnium	.03	.3	Chromium	3.97	.2
Holmium	.03	0	Stibium	.41	.2
Indium	6.59	0	Tellurium	.04	.1
Lanthanum	1.36	-.1	Thulium	.01	.1
Lithium	2.76	-.3	Lutetium	.01	.1
Lutetium	.01	.1	Thallium	.17	.1
Magnesium	150000	2.1	Uranium	.16	.1
Manganum	357.79	-.2	Borium	379.83	0
Molybdenum	72	.9	Indium	6.59	0
Natrium	1926.66	-.5	Neodymium	1.12	0
Neodymium	1.12	0	Gadolinium	.32	0
Niccolum	6.1	-.3	Terbium	.04	0
Niobium	.27	1.2	Dysprosium	.16	0
Plumbum	13.39	.3	Holmium	.03	0
Praseodymium	.3	-.1	Erbium	.08	0
Rhenium	13.74	.8	Ytterbium	.06	0
Rubidium	34.19	-.4	Thorium	.31	0
Samarium	.26	-.1	Beryllium	.12	-.1
Scandium	1.8	.6	Vanadium	5.69	-.1
Selenium	1.83	-.2	Ferrum	3967.03	-.1
Silicium	2478.49	.6	Cuprum	64.06	-.1
Stibium	.41	.2	Gallium	.94	-.1
Strontium	1140	.3	Yttrium	.79	-.1
Tantalum	.02	2	Lanthanum	1.36	-.1
Tellurium	.04	.1	Cerium	3.02	-.1
Terbium	.04	0	Praseodymium	.3	-.1
Thallium	.17	.1	Samarium	.26	-.1
Thorium	.31	0	Manganum	357.79	-.2
Thulium	.01	.1	Selenium	1.83	-.2
Titanium	311.65	.8	Argentum	0	-.2
Tungstenium	.54	1.2	Lithium	2.76	-.3
Uranium	.16	.1	Niccolum	6.1	-.3
Vanadium	5.69	-.1	Caesium	.23	-.3
Ytterbium	.06	0	Cobaltum	1.29	-.4
Yttrium	.79	-.1	Rubidium	34.19	-.4
Zincum	1130	.6	Natrium	1926.66	-.5
Zirconium	.93	.5	Cadmium	.47	-.5

Mentha arvensis

Name	Value	Deviation	Name	Value	Deviation
Aluminium	1960	-.4	Bismuthum	.07	2.3
Argentum	0	-.2	Selenium	4.91	1.7
Arsenicum	1.37	-.7	Tungstenium	.58	1.4
Barium	270.26	1	Barium	270.26	1
Beryllium	.06	-.4	Calcium	387424.19	.7
Bismuthum	.07	2.3	Stibium	.55	.7
Borium	325.86	-.3	Tantalum	.01	.6
Cadmium	.33	-.6	Strontium	1210	.5
Caesium	.26	-.3	Tellurium	.05	.5
Calcium	387424.19	.7	Zirconium	.87	.4
Cerium	2.21	-.4	Niobium	.17	.3
Chromium	0	-.5	Molybdenum	30.66	.2
Cobaltum	1.27	-.4	Indium	8.56	.2
Cuprum	52.39	-.3	Thulium	.01	.1
Dysprosium	.08	-.4	Ferrum	6749.6	0
Erbium	.04	-.3	Europium	.08	0
Europium	.08	0	Hafnium	.02	0
Ferrum	6749.6	0	Plumbum	10.67	0
Gadolinium	.18	-.3	Scandium	1.12	-.1
Gallium	.55	-.6	Natrium	5974.74	-.2
Germanium	0	-.3	Titanium	226.66	-.2
Hafnium	.02	0	Vanadium	2.59	-.2
Holmium	.02	-.2	Manganum	439.82	-.2
Indium	8.56	.2	Argentum	0	-.2
Lanthanum	.94	-.3	Holmium	.02	-.2
Lithium	1.95	-.5	Borium	325.86	-.3
Lutetium	0	-.6	Cuprum	52.39	-.3
Magnesium	34978.45	-.4	Germanium	0	-.3
Manganum	439.82	-.2	Caesium	.26	-.3
Molybdenum	30.66	.2	Lanthanum	.94	-.3
Natrium	5974.74	-.2	Praseodymium	.19	-.3
Neodymium	.68	-.3	Neodymium	.68	-.3
Niccolum	0	-.8	Samarium	.15	-.3
Niobium	.17	.3	Gadolinium	.18	-.3
Plumbum	10.67	0	Erbium	.04	-.3
Praseodymium	.19	-.3	Ytterbium	.04	-.3
Rhenium	3.42	-.4	Thorium	.18	-.3
Rubidium	28.79	-.4	Uranium	.08	-.3
Samarium	.15	-.3	Beryllium	.06	-.4
Scandium	1.12	-.1	Magnesium	34978.45	-.4
Selenium	4.91	1.7	Aluminium	1960	-.4
Silicium	0	-.4	Silicium	0	-.4
Stibium	.55	.7	Cobaltum	1.27	-.4
Strontium	1210	.5	Zincum	385.47	-.4
Tantalum	.01	.6	Rubidium	28.79	-.4
Tellurium	.05	.5	Yttrium	.46	-.4
Terbium	.02	-.4	Cerium	2.21	-.4
Thallium	.03	-.4	Terbium	.02	-.4
Thorium	.18	-.3	Dysprosium	.08	-.4
Thulium	.01	.1	Rhenium	3.42	-.4
Titanium	226.66	-.2	Thallium	.03	-.4
Tungstenium	.58	1.4	Lithium	1.95	-.5
Uranium	.08	-.3	Chromium	0	-.5
Vanadium	2.59	-.2	Gallium	.55	-.6
Ytterbium	.04	-.3	Cadmium	.33	-.6
Yttrium	.46	-.4	Lutetium	0	-.6
Zincum	385.47	-.4	Arsenicum	1.37	-.7
Zirconium	.87	.4	Niccolum	0	-.8

Mercurialis perennis

Name	Value	Deviation	Name	Value	Deviation
Aluminium	5340	.3	Tantalum	.02	2
Argentum	0	-.2	Tungstenium	.57	1.3
Arsenicum	3.58	-.1	Lithium	9.31	1.2
Barium	98.96	-.4	Selenium	3.91	1.1
Beryllium	.2	.3	Zincum	1400	1
Bismuthum	0	-.4	Tellurium	.06	1
Borium	294.96	-.5	Niobium	.24	.9
Cadmium	1.39	-.1	Niccolum	17.98	.8
Caesium	.61	.3	Hafnium	.05	.8
Calcium	326193.11	.3	Titanium	301.7	.7
Cerium	4.52	.3	Zirconium	1.13	.7
Chromium	0	-.5	Thorium	.53	.5
Cobaltum	2.54	.1	Scandium	1.58	.4
Cuprum	53.82	-.3	Lanthanum	2.18	.4
Dysprosium	.22	.2	Neodymium	1.74	.4
Erbium	.12	.3	Beryllium	.2	.3
Europium	.1	.2	Aluminium	5340	.3
Ferrum	7855.42	0	Calcium	326193.11	.3
Gadolinium	.46	.3	Caesium	.61	.3
Gallium	1.08	0	Cerium	4.52	.3
Germanium	0	-.3	Praseodymium	.46	.3
Hafnium	.05	.8	Samarium	.38	.3
Holmium	.04	.3	Gadolinium	.46	.3
Indium	5.5	-.1	Holmium	.04	.3
Lanthanum	2.18	.4	Erbium	.12	.3
Lithium	9.31	1.2	Ytterbium	.09	.3
Lutetium	.01	.1	Rubidium	254.82	.2
Magnesium	40372.82	-.2	Strontium	1050	.2
Manganum	996.08	-.1	Yttrium	1.26	.2
Molybdenum	10.88	-.2	Europium	.1	.2
Natrium	8080.44	-.1	Terbium	.05	.2
Neodymium	1.74	.4	Dysprosium	.22	.2
Niccolum	17.98	.8	Rhenium	8.52	.2
Niobium	.24	.9	Cobaltum	2.54	.1
Plumbum	11.37	.1	Thulium	.01	.1
Praseodymium	.46	.3	Lutetium	.01	.1
Rhenium	8.52	.2	Plumbum	11.37	.1
Rubidium	254.82	.2	Uranium	.17	.1
Samarium	.38	.3	Ferrum	7855.42	0
Scandium	1.58	.4	Gallium	1.08	0
Selenium	3.91	1.1	Natrium	8080.44	-.1
Silicium	0	-.4	Vanadium	5.76	-.1
Stibium	.23	-.3	Manganum	996.08	-.1
Strontium	1050	.2	Arsenicum	3.58	-.1
Tantalum	.02	2	Cadmium	1.39	-.1
Tellurium	.06	1	Indium	5.5	-.1
Terbium	.05	.2	Thallium	.12	-.1
Thallium	.12	-.1	Magnesium	40372.82	-.2
Thorium	.53	.5	Molybdenum	10.88	-.2
Thulium	.01	.1	Argentum	0	-.2
Titanium	301.7	.7	Cuprum	53.82	-.3
Tungstenium	.57	1.3	Germanium	0	-.3
Uranium	.17	.1	Stibium	.23	-.3
Vanadium	5.76	-.1	Silicium	0	-.4
Ytterbium	.09	.3	Barium	98.96	-.4
Yttrium	1.26	.2	Bismuthum	0	-.4
Zincum	1400	1	Borium	294.96	-.5
Zirconium	1.13	.7	Chromium	0	-.5

Milium solis-Lithospermum officinalis

Name	Value	Deviation	Name	Value	Deviation
Aluminium	1500	-.5	Tungstenium	.52	1.2
Argentum	0	-.2	Arsenicum	5.93	.7
Arsenicum	5.93	.7	Borium	506.12	.6
Barium	71.08	-.6	Tantalum	.01	.6
Beryllium	.07	-.3	Tellurium	.05	.5
Bismuthum	0	-.4	Niccolum	12.63	.3
Borium	506.12	.6	Niobium	.16	.2
Cadmium	.29	-.6	Stibium	.29	-.1
Caesium	.1	-.5	Titanium	224.6	-.2
Calcium	158249.57	-.8	Manganum	465.52	-.2
Cerium	2.02	-.4	Ferrum	1533.16	-.2
Chromium	0	-.5	Selenium	1.8	-.2
Cobaltum	1.04	-.5	Rubidium	113.92	-.2
Cuprum	37.34	-.6	Argentum	0	-.2
Dysprosium	.07	-.4	Beryllium	.07	-.3
Erbium	.03	-.4	Scandium	.86	-.3
Europium	.04	-.5	Vanadium	1.7	-.3
Ferrum	1533.16	-.2	Zincum	477.01	-.3
Gadolinium	.15	-.4	Germanium	0	-.3
Gallium	.47	-.7	Molybdenum	6.95	-.3
Germanium	0	-.3	Indium	3.95	-.3
Hafnium	.01	-.3	Hafnium	.01	-.3
Holmium	.01	-.5	Thallium	.07	-.3
Indium	3.95	-.3	Magnesium	31279.83	-.4
Lanthanum	.69	-.4	Silicium	0	-.4
Lithium	1.86	-.5	Yttrium	.37	-.4
Lutetium	0	-.6	Lanthanum	.69	-.4
Magnesium	31279.83	-.4	Cerium	2.02	-.4
Manganum	465.52	-.2	Neodymium	.54	-.4
Molybdenum	6.95	-.3	Samarium	.12	-.4
Natrium	1021.9	-.6	Gadolinium	.15	-.4
Neodymium	.54	-.4	Terbium	.02	-.4
Niccolum	12.63	.3	Dysprosium	.07	-.4
Niobium	.16	.2	Erbium	.03	-.4
Plumbum	6.11	-.4	Ytterbium	.03	-.4
Praseodymium	.14	-.5	Plumbum	6.11	-.4
Rhenium	1.85	-.6	Bismuthum	0	-.4
Rubidium	113.92	-.2	Uranium	.06	-.4
Samarium	.12	-.4	Lithium	1.86	-.5
Scandium	.86	-.3	Aluminium	1500	-.5
Selenium	1.8	-.2	Chromium	0	-.5
Silicium	0	-.4	Cobaltum	1.04	-.5
Stibium	.29	-.1	Zirconium	0	-.5
Strontium	511	-.8	Caesium	.1	-.5
Tantalum	.01	.6	Praseodymium	.14	-.5
Tellurium	.05	.5	Europium	.04	-.5
Terbium	.02	-.4	Holmium	.01	-.5
Thallium	.07	-.3	Thorium	.13	-.5
Thorium	.13	-.5	Natrium	1021.9	-.6
Thulium	0	-.6	Cuprum	37.34	-.6
Titanium	224.6	-.2	Cadmium	.29	-.6
Tungstenium	.52	1.2	Barium	71.08	-.6
Uranium	.06	-.4	Thulium	0	-.6
Vanadium	1.7	-.3	Lutetium	0	-.6
Ytterbium	.03	-.4	Rhenium	1.85	-.6
Yttrium	.37	-.4	Gallium	.47	-.7
Zincum	477.01	-.3	Calcium	158249.57	-.8
Zirconium	0	-.5	Strontium	511	-.8

Ocimum canum

Name	Value	Deviation	Name	Value	Deviation
Aluminium	7600	.8	Magnesium	150000	2.1
Argentum	0	-.2	Selenium	4.09	1.2
Arsenicum	7.35	1.1	Arsenicum	7.35	1.1
Barium	199.14	.4	Silicium	3283.17	1
Beryllium	.1	-.2	Scandium	2.19	1
Bismuthum	0	-.4	Calcium	413067.18	.9
Borium	194.41	-1	Aluminium	7600	.8
Cadmium	.17	-.6	Zirconium	1.09	.7
Caesium	.21	-.4	Niobium	.22	.7
Calcium	413067.18	.9	Tantalum	.01	.6
Cerium	3.01	-.1	Rhenium	11.87	.6
Chromium	0	-.5	Strontium	1230	.5
Cobaltum	1.38	-.4	Tellurium	.05	.5
Cuprum	33.26	-.6	Barium	199.14	.4
Dysprosium	.19	.1	Hafnium	.03	.3
Erbium	.08	0	Gallium	1.19	.2
Europium	.1	.2	Europium	.1	.2
Ferrum	2803.07	-.2	Samarium	.3	.1
Gadolinium	.36	.1	Gadolinium	.36	.1
Gallium	1.19	.2	Dysprosium	.19	.1
Germanium	0	-.3	Thulium	.01	.1
Hafnium	.03	.3	Lutetium	.01	.1
Holmium	.03	0	Neodymium	1.12	0
Indium	3.86	-.3	Terbium	.04	0
Lanthanum	1.26	-.1	Holmium	.03	0
Lithium	2.85	-.3	Erbium	.08	0
Lutetium	.01	.1	Ytterbium	.06	0
Magnesium	150000	2.1	Thorium	.33	0
Manganum	314.94	-.2	Titanium	233.41	-.1
Molybdenum	8.89	-.2	Yttrium	.86	-.1
Natrium	1828.62	-.5	Lanthanum	1.26	-.1
Neodymium	1.12	0	Cerium	3.01	-.1
Niccolum	0	-.8	Praseodymium	.3	-.1
Niobium	.22	.7	Uranium	.13	-.1
Plumbum	4.62	-.5	Beryllium	.1	-.2
Praseodymium	.3	-.1	Vanadium	4.12	-.2
Rhenium	11.87	.6	Manganum	314.94	-.2
Rubidium	14.56	-.4	Ferrum	2803.07	-.2
Samarium	.3	.1	Molybdenum	8.89	-.2
Scandium	2.19	1	Argentum	0	-.2
Selenium	4.09	1.2	Thallium	.08	-.2
Silicium	3283.17	1	Lithium	2.85	-.3
Stibium	.2	-.4	Germanium	0	-.3
Strontium	1230	.5	Indium	3.86	-.3
Tantalum	.01	.6	Cobaltum	1.38	-.4
Tellurium	.05	.5	Rubidium	14.56	-.4
Terbium	.04	0	Stibium	.2	-.4
Thallium	.08	-.2	Caesium	.21	-.4
Thorium	.33	0	Bismuthum	0	-.4
Thulium	.01	.1	Natrium	1828.62	-.5
Titanium	233.41	-.1	Chromium	0	-.5
Tungstenium	0	-.6	Plumbum	4.62	-.5
Uranium	.13	-.1	Cuprum	33.26	-.6
Vanadium	4.12	-.2	Zincum	280.05	-.6
Ytterbium	.06	0	Cadmium	.17	-.6
Yttrium	.86	-.1	Tungstenium	0	-.6
Zincum	280.05	-.6	Niccolum	0	-.8
Zirconium	1.09	.7	Borium	194.41	-1

Ocimum canum R

Name	Value	Deviation	Name	Value	Deviation
Aluminium	6430	.6	Magnesium	150000	2.1
Argentum	0	-.2	Selenium	4.47	1.4
Arsenicum	7.89	1.3	Arsenicum	7.89	1.3
Barium	208.45	.5	Calcium	430962.07	1
Beryllium	.12	-.1	Tellurium	.06	1
Bismuthum	0	-.4	Silicium	3135.55	.9
Borium	211.84	-.9	Scandium	2.12	.9
Cadmium	.18	-.6	Rhenium	13.27	.8
Caesium	.24	-.3	Aluminium	6430	.6
Calcium	430962.07	1	Strontium	1280	.6
Cerium	3.01	-.1	Tantalum	.01	.6
Chromium	0	-.5	Barium	208.45	.5
Cobaltum	1.53	-.3	Niobium	.18	.4
Cuprum	35.6	-.6	Zirconium	.65	.2
Dysprosium	.17	0	Europium	.09	.1
Erbium	.08	0	Thulium	.01	.1
Europium	.09	.1	Ytterbium	.07	.1
Ferrum	4126.46	-.1	Lutetium	.01	.1
Gadolinium	.32	0	Titanium	239.31	0
Gallium	1.06	0	Gallium	1.06	0
Germanium	0	-.3	Samarium	.27	0
Hafnium	.01	-.3	Gadolinium	.32	0
Holmium	.03	0	Terbium	.04	0
Indium	5.64	-.1	Dysprosium	.17	0
Lanthanum	1.24	-.1	Holmium	.03	0
Lithium	2.84	-.3	Erbium	.08	0
Lutetium	.01	.1	Thorium	.34	0
Magnesium	150000	2.1	Beryllium	.12	-.1
Manganum	354.4	-.2	Ferrum	4126.46	-.1
Molybdenum	9.55	-.2	Yttrium	.81	-.1
Natrium	2082.08	-.5	Indium	5.64	-.1
Neodymium	1.09	-.1	Lanthanum	1.24	-.1
Niccolum	0	-.8	Cerium	3.01	-.1
Niobium	.18	.4	Praseodymium	.3	-.1
Plumbum	5.09	-.5	Neodymium	1.09	-.1
Praseodymium	.3	-.1	Uranium	.13	-.1
Rhenium	13.27	.8	Vanadium	4.27	-.2
Rubidium	15.19	-.4	Manganum	354.4	-.2
Samarium	.27	0	Molybdenum	9.55	-.2
Scandium	2.12	.9	Argentum	0	-.2
Selenium	4.47	1.4	Thallium	.08	-.2
Silicium	3135.55	.9	Lithium	2.84	-.3
Stibium	.18	-.4	Cobaltum	1.53	-.3
Strontium	1280	.6	Germanium	0	-.3
Tantalum	.01	.6	Caesium	.24	-.3
Tellurium	.06	1	Hafnium	.01	-.3
Terbium	.04	0	Rubidium	15.19	-.4
Thallium	.08	-.2	Stibium	.18	-.4
Thorium	.34	0	Bismuthum	0	-.4
Thulium	.01	.1	Natrium	2082.08	-.5
Titanium	239.31	0	Chromium	0	-.5
Tungstenium	0	-.6	Plumbum	5.09	-.5
Uranium	.13	-.1	Cuprum	35.6	-.6
Vanadium	4.27	-.2	Zincum	287.23	-.6
Ytterbium	.07	.1	Cadmium	.18	-.6
Yttrium	.81	-.1	Tungstenium	0	-.6
Zincum	287.23	-.6	Niccolum	0	-.8
Zirconium	.65	.2	Borium	211.84	-.9

Oenanthe crocata

Name	Value	Deviation	Name	Value	Deviation
Aluminium	5550	.4	Titanium	371.37	1.5
Argentum	0	-.2	Tungstenium	.5	1.1
Arsenicum	5.05	.4	Niobium	.24	.9
Barium	24.47	-1	Natrium	19865.9	.7
Beryllium	.19	.3	Silicium	2404.79	.6
Bismuthum	0	-.4	Scandium	1.78	.6
Borium	215.3	-.9	Tantalum	.01	.6
Cadmium	.22	-.6	Aluminium	5550	.4
Caesium	.48	.1	Arsenicum	5.05	.4
Calcium	86363.63	-1.3	Uranium	.22	.4
Cerium	4.11	.2	Beryllium	.19	.3
Chromium	2.67	-.1	Lanthanum	1.94	.3
Cobaltum	1.3	-.4	Holmium	.04	.3
Cuprum	62.54	-.1	Erbium	.12	.3
Dysprosium	.21	.2	Ytterbium	.09	.3
Erbium	.12	.3	Hafnium	.03	.3
Europium	.08	0	Thorium	.43	.3
Ferrum	2315.29	-.2	Yttrium	1.3	.2
Gadolinium	.42	.2	Cerium	4.11	.2
Gallium	1.14	.1	Praseodymium	.43	.2
Germanium	0	-.3	Neodymium	1.54	.2
Hafnium	.03	.3	Samarium	.35	.2
Holmium	.04	.3	Gadolinium	.42	.2
Indium	3.7	-.3	Terbium	.05	.2
Lanthanum	1.94	.3	Dysprosium	.21	.2
Lithium	4.03	0	Vanadium	9.7	.1
Lutetium	.01	.1	Gallium	1.14	.1
Magnesium	35550.85	-.3	Selenium	2.21	.1
Manganum	317.94	-.2	Caesium	.48	.1
Molybdenum	3.64	-.3	Thulium	.01	.1
Natrium	19865.9	.7	Lutetium	.01	.1
Neodymium	1.54	.2	Lithium	4.03	0
Niccolum	8.39	-.1	Europium	.08	0
Niobium	.24	.9	Chromium	2.67	-.1
Plumbum	5.49	-.4	Niccolum	8.39	-.1
Praseodymium	.43	.2	Cuprum	62.54	-.1
Rhenium	.7	-.7	Manganum	317.94	-.2
Rubidium	56.39	-.3	Ferrum	2315.29	-.2
Samarium	.35	.2	Argentum	0	-.2
Scandium	1.78	.6	Magnesium	35550.85	-.3
Selenium	2.21	.1	Germanium	0	-.3
Silicium	2404.79	.6	Rubidium	56.39	-.3
Stibium	.1	-.7	Molybdenum	3.64	-.3
Strontium	414	-1	Indium	3.7	-.3
Tantalum	.01	.6	Cobaltum	1.3	-.4
Tellurium	.02	-.8	Thallium	.03	-.4
Terbium	.05	.2	Plumbum	5.49	-.4
Thallium	.03	-.4	Bismuthum	0	-.4
Thorium	.43	.3	Zirconium	0	-.5
Thulium	.01	.1	Cadmium	.22	-.6
Titanium	371.37	1.5	Zincum	234.5	-.7
Tungstenium	.5	1.1	Stibium	.1	-.7
Uranium	.22	.4	Rhenium	.7	-.7
Vanadium	9.7	.1	Tellurium	.02	-.8
Ytterbium	.09	.3	Borium	215.3	-.9
Yttrium	1.3	.2	Strontium	414	-1
Zincum	234.5	-.7	Barium	24.47	-1
Zirconium	0	-.5	Calcium	86363.63	-1.3

Ononis spinosa

Name	Value	Deviation	Name	Value	Deviation
Aluminium	3020	-.2	Bismuthum	.08	2.7
Argentum	0	-.2	Arsenicum	6.38	.8
Arsenicum	6.38	.8	Borium	493.5	.6
Barium	64.37	-.7	Selenium	3.07	.6
Beryllium	.12	-.1	Molybdenum	48.76	.5
Bismuthum	.08	2.7	Plumbum	15.08	.4
Borium	493.5	.6	Hafnium	.03	.3
Cadmium	.36	-.5	Uranium	.18	.2
Caesium	.35	-.1	Strontium	1020	.1
Calcium	285531.79	0	Zirconium	.56	.1
Cerium	3.08	-.1	Tellurium	.04	.1
Chromium	0	-.5	Erbium	.09	.1
Cobaltum	1.5	-.3	Thulium	.01	.1
Cuprum	30.21	-.7	Ytterbium	.07	.1
Dysprosium	.17	0	Lutetium	.01	.1
Erbium	.09	.1	Calcium	285531.79	0
Europium	.08	0	Scandium	1.23	0
Ferrum	6226.23	0	Ferrum	6226.23	0
Gadolinium	.34	0	Yttrium	.99	0
Gallium	.61	-.5	Indium	6.53	0
Germanium	0	-.3	Neodymium	1.14	0
Hafnium	.03	.3	Samarium	.29	0
Holmium	.03	0	Europium	.08	0
Indium	6.53	0	Gadolinium	.34	0
Lanthanum	1.25	-.1	Terbium	.04	0
Lithium	2.32	-.4	Dysprosium	.17	0
Lutetium	.01	.1	Holmium	.03	0
Magnesium	26426.59	-.5	Beryllium	.12	-.1
Manganum	272.24	-.2	Stibium	.29	-.1
Molybdenum	48.76	.5	Caesium	.35	-.1
Natrium	3534.07	-.4	Lanthanum	1.25	-.1
Neodymium	1.14	0	Cerium	3.08	-.1
Niccolum	6	-.3	Praseodymium	.3	-.1
Niobium	.11	-.2	Thorium	.28	-.1
Plumbum	15.08	.4	Aluminium	3020	-.2
Praseodymium	.3	-.1	Vanadium	3.8	-.2
Rhenium	3.88	-.3	Manganum	272.24	-.2
Rubidium	49.76	-.3	Niobium	.11	-.2
Samarium	.29	0	Argentum	0	-.2
Scandium	1.23	0	Thallium	.09	-.2
Selenium	3.07	.6	Cobaltum	1.5	-.3
Silicium	0	-.4	Niccolum	6	-.3
Stibium	.29	-.1	Germanium	0	-.3
Strontium	1020	.1	Rubidium	49.76	-.3
Tantalum	0	-.7	Rhenium	3.88	-.3
Tellurium	.04	.1	Lithium	2.32	-.4
Terbium	.04	0	Natrium	3534.07	-.4
Thallium	.09	-.2	Silicium	0	-.4
Thorium	.28	-.1	Magnesium	26426.59	-.5
Thulium	.01	.1	Chromium	0	-.5
Titanium	160.56	-.9	Gallium	.61	-.5
Tungstenium	0	-.6	Cadmium	.36	-.5
Uranium	.18	.2	Zincum	287.71	-.6
Vanadium	3.8	-.2	Tungstenium	0	-.6
Ytterbium	.07	.1	Cuprum	30.21	-.7
Yttrium	.99	0	Barium	64.37	-.7
Zincum	287.71	-.6	Tantalum	0	-.7
Zirconium	.56	.1	Titanium	160.56	-.9

Osmundo regalis 1

Name	Value	Deviation	Name	Value	Deviation
Aluminium	35500	6.8	Vanadium	212.16	9
Argentum	.22	4.4	Beryllium	1.56	7.5
Arsenicum	11.29	2.3	Hafnium	.28	7.5
Barium	292.66	1.2	Erbium	.99	7.3
Beryllium	1.56	7.5	Zirconium	7.35	7.2
Bismuthum	.1	3.5	Yttrium	11.06	7.1
Borium	322.98	-.3	Dysprosium	1.82	7.1
Cadmium	7.39	2.8	Gadolinium	3.39	7
Caesium	2.15	2.7	Terbium	.4	7
Calcium	194368.22	-.6	Ytterbium	.72	6.9
Cerium	23.37	6.1	Aluminium	35500	6.8
Chromium	25.56	3.8	Holmium	.31	6.8
Cobaltum	16.59	5.7	Thulium	.11	6.8
Cuprum	343.14	5	Plumbum	87.09	6.8
Dysprosium	1.82	7.1	Silicium	17158.57	6.7
Erbium	.99	7.3	Neodymium	11.54	6.6
Europium	.67	6.6	Samarium	2.79	6.6
Ferrum	22331.96	.5	Europium	.67	6.6
Gadolinium	3.39	7	Lutetium	.1	6.3
Gallium	3.94	3.5	Uranium	1.46	6.3
Germanium	.32	1.2	Scandium	7.42	6.2
Hafnium	.28	7.5	Lanthanum	12.59	6.2
Holmium	.31	6.8	Cerium	23.37	6.1
Indium	25.07	2	Praseodymium	2.89	6.1
Lanthanum	12.59	6.2	Cobaltum	16.59	5.7
Lutetium	.1	6.3	Cuprum	343.14	5
Magnesium	150000	2.1	Thorium	2.34	4.9
Manganum	2480	0	Niobium	.68	4.7
Molybdenum	21.65	0	Tantalum	.04	4.7
Natrium	22729.95	.9	Argentum	.22	4.4
Neodymium	11.54	6.6	Stibium	1.79	4.4
Niccolum	33.21	2.2	Chromium	25.56	3.8
Niobium	.68	4.7	Gallium	3.94	3.5
Plumbum	87.09	6.8	Bismuthum	.1	3.5
Praseodymium	2.89	6.1	Titanium	484.69	2.8
Rhenium	2.32	-.5	Cadmium	7.39	2.8
Rubidium	265.49	.2	Caesium	2.15	2.7
Samarium	2.79	6.6	Arsenicum	11.29	2.3
Scandium	7.42	6.2	Niccolum	33.21	2.2
Selenium	2.7	.4	Magnesium	150000	2.1
Silicium	17158.57	6.7	Indium	25.07	2
Stibium	1.79	4.4	Tellurium	.08	1.8
Strontium	892	-.1	Tungstenium	.7	1.8
Tantalum	.04	4.7	Zincum	1590	1.3
Tellurium	.08	1.8	Germanium	.32	1.2
Terbium	.4	7	Barium	292.66	1.2
Thallium	.13	0	Natrium	22729.95	.9
Thorium	2.34	4.9	Ferrum	22331.96	.5
Thulium	.11	6.8	Selenium	2.7	.4
Titanium	484.69	2.8	Rubidium	265.49	.2
Tungstenium	.7	1.8	Manganum	2480	0
Uranium	1.46	6.3	Molybdenum	21.65	0
Vanadium	212.16	9	Thallium	.13	0
Ytterbium	.72	6.9	Strontium	892	-.1
Yttrium	11.06	7.1	Borium	322.98	-.3
Zincum	1590	1.3	Rhenium	2.32	-.5
Zirconium	7.35	7.2	Calcium	194368.22	-.6
Lithium	15.92	2.7	Lithium	15.92	2.7

Osmundo regalis 2

Name	Value	Deviation	Name	Value	Deviation
Aluminium	427	-.7	Cadmium	7.27	2.8
Argentum	0	-.2	Caesium	1.82	2.2
Arsenicum	1.61	-.7	Rubidium	777.2	1.6
Barium	84.69	-.5	Cuprum	130.15	1.1
Beryllium	.01	-.7	Borium	550.83	.9
Bismuthum	0	-.4	Silicium	2552.63	.7
Borium	550.83	.9	Scandium	1.63	.4
Cadmium	7.27	2.8	Natrium	11166.67	.1
Caesium	1.82	2.2	Manganum	877.65	-.1
Calcium	35926.12	-1.7	Ferrum	612.64	-.2
Cerium	.93	-.8	Molybdenum	13.11	-.2
Chromium	0	-.5	Argentum	0	-.2
Cobaltum	.26	-.8	Zincum	501.17	-.3
Cuprum	130.15	1.1	Germanium	0	-.3
Dysprosium	.01	-.7	Vanadium	0	-.4
Erbium	0	-.7	Bismuthum	0	-.4
Europium	.02	-.7	Magnesium	28954.08	-.5
Ferrum	612.64	-.2	Titanium	198.23	-.5
Gadolinium	.03	-.7	Chromium	0	-.5
Gallium	.46	-.7	Selenium	1.3	-.5
Germanium	0	-.3	Zirconium	0	-.5
Hafnium	0	-.6	Indium	1.59	-.5
Holmium	0	-.7	Barium	84.69	-.5
Indium	1.59	-.5	Thallium	.02	-.5
Lanthanum	.09	-.8	Yttrium	.05	-.6
Lithium	.81	-.8	Thulium	0	-.6
Lutetium	0	-.6	Lutetium	0	-.6
Magnesium	28954.08	-.5	Hafnium	0	-.6
Manganum	877.65	-.1	Tungstenium	0	-.6
Molybdenum	13.11	-.2	Uranium	.02	-.6
Natrium	11166.67	.1	Beryllium	.01	-.7
Neodymium	.06	-.7	Aluminium	427	-.7
Niccolum	0	-.8	Gallium	.46	-.7
Niobium	.02	-1	Arsenicum	1.61	-.7
Plumbum	2.19	-.7	Praseodymium	.02	-.7
Praseodymium	.02	-.7	Neodymium	.06	-.7
Rhenium	.62	-.7	Samarium	.02	-.7
Rubidium	777.2	1.6	Europium	.02	-.7
Samarium	.02	-.7	Gadolinium	.03	-.7
Scandium	1.63	.4	Terbium	0	-.7
Selenium	1.3	-.5	Dysprosium	.01	-.7
Silicium	2552.63	.7	Holmium	0	-.7
Stibium	.04	-.9	Erbium	0	-.7
Strontium	127	-1.5	Ytterbium	0	-.7
Tantalum	0	-.7	Tantalum	0	-.7
Tellurium	0	-1.7	Rhenium	.62	-.7
Terbium	0	-.7	Plumbum	2.19	-.7
Thallium	.02	-.5	Thorium	.05	-.7
Thorium	.05	-.7	Lithium	.81	-.8
Thulium	0	-.6	Cobaltum	.26	-.8
Titanium	198.23	-.5	Niccolum	0	-.8
Tungstenium	0	-.6	Lanthanum	.09	-.8
Uranium	.02	-.6	Cerium	.93	-.8
Vanadium	0	-.4	Stibium	.04	-.9
Ytterbium	0	-.7	Niobium	.02	-1
Yttrium	.05	-.6	Strontium	127	-1.5
Zincum	501.17	-.3	Calcium	35926.12	-1.7
Zirconium	0	-.5	Tellurium	0	-1.7

Paeonia officinalis

Name	Value	Deviation	Name	Value	Deviation
Aluminium	3410	-.1	Natrium	60000	3.6
Argentum	0	-.2	Strontium	2660	3.1
Arsenicum	12.82	2.8	Arsenicum	12.82	2.8
Barium	161.39	.1	Lithium	13.73	2.2
Beryllium	.29	.8	Calcium	586652.39	2
Bismuthum	0	-.4	Titanium	405.86	1.9
Borium	368.98	-.1	Cadmium	3.55	1
Cadmium	3.55	1	Beryllium	.29	.8
Caesium	.36	-.1	Uranium	.3	.8
Calcium	586652.39	2	Zincum	1180	.7
Cerium	3.37	0	Tantalum	.01	.6
Chromium	3.64	.1	Thallium	.29	.6
Cobaltum	2.53	.1	Zirconium	.92	.5
Cuprum	46.96	-.4	Hafnium	.04	.5
Dysprosium	.18	.1	Scandium	1.38	.2
Erbium	.09	.1	Germanium	.1	.2
Europium	.09	.1	Niobium	.16	.2
Ferrum	9069.07	0	Vanadium	9.41	.1
Gadolinium	.35	.1	Chromium	3.64	.1
Gallium	1.01	0	Cobaltum	2.53	.1
Germanium	.1	.2	Yttrium	1.1	.1
Hafnium	.04	.5	Barium	161.39	.1
Holmium	.03	0	Samarium	.3	.1
Indium	4.05	-.3	Europium	.09	.1
Lanthanum	1.52	0	Gadolinium	.35	.1
Lithium	13.73	2.2	Dysprosium	.18	.1
Lutetium	.01	.1	Erbium	.09	.1
Magnesium	29100.22	-.5	Thulium	.01	.1
Manganum	346.45	-.2	Ytterbium	.07	.1
Molybdenum	5.65	-.3	Lutetium	.01	.1
Natrium	60000	3.6	Ferrum	9069.07	0
Neodymium	1.25	0	Gallium	1.01	0
Niccolum	5.98	-.3	Lanthanum	1.52	0
Niobium	.16	.2	Cerium	3.37	0
Plumbum	8.37	-.2	Praseodymium	.32	0
Praseodymium	.32	0	Neodymium	1.25	0
Rhenium	.25	-.8	Terbium	.04	0
Rubidium	52.99	-.3	Holmium	.03	0
Samarium	.3	.1	Borium	368.98	-.1
Scandium	1.38	.2	Aluminium	3410	-.1
Selenium	1.47	-.4	Caesium	.36	-.1
Silicium	0	-.4	Thorium	.28	-.1
Stibium	.16	-.5	Manganum	346.45	-.2
Strontium	2660	3.1	Argentum	0	-.2
Tantalum	.01	.6	Plumbum	8.37	-.2
Tellurium	.03	-.4	Niccolum	5.98	-.3
Terbium	.04	0	Rubidium	52.99	-.3
Thallium	.29	.6	Molybdenum	5.65	-.3
Thorium	.28	-.1	Indium	4.05	-.3
Thulium	.01	.1	Silicium	0	-.4
Titanium	405.86	1.9	Cuprum	46.96	-.4
Tungstenium	0	-.6	Selenium	1.47	-.4
Uranium	.3	.8	Tellurium	.03	-.4
Vanadium	9.41	.1	Bismuthum	0	-.4
Ytterbium	.07	.1	Magnesium	29100.22	-.5
Yttrium	1.1	.1	Stibium	.16	-.5
Zincum	1180	.7	Tungstenium	0	-.6
Zirconium	.92	.5	Rhenium	.25	-.8

Petroselinum crispum

Name	Value	Deviation	Name	Value	Deviation
Aluminium	6770	.6	Tungstenium	.55	1.3
Argentum	0	-.2	Lutetium	.02	.8
Arsenicum	2.07	-.5	Silicium	2548.97	.7
Barium	144.13	0	Scandium	1.96	.7
Beryllium	.21	.4	Samarium	.53	.7
Bismuthum	0	-.4	Thulium	.02	.7
Borium	497.24	.6	Plumbum	18.32	.7
Cadmium	.4	-.5	Borium	497.24	.6
Caesium	.54	.2	Aluminium	6770	.6
Calcium	241300.35	-.3	Lanthanum	2.48	.6
Cerium	5.09	.5	Neodymium	2.13	.6
Chromium	3.41	.1	Gadolinium	.57	.6
Cobaltum	1.93	-.1	Terbium	.07	.6
Cuprum	47.56	-.4	Tantalum	.01	.6
Dysprosium	.29	.5	Thorium	.56	.6
Erbium	.14	.5	Cerium	5.09	.5
Europium	.13	.5	Praseodymium	.55	.5
Ferrum	8787.09	0	Europium	.13	.5
Gadolinium	.57	.6	Dysprosium	.29	.5
Gallium	1.11	.1	Holmium	.05	.5
Germanium	.11	.2	Erbium	.14	.5
Hafnium	.02	0	Ytterbium	.11	.5
Holmium	.05	.5	Beryllium	.21	.4
Indium	5.61	-.1	Yttrium	1.57	.4
Lanthanum	2.48	.6	Niobium	.18	.4
Lithium	5.42	.3	Lithium	5.42	.3
Lutetium	.02	.8	Natrium	14041.81	.3
Magnesium	17536.21	-.7	Germanium	.11	.2
Manganum	337.43	-.2	Caesium	.54	.2
Molybdenum	13.05	-.2	Uranium	.18	.2
Natrium	14041.81	.3	Chromium	3.41	.1
Neodymium	2.13	.6	Niccolum	10.32	.1
Niccolum	10.32	.1	Gallium	1.11	.1
Niobium	.18	.4	Vanadium	8.54	0
Plumbum	18.32	.7	Ferrum	8787.09	0
Praseodymium	.55	.5	Selenium	2.13	0
Rhenium	2.18	-.5	Barium	144.13	0
Rubidium	42.67	-.4	Hafnium	.02	0
Samarium	.53	.7	Titanium	228.5	-.1
Scandium	1.96	.7	Cobaltum	1.93	-.1
Selenium	2.13	0	Indium	5.61	-.1
Silicium	2548.97	.7	Manganum	337.43	-.2
Stibium	.14	-.6	Molybdenum	13.05	-.2
Strontium	659	-.5	Argentum	0	-.2
Tantalum	.01	.6	Calcium	241300.35	-.3
Tellurium	.02	-.8	Thallium	.06	-.3
Terbium	.07	.6	Cuprum	47.56	-.4
Thallium	.06	-.3	Rubidium	42.67	-.4
Thorium	.56	.6	Bismuthum	0	-.4
Thulium	.02	.7	Arsenicum	2.07	-.5
Titanium	228.5	-.1	Strontium	659	-.5
Tungstenium	.55	1.3	Zirconium	0	-.5
Uranium	.18	.2	Cadmium	.4	-.5
Vanadium	8.54	0	Rhenium	2.18	-.5
Ytterbium	.11	.5	Zincum	293.85	-.6
Yttrium	1.57	.4	Stibium	.14	-.6
Zincum	293.85	-.6	Magnesium	17536.21	-.7
Zirconium	0	-.5	Tellurium	.02	-.8

Plantago major

Name	Value	Deviation	Name	Value	Deviation
Aluminium	3410	-.1	Calcium	380050.74	.6
Argentum	0	-.2	Germanium	.18	.6
Arsenicum	2.1	-.5	Tantalum	.01	.6
Barium	120.65	-.2	Zirconium	.95	.5
Beryllium	.14	0	Niobium	.18	.4
Bismuthum	0	-.4	Lanthanum	2.14	.4
Borium	257.41	-.7	Cerium	4.14	.2
Cadmium	.87	-.3	Tellurium	.04	.1
Caesium	.23	-.3	Praseodymium	.38	.1
Calcium	380050.74	.6	Neodymium	1.4	.1
Cerium	4.14	.2	Samarium	.33	.1
Chromium	1.85	-.2	Thulium	.01	.1
Cobaltum	1.65	-.3	Lutetium	.01	.1
Cuprum	21.12	-.9	Thorium	.35	.1
Dysprosium	.13	-.1	Beryllium	.14	0
Erbium	.07	-.1	Strontium	972	0
Europium	.08	0	Europium	.08	0
Ferrum	3010.13	-.2	Gadolinium	.34	0
Gadolinium	.34	0	Hafnium	.02	0
Gallium	.61	-.5	Aluminium	3410	-.1
Germanium	.18	.6	Scandium	1.07	-.1
Hafnium	.02	0	Molybdenum	13.81	-.1
Holmium	.02	-.2	Dysprosium	.13	-.1
Indium	1.86	-.5	Erbium	.07	-.1
Lanthanum	2.14	.4	Ytterbium	.05	-.1
Lithium	2.46	-.4	Uranium	.12	-.1
Lutetium	.01	.1	Vanadium	3.96	-.2
Magnesium	19147.11	-.7	Chromium	1.85	-.2
Manganum	154.82	-.2	Manganum	154.82	-.2
Molybdenum	13.81	-.1	Ferrum	3010.13	-.2
Natrium	1553.84	-.6	Selenium	1.73	-.2
Neodymium	1.4	.1	Yttrium	.7	-.2
Niccolum	0	-.8	Argentum	0	-.2
Niobium	.18	.4	Barium	120.65	-.2
Plumbum	5.01	-.5	Terbium	.03	-.2
Praseodymium	.38	.1	Holmium	.02	-.2
Rhenium	4.28	-.3	Cobaltum	1.65	-.3
Rubidium	20.87	-.4	Cadmium	.87	-.3
Samarium	.33	.1	Caesium	.23	-.3
Scandium	1.07	-.1	Rhenium	4.28	-.3
Selenium	1.73	-.2	Lithium	2.46	-.4
Silicium	0	-.4	Silicium	0	-.4
Stibium	.09	-.7	Rubidium	20.87	-.4
Strontium	972	0	Thallium	.03	-.4
Tantalum	.01	.6	Bismuthum	0	-.4
Tellurium	.04	.1	Gallium	.61	-.5
Terbium	.03	-.2	Arsenicum	2.1	-.5
Thallium	.03	-.4	Indium	1.86	-.5
Thorium	.35	.1	Plumbum	5.01	-.5
Thulium	.01	.1	Natrium	1553.84	-.6
Titanium	183.84	-.7	Tungstenium	0	-.6
Tungstenium	0	-.6	Borium	257.41	-.7
Uranium	.12	-.1	Magnesium	19147.11	-.7
Vanadium	3.96	-.2	Titanium	183.84	-.7
Ytterbium	.05	-.1	Stibium	.09	-.7
Yttrium	.7	-.2	Niccolum	0	-.8
Zincum	167.04	-.8	Zincum	167.04	-.8
Zirconium	.95	.5	Cuprum	21.12	-.9

Polygonum aviculare

Name	Value	Deviation	Name	Value	Deviation
Aluminium	3130	-.1	Titanium	308.23	.8
Argentum	0	-.2	Zincum	1200	.7
Arsenicum	0	-1.1	Niobium	.21	.6
Barium	32.42	-.9	Tantalum	.01	.6
Beryllium	.07	-.3	Selenium	2.57	.3
Bismuthum	0	-.4	Hafnium	.03	.3
Borium	279.1	-.6	Scandium	1.46	.2
Cadmium	.62	-.4	Zirconium	.66	.2
Caesium	.24	-.3	Magnesium	56810.42	.1
Calcium	302427.37	.1	Calcium	302427.37	.1
Cerium	2.51	-.3	Molybdenum	25.39	.1
Chromium	0	-.5	Thulium	.01	.1
Cobaltum	1.26	-.4	Lutetium	.01	.1
Cuprum	63.58	-.1	Rhenium	7.07	0
Dysprosium	.11	-.2	Aluminium	3130	-.1
Erbium	.06	-.2	Cuprum	63.58	-.1
Europium	.05	-.4	Strontium	873	-.1
Ferrum	2767.36	-.2	Ytterbium	.05	-.1
Gadolinium	.22	-.2	Thorium	.27	-.1
Gallium	.72	-.4	Vanadium	2.83	-.2
Germanium	0	-.3	Manganum	217.98	-.2
Hafnium	.03	.3	Ferrum	2767.36	-.2
Holmium	.02	-.2	Yttrium	.63	-.2
Indium	3.29	-.4	Argentum	0	-.2
Lanthanum	.98	-.3	Neodymium	.84	-.2
Lithium	2.72	-.3	Samarium	.19	-.2
Lutetium	.01	.1	Gadolinium	.22	-.2
Magnesium	56810.42	.1	Terbium	.03	-.2
Manganum	217.98	-.2	Dysprosium	.11	-.2
Molybdenum	25.39	.1	Holmium	.02	-.2
Natrium	1626.58	-.5	Erbium	.06	-.2
Neodymium	.84	-.2	Plumbum	7.77	-.2
Niccolum	0	-.8	Lithium	2.72	-.3
Niobium	.21	.6	Beryllium	.07	-.3
Plumbum	7.77	-.2	Germanium	0	-.3
Praseodymium	.22	-.3	Caesium	.24	-.3
Rhenium	7.07	0	Lanthanum	.98	-.3
Rubidium	31.9	-.4	Cerium	2.51	-.3
Samarium	.19	-.2	Praseodymium	.22	-.3
Scandium	1.46	.2	Uranium	.08	-.3
Selenium	2.57	.3	Silicium	0	-.4
Silicium	0	-.4	Cobaltum	1.26	-.4
Stibium	.09	-.7	Gallium	.72	-.4
Strontium	873	-.1	Rubidium	31.9	-.4
Tantalum	.01	.6	Cadmium	.62	-.4
Tellurium	.02	-.8	Indium	3.29	-.4
Terbium	.03	-.2	Europium	.05	-.4
Thallium	.03	-.4	Thallium	.03	-.4
Thorium	.27	-.1	Bismuthum	0	-.4
Thulium	.01	.1	Natrium	1626.58	-.5
Titanium	308.23	.8	Chromium	0	-.5
Tungstenium	0	-.6	Borium	279.1	-.6
Uranium	.08	-.3	Tungstenium	0	-.6
Vanadium	2.83	-.2	Stibium	.09	-.7
Ytterbium	.05	-.1	Niccolum	0	-.8
Yttrium	.63	-.2	Tellurium	.02	-.8
Zincum	1200	.7	Barium	32.42	-.9
Zirconium	.66	.2	Arsenicum	0	-1.1

Primula veris

Name	Value	Deviation	Name	Value	Deviation
Aluminium	4820	.2	Tantalum	.02	2
Argentum	0	-.2	Niobium	.32	1.6
Arsenicum	2.97	-.2	Tungstenium	.54	1.2
Barium	26.63	-1	Zirconium	1.11	.7
Beryllium	.12	-.1	Silicium	2020.66	.5
Bismuthum	0	-.4	Selenium	2.96	.5
Borium	416.31	.2	Titanium	268.6	.3
Cadmium	.25	-.6	Stibium	.42	.3
Caesium	.25	-.3	Hafnium	.03	.3
Calcium	145533.2	-.9	Borium	416.31	.2
Cerium	3.17	-.1	Aluminium	4820	.2
Chromium	0	-.5	Scandium	1.46	.2
Cobaltum	1.02	-.5	Tellurium	.04	.1
Cuprum	18.54	-.9	Lanthanum	1.58	.1
Dysprosium	.16	0	Gadolinium	.35	.1
Erbium	.08	0	Thulium	.01	.1
Europium	.07	-.2	Lutetium	.01	.1
Ferrum	2148.48	-.2	Thorium	.36	.1
Gadolinium	.35	.1	Uranium	.16	.1
Gallium	.9	-.2	Indium	6.63	0
Germanium	0	-.3	Praseodymium	.33	0
Hafnium	.03	.3	Neodymium	1.18	0
Holmium	.03	0	Samarium	.29	0
Indium	6.63	0	Terbium	.04	0
Lanthanum	1.58	.1	Dysprosium	.16	0
Lithium	2.55	-.4	Holmium	.03	0
Lutetium	.01	.1	Erbium	.08	0
Magnesium	31921.31	-.4	Ytterbium	.06	0
Manganum	418.96	-.2	Beryllium	.12	-.1
Molybdenum	15.24	-.1	Vanadium	4.76	-.1
Natrium	2498.95	-.5	Yttrium	.84	-.1
Neodymium	1.18	0	Molybdenum	15.24	-.1
Niccolum	5.78	-.3	Cerium	3.17	-.1
Niobium	.32	1.6	Plumbum	9.16	-.1
Plumbum	9.16	-.1	Manganum	418.96	-.2
Praseodymium	.33	0	Ferrum	2148.48	-.2
Rhenium	2.07	-.6	Gallium	.9	-.2
Rubidium	77.46	-.3	Arsenicum	2.97	-.2
Samarium	.29	0	Argentum	0	-.2
Scandium	1.46	.2	Europium	.07	-.2
Selenium	2.96	.5	Niccolum	5.78	-.3
Silicium	2020.66	.5	Germanium	0	-.3
Stibium	.42	.3	Rubidium	77.46	-.3
Strontium	349	-1.1	Caesium	.25	-.3
Tantalum	.02	2	Lithium	2.55	-.4
Tellurium	.04	.1	Magnesium	31921.31	-.4
Terbium	.04	0	Thallium	.05	-.4
Thallium	.05	-.4	Bismuthum	0	-.4
Thorium	.36	.1	Natrium	2498.95	-.5
Thulium	.01	.1	Chromium	0	-.5
Titanium	268.6	.3	Cobaltum	1.02	-.5
Tungstenium	.54	1.2	Zincum	292.01	-.6
Uranium	.16	.1	Cadmium	.25	-.6
Vanadium	4.76	-.1	Rhenium	2.07	-.6
Ytterbium	.06	0	Calcium	145533.2	-.9
Yttrium	.84	-.1	Cuprum	18.54	-.9
Zincum	292.01	-.6	Barium	26.63	-1
Zirconium	1.11	.7	Strontium	349	-1.1

Prunus padus

Name	Value	Deviation	Name	Value	Deviation
Aluminium	1910	-.4	Borium	1170	4.2
Argentum	0	-.2	Stibium	1	2
Arsenicum	8.93	1.6	Arsenicum	8.93	1.6
Barium	106.23	-.3	Tellurium	.06	1
Beryllium	.06	-.4	Calcium	391988.56	.7
Bismuthum	0	-.4	Cuprum	109.32	.7
Borium	1170	4.2	Zirconium	1.04	.6
Cadmium	.44	-.5	Tantalum	.01	.6
Caesium	.07	-.6	Titanium	280.7	.5
Calcium	391988.56	.7	Strontium	1230	.5
Cerium	2.05	-.4	Selenium	2.74	.4
Chromium	0	-.5	Magnesium	59863.47	.2
Cobaltum	1.46	-.3	Niobium	.15	.1
Cuprum	109.32	.7	Thulium	.01	.1
Dysprosium	.07	-.4	Hafnium	.02	0
Erbium	.04	-.3	Manganum	2040	-.1
Europium	.05	-.4	Rubidium	130.3	-.1
Ferrum	2971.75	-.2	Rhenium	6.26	-.1
Gadolinium	.17	-.4	Thallium	.11	-.1
Gallium	.74	-.4	Vanadium	2.64	-.2
Germanium	0	-.3	Ferrum	2971.75	-.2
Hafnium	.02	0	Argentum	0	-.2
Holmium	.01	-.5	Indium	4.42	-.2
Indium	4.42	-.2	Plumbum	8.26	-.2
Lanthanum	.8	-.4	Natrium	5431.5	-.3
Lithium	1.93	-.5	Scandium	.93	-.3
Lutetium	0	-.6	Cobaltum	1.46	-.3
Magnesium	59863.47	.2	Germanium	0	-.3
Manganum	2040	-.1	Molybdenum	2.65	-.3
Molybdenum	2.65	-.3	Barium	106.23	-.3
Natrium	5431.5	-.3	Erbium	.04	-.3
Neodymium	.57	-.4	Beryllium	.06	-.4
Niccolum	0	-.8	Aluminium	1910	-.4
Niobium	.15	.1	Silicium	0	-.4
Plumbum	8.26	-.2	Zincum	413.62	-.4
Praseodymium	.16	-.4	Gallium	.74	-.4
Rhenium	6.26	-.1	Yttrium	.43	-.4
Rubidium	130.3	-.1	Lanthanum	.8	-.4
Samarium	.14	-.4	Cerium	2.05	-.4
Scandium	.93	-.3	Praseodymium	.16	-.4
Selenium	2.74	.4	Neodymium	.57	-.4
Silicium	0	-.4	Samarium	.14	-.4
Stibium	1	2	Europium	.05	-.4
Strontium	1230	.5	Gadolinium	.17	-.4
Tantalum	.01	.6	Terbium	.02	-.4
Tellurium	.06	1	Dysprosium	.07	-.4
Terbium	.02	-.4	Ytterbium	.03	-.4
Thallium	.11	-.1	Bismuthum	0	-.4
Thorium	.17	-.4	Thorium	.17	-.4
Thulium	.01	.1	Uranium	.06	-.4
Titanium	280.7	.5	Lithium	1.93	-.5
Tungstenium	0	-.6	Chromium	0	-.5
Uranium	.06	-.4	Cadmium	.44	-.5
Vanadium	2.64	-.2	Holmium	.01	-.5
Ytterbium	.03	-.4	Caesium	.07	-.6
Yttrium	.43	-.4	Lutetium	0	-.6
Zincum	413.62	-.4	Tungstenium	0	-.6
Zirconium	1.04	.6	Niccolum	0	-.8

Psoralea bituminosa

Name	Value	Deviation	Name	Value	Deviation
Aluminium	6330	.5	Bismuthum	.07	2.3
Argentum	0	-.2	Rhenium	25.82	2.2
Arsenicum	3.53	-.1	Magnesium	150000	2.1
Barium	246.1	.8	Molybdenum	94.55	1.3
Beryllium	.13	0	Cobaltum	4.93	1.1
Bismuthum	.07	2.3	Calcium	436703.3	1
Borium	441.84	.3	Cadmium	3.36	.9
Cadmium	3.36	.9	Zincum	1250	.8
Caesium	.24	-.3	Strontium	1420	.8
Calcium	436703.3	1	Barium	246.1	.8
Cerium	2.56	-.3	Niccolum	16.75	.7
Chromium	0	-.5	Lithium	6.83	.6
Cobaltum	4.93	1.1	Aluminium	6330	.5
Cuprum	39.13	-.5	Tellurium	.05	.5
Dysprosium	.16	0	Borium	441.84	.3
Erbium	.06	-.2	Zirconium	.78	.3
Europium	.09	.1	Gallium	1.2	.2
Ferrum	5709.08	-.1	Selenium	2.38	.2
Gadolinium	.28	-.1	Thallium	.2	.2
Gallium	1.2	.2	Europium	.09	.1
Germanium	0	-.3	Thulium	.01	.1
Hafnium	.02	0	Beryllium	.13	0
Holmium	.02	-.2	Niobium	.14	0
Indium	3.01	-.4	Dysprosium	.16	0
Lanthanum	.99	-.3	Hafnium	.02	0
Lithium	6.83	.6	Natrium	7969.2	-.1
Lutetium	0	-.6	Manganum	1860	-.1
Magnesium	150000	2.1	Ferrum	5709.08	-.1
Manganum	1860	-.1	Arsenicum	3.53	-.1
Molybdenum	94.55	1.3	Gadolinium	.28	-.1
Natrium	7969.2	-.1	Scandium	.96	-.2
Neodymium	.87	-.2	Titanium	226.74	-.2
Niccolum	16.75	.7	Vanadium	3.88	-.2
Niobium	.14	0	Yttrium	.66	-.2
Plumbum	5.54	-.4	Argentum	0	-.2
Praseodymium	.22	-.3	Stibium	.26	-.2
Rhenium	25.82	2.2	Neodymium	.87	-.2
Rubidium	45.37	-.3	Samarium	.22	-.2
Samarium	.22	-.2	Terbium	.03	-.2
Scandium	.96	-.2	Holmium	.02	-.2
Selenium	2.38	.2	Erbium	.06	-.2
Silicium	0	-.4	Thorium	.23	-.2
Stibium	.26	-.2	Uranium	.1	-.2
Strontium	1420	.8	Germanium	0	-.3
Tantalum	0	-.7	Rubidium	45.37	-.3
Tellurium	.05	.5	Caesium	.24	-.3
Terbium	.03	-.2	Lanthanum	.99	-.3
Thallium	.2	.2	Cerium	2.56	-.3
Thorium	.23	-.2	Praseodymium	.22	-.3
Thulium	.01	.1	Silicium	0	-.4
Titanium	226.74	-.2	Indium	3.01	-.4
Tungstenium	0	-.6	Ytterbium	.03	-.4
Uranium	.1	-.2	Plumbum	5.54	-.4
Vanadium	3.88	-.2	Chromium	0	-.5
Ytterbium	.03	-.4	Cuprum	39.13	-.5
Yttrium	.66	-.2	Lutetium	0	-.6
Zincum	1250	.8	Tungstenium	0	-.6
Zirconium	.78	.3	Tantalum	0	-.7

Rhus toxicodendron

Name	Value	Deviation	Name	Value	Deviation
Aluminium	1320	-.5	Barium	398.78	2
Argentum	0	-.2	Calcium	536080.95	1.7
Arsenicum	2.85	-.3	Strontium	1830	1.6
Barium	398.78	2	Borium	505.67	.6
Beryllium	.1	-.2	Thallium	.23	.4
Bismuthum	0	-.4	Manganum	7500	.3
Borium	505.67	.6	Gallium	1.33	.3
Cadmium	.9	-.3	Europium	.09	.1
Caesium	.05	-.6	Titanium	236.66	0
Calcium	536080.95	1.7	Ferrum	7549.41	0
Cerium	1.78	-.5	Chromium	2.68	-.1
Chromium	2.68	-.1	Niccolum	8.37	-.1
Cobaltum	1.45	-.3	Cuprum	63.85	-.1
Cuprum	63.85	-.1	Stibium	.3	-.1
Dysprosium	.06	-.4	Lithium	3.2	-.2
Erbium	.03	-.4	Beryllium	.1	-.2
Europium	.09	.1	Argentum	0	-.2
Ferrum	7549.41	0	Magnesium	37166.07	-.3
Gadolinium	.13	-.4	Vanadium	1.67	-.3
Gallium	1.33	.3	Cobaltum	1.45	-.3
Germanium	0	-.3	Germanium	0	-.3
Hafnium	0	-.6	Arsenicum	2.85	-.3
Holmium	.01	-.5	Selenium	1.68	-.3
Indium	3.48	-.3	Molybdenum	3.99	-.3
Lanthanum	.6	-.5	Cadmium	.9	-.3
Lithium	3.2	-.2	Indium	3.48	-.3
Lutetium	0	-.6	Plumbum	7.36	-.3
Magnesium	37166.07	-.3	Natrium	3151.79	-.4
Manganum	7500	.3	Silicium	0	-.4
Molybdenum	3.99	-.3	Zincum	399.98	-.4
Natrium	3151.79	-.4	Rubidium	32.51	-.4
Neodymium	.39	-.5	Yttrium	.39	-.4
Niccolum	8.37	-.1	Gadolinium	.13	-.4
Niobium	.06	-.7	Dysprosium	.06	-.4
Plumbum	7.36	-.3	Erbium	.03	-.4
Praseodymium	.11	-.5	Bismuthum	0	-.4
Rhenium	1.94	-.6	Aluminium	1320	-.5
Rubidium	32.51	-.4	Scandium	.73	-.5
Samarium	.09	-.5	Zirconium	0	-.5
Scandium	.73	-.5	Lanthanum	.6	-.5
Selenium	1.68	-.3	Cerium	1.78	-.5
Silicium	0	-.4	Praseodymium	.11	-.5
Stibium	.3	-.1	Neodymium	.39	-.5
Strontium	1830	1.6	Samarium	.09	-.5
Tantalum	0	-.7	Terbium	.01	-.5
Tellurium	.01	-1.3	Holmium	.01	-.5
Terbium	.01	-.5	Ytterbium	.02	-.5
Thallium	.23	.4	Uranium	.03	-.5
Thorium	.09	-.6	Caesium	.05	-.6
Thulium	0	-.6	Thulium	0	-.6
Titanium	236.66	0	Lutetium	0	-.6
Tungstenium	0	-.6	Hafnium	0	-.6
Uranium	.03	-.5	Tungstenium	0	-.6
Vanadium	1.67	-.3	Rhenium	1.94	-.6
Ytterbium	.02	-.5	Thorium	.09	-.6
Yttrium	.39	-.4	Niobium	.06	-.7
Zincum	399.98	-.4	Tantalum	0	-.7
Zirconium	0	-.5	Tellurium	.01	-1.3

Rumex acetosa / acetosella

Name	Value	Deviation	Name	Value	Deviation
Aluminium	16900	2.8	Uranium	1.1	4.6
Argentum	0	-.2	Niccolum	48.7	3.7
Arsenicum	5.38	.5	Hafnium	.15	3.7
Barium	379.28	1.9	Zirconium	3.69	3.4
Beryllium	.73	3.1	Europium	.38	3.3
Bismuthum	0	-.4	Beryllium	.73	3.1
Borium	436.84	.3	Caesium	2.38	3.1
Cadmium	3.38	.9	Neodymium	5.92	3
Caesium	2.38	3.1	Terbium	.19	2.9
Calcium	293406.38	.1	Dysprosium	.85	2.9
Cerium	12.4	2.8	Holmium	.15	2.9
Chromium	18.24	2.5	Erbium	.44	2.9
Cobaltum	6.9	1.8	Aluminium	16900	2.8
Cuprum	156.97	1.6	Lanthanum	6.52	2.8
Dysprosium	.85	2.9	Cerium	12.4	2.8
Erbium	.44	2.9	Samarium	1.36	2.8
Europium	.38	3.3	Gadolinium	1.55	2.8
Ferrum	15186.62	.2	Thulium	.05	2.8
Gadolinium	1.55	2.8	Ytterbium	.33	2.8
Gallium	3.17	2.6	Lutetium	.05	2.8
Germanium	.12	.3	Praseodymium	1.44	2.7
Hafnium	.15	3.7	Gallium	3.17	2.6
Holmium	.15	2.9	Yttrium	4.67	2.6
Indium	13.12	.7	Thorium	1.4	2.6
Lanthanum	6.52	2.8	Chromium	18.24	2.5
Lithium	12.52	1.9	Thallium	.71	2.3
Lutetium	.05	2.8	Strontium	2110	2.1
Magnesium	33408.55	-.4	Lithium	12.52	1.9
Manganum	3620	0	Barium	379.28	1.9
Molybdenum	33.27	.2	Cobaltum	6.9	1.8
Natrium	12337.57	.2	Cuprum	156.97	1.6
Neodymium	5.92	3	Vanadium	40.52	1.4
Niccolum	48.7	3.7	Plumbum	26.5	1.4
Niobium	.28	1.2	Scandium	2.39	1.2
Plumbum	26.5	1.4	Niobium	.28	1.2
Praseodymium	1.44	2.7	Titanium	335.91	1.1
Rhenium	3.61	-.4	Cadmium	3.38	.9
Rubidium	297.03	.3	Indium	13.12	.7
Samarium	1.36	2.8	Tantalum	.01	.6
Scandium	2.39	1.2	Arsenicum	5.38	.5
Selenium	1.36	-.5	Borium	436.84	.3
Silicium	0	-.4	Germanium	.12	.3
Stibium	.15	-.5	Rubidium	297.03	.3
Strontium	2110	2.1	Natrium	12337.57	.2
Tantalum	.01	.6	Ferrum	15186.62	.2
Tellurium	.03	-.4	Molybdenum	33.27	.2
Terbium	.19	2.9	Calcium	293406.38	.1
Thallium	.71	2.3	Manganum	3620	0
Thorium	1.4	2.6	Argentum	0	-.2
Thulium	.05	2.8	Magnesium	33408.55	-.4
Titanium	335.91	1.1	Silicium	0	-.4
Tungstenium	0	-.6	Zincum	441.65	-.4
Uranium	1.1	4.6	Tellurium	.03	-.4
Vanadium	40.52	1.4	Rhenium	3.61	-.4
Ytterbium	.33	2.8	Bismuthum	0	-.4
Yttrium	4.67	2.6	Selenium	1.36	-.5
Zincum	441.65	-.4	Stibium	.15	-.5
Zirconium	3.69	3.4	Tungstenium	0	-.6

Ruta graveolens

Name	Value	Deviation	Name	Value	Deviation
Aluminium	1310	-.5	Lithium	15.9	2.7
Argentum	0	-.2	Niccolum	26.52	1.6
Arsenicum	3.04	-.2	Borium	660.93	1.5
Barium	109.52	-.3	Strontium	1200	.5
Beryllium	.19	.3	Cadmium	2.61	.5
Bismuthum	0	-.4	Calcium	343347.9	.4
Borium	660.93	1.5	Beryllium	.19	.3
Cadmium	2.61	.5	Cobaltum	2.97	.3
Caesium	.06	-.6	Selenium	2.43	.2
Calcium	343347.9	.4	Rhenium	8.73	.2
Cerium	1.61	-.5	Natrium	10369.96	.1
Chromium	0	-.5	Manganum	4740	.1
Cobaltum	2.97	.3	Gallium	1.12	.1
Cuprum	68.53	0	Stibium	.35	.1
Dysprosium	.06	-.4	Tellurium	.04	.1
Erbium	.03	-.4	Titanium	238.14	0
Europium	.04	-.5	Cuprum	68.53	0
Ferrum	2296.9	-.2	Magnesium	41143.67	-.2
Gadolinium	.12	-.5	Ferrum	2296.9	-.2
Gallium	1.12	.1	Zincum	542.87	-.2
Germanium	0	-.3	Arsenicum	3.04	-.2
Hafnium	0	-.6	Argentum	0	-.2
Holmium	.01	-.5	Indium	4.82	-.2
Indium	4.82	-.2	Germanium	0	-.3
Lanthanum	.53	-.5	Rubidium	45.25	-.3
Lithium	15.9	2.7	Molybdenum	4.91	-.3
Lutetium	0	-.6	Barium	109.52	-.3
Magnesium	41143.67	-.2	Silicium	0	-.4
Manganum	4740	.1	Vanadium	0	-.4
Molybdenum	4.91	-.3	Yttrium	.45	-.4
Natrium	10369.96	.1	Dysprosium	.06	-.4
Neodymium	.35	-.5	Erbium	.03	-.4
Niccolum	26.52	1.6	Thallium	.04	-.4
Niobium	.04	-.8	Bismuthum	0	-.4
Plumbum	5.07	-.5	Aluminium	1310	-.5
Praseodymium	.09	-.6	Scandium	.72	-.5
Rhenium	8.73	.2	Chromium	0	-.5
Rubidium	45.25	-.3	Zirconium	0	-.5
Samarium	.08	-.5	Lanthanum	.53	-.5
Scandium	.72	-.5	Cerium	1.61	-.5
Selenium	2.43	.2	Neodymium	.35	-.5
Silicium	0	-.4	Samarium	.08	-.5
Stibium	.35	.1	Europium	.04	-.5
Strontium	1200	.5	Gadolinium	.12	-.5
Tantalum	0	-.7	Terbium	.01	-.5
Tellurium	.04	.1	Holmium	.01	-.5
Terbium	.01	-.5	Ytterbium	.02	-.5
Thallium	.04	-.4	Plumbum	5.07	-.5
Thorium	.1	-.5	Thorium	.1	-.5
Thulium	0	-.6	Uranium	.04	-.5
Titanium	238.14	0	Caesium	.06	-.6
Tungstenium	0	-.6	Praseodymium	.09	-.6
Uranium	.04	-.5	Thulium	0	-.6
Vanadium	0	-.4	Lutetium	0	-.6
Ytterbium	.02	-.5	Hafnium	0	-.6
Yttrium	.45	-.4	Tungstenium	0	-.6
Zincum	542.87	-.2	Tantalum	0	-.7
Zirconium	0	-.5	Niobium	.04	-.8

Salicaria purpurea

Name	Value	Deviation	Name	Value	Deviation
Aluminium	2050	-.4	Lithium	14.21	2.3
Argentum	0	-.2	Bismuthum	.06	1.9
Arsenicum	1.31	-.7	Natrium	29136.5	1.4
Barium	70.94	-.6	Tungstenium	.59	1.4
Beryllium	.06	-.4	Cobaltum	5.21	1.2
Bismuthum	.06	1.9	Zincum	1540	1.2
Borium	358.33	-.1	Thallium	.39	1
Cadmium	.78	-.3	Gallium	1.69	.8
Caesium	.2	-.4	Selenium	3.23	.7
Calcium	337844.41	.4	Strontium	1310	.7
Cerium	1.69	-.5	Tellurium	.05	.5
Chromium	0	-.5	Calcium	337844.41	.4
Cobaltum	5.21	1.2	Manganum	8460	.4
Cuprum	48.25	-.4	Rubidium	205.49	.1
Dysprosium	.07	-.4	Niccolum	8.83	0
Erbium	.03	-.4	Borium	358.33	-.1
Europium	.04	-.5	Magnesium	45003.1	-.1
Ferrum	3960.36	-.1	Ferrum	3960.36	-.1
Gadolinium	.14	-.4	Plumbum	9.21	-.1
Gallium	1.69	.8	Titanium	223.67	-.2
Germanium	0	-.3	Argentum	0	-.2
Hafnium	.01	-.3	Indium	4.6	-.2
Holmium	.01	-.5	Stibium	.27	-.2
Indium	4.6	-.2	Vanadium	1.04	-.3
Lanthanum	.54	-.5	Germanium	0	-.3
Lithium	14.21	2.3	Molybdenum	7.34	-.3
Lutetium	0	-.6	Cadmium	.78	-.3
Magnesium	45003.1	-.1	Hafnium	.01	-.3
Manganum	8460	.4	Beryllium	.06	-.4
Molybdenum	7.34	-.3	Aluminium	2050	-.4
Natrium	29136.5	1.4	Silicium	0	-.4
Neodymium	.42	-.5	Cuprum	48.25	-.4
Niccolum	8.83	0	Caesium	.2	-.4
Niobium	.05	-.7	Gadolinium	.14	-.4
Plumbum	9.21	-.1	Terbium	.02	-.4
Praseodymium	.11	-.5	Dysprosium	.07	-.4
Rhenium	1.79	-.6	Erbium	.03	-.4
Rubidium	205.49	.1	Chromium	0	-.5
Samarium	.1	-.5	Yttrium	.31	-.5
Scandium	.56	-.6	Zirconium	0	-.5
Selenium	3.23	.7	Lanthanum	.54	-.5
Silicium	0	-.4	Cerium	1.69	-.5
Stibium	.27	-.2	Praseodymium	.11	-.5
Strontium	1310	.7	Neodymium	.42	-.5
Tantalum	0	-.7	Samarium	.1	-.5
Tellurium	.05	.5	Europium	.04	-.5
Terbium	.02	-.4	Holmium	.01	-.5
Thallium	.39	1	Ytterbium	.02	-.5
Thorium	.09	-.6	Uranium	.04	-.5
Thulium	0	-.6	Scandium	.56	-.6
Titanium	223.67	-.2	Barium	70.94	-.6
Tungstenium	.59	1.4	Thulium	0	-.6
Uranium	.04	-.5	Lutetium	0	-.6
Vanadium	1.04	-.3	Rhenium	1.79	-.6
Ytterbium	.02	-.5	Thorium	.09	-.6
Yttrium	.31	-.5	Arsenicum	1.31	-.7
Zincum	1540	1.2	Niobium	.05	-.7
Zirconium	0	-.5	Tantalum	0	-.7

Salix purpurea

Name	Value	Deviation	Name	Value	Deviation
Aluminium	1580	-.5	Cadmium	13.03	5.5
Argentum	0	-.2	Zincum	3000	3.4
Arsenicum	2.05	-.5	Calcium	765546.53	3.2
Barium	183.52	.3	Strontium	2180	2.2
Beryllium	.07	-.3	Cobaltum	5.29	1.2
Bismuthum	0	-.4	Stibium	.64	.9
Borium	417.08	.2	Gallium	1.5	.5
Cadmium	13.03	5.5	Tellurium	.05	.5
Caesium	.06	-.6	Indium	9.06	.3
Calcium	765546.53	3.2	Barium	183.52	.3
Cerium	2.16	-.4	Borium	417.08	.2
Chromium	3.9	.1	Manganum	6030	.2
Cobaltum	5.29	1.2	Plumbum	13.09	.2
Cuprum	36.61	-.6	Chromium	3.9	.1
Dysprosium	.06	-.4	Ferrum	9877.26	.1
Erbium	.03	-.4	Vanadium	5.62	-.1
Europium	.06	-.3	Niccolum	6.59	-.2
Ferrum	9877.26	.1	Argentum	0	-.2
Gadolinium	.15	-.4	Lithium	2.71	-.3
Gallium	1.5	.5	Beryllium	.07	-.3
Germanium	0	-.3	Natrium	5277.95	-.3
Hafnium	0	-.6	Germanium	0	-.3
Holmium	.01	-.5	Lanthanum	.9	-.3
Indium	9.06	.3	Europium	.06	-.3
Lanthanum	.9	-.3	Silicium	0	-.4
Lithium	2.71	-.3	Rubidium	26.99	-.4
Lutetium	0	-.6	Yttrium	.36	-.4
Magnesium	15658.32	-.8	Niobium	.09	-.4
Manganum	6030	.2	Molybdenum	.93	-.4
Molybdenum	.93	-.4	Cerium	2.16	-.4
Natrium	5277.95	-.3	Praseodymium	.17	-.4
Neodymium	.55	-.4	Neodymium	.55	-.4
Niccolum	6.59	-.2	Samarium	.12	-.4
Niobium	.09	-.4	Gadolinium	.15	-.4
Plumbum	13.09	.2	Terbium	.02	-.4
Praseodymium	.17	-.4	Dysprosium	.06	-.4
Rhenium	.58	-.7	Erbium	.03	-.4
Rubidium	26.99	-.4	Ytterbium	.03	-.4
Samarium	.12	-.4	Bismuthum	0	-.4
Scandium	.66	-.5	Uranium	.06	-.4
Selenium	1.04	-.7	Aluminium	1580	-.5
Silicium	0	-.4	Scandium	.66	-.5
Stibium	.64	.9	Arsenicum	2.05	-.5
Strontium	2180	2.2	Zirconium	0	-.5
Tantalum	0	-.7	Holmium	.01	-.5
Tellurium	.05	.5	Thallium	.02	-.5
Terbium	.02	-.4	Thorium	.11	-.5
Thallium	.02	-.5	Cuprum	36.61	-.6
Thorium	.11	-.5	Caesium	.06	-.6
Thulium	0	-.6	Thulium	0	-.6
Titanium	155.64	-1	Lutetium	0	-.6
Tungstenium	0	-.6	Hafnium	0	-.6
Uranium	.06	-.4	Tungstenium	0	-.6
Vanadium	5.62	-.1	Selenium	1.04	-.7
Ytterbium	.03	-.4	Tantalum	0	-.7
Yttrium	.36	-.4	Rhenium	.58	-.7
Zincum	3000	3.4	Magnesium	15658.32	-.8
Zirconium	0	-.5	Titanium	155.64	-1

Salvia sclarea

Name	Value	Deviation	Name	Value	Deviation
Aluminium	1230	-.6	Lithium	20.48	3.7
Argentum	0	-.2	Rhenium	20.54	1.6
Arsenicum	2.58	-.4	Strontium	1360	.7
Barium	192.77	.4	Beryllium	.23	.5
Beryllium	.23	.5	Barium	192.77	.4
Bismuthum	0	-.4	Calcium	328233.78	.3
Borium	279.77	-.6	Natrium	12278.88	.2
Cadmium	.3	-.6	Manganum	1100	-.1
Caesium	.1	-.5	Thallium	.12	-.1
Calcium	328233.78	.3	Ferrum	2321	-.2
Cerium	1.68	-.5	Argentum	0	-.2
Chromium	1.2	-.3	Magnesium	39324.53	-.3
Cobaltum	1.05	-.5	Vanadium	2.03	-.3
Cuprum	16.87	-.9	Chromium	1.2	-.3
Dysprosium	.05	-.5	Niccolum	5.49	-.3
Erbium	.03	-.4	Gallium	.78	-.3
Europium	.05	-.4	Germanium	0	-.3
Ferrum	2321	-.2	Molybdenum	4.35	-.3
Gadolinium	.1	-.5	Indium	4.23	-.3
Gallium	.78	-.3	Hafnium	.01	-.3
Germanium	0	-.3	Plumbum	7.12	-.3
Hafnium	.01	-.3	Silicium	0	-.4
Holmium	.01	-.5	Arsenicum	2.58	-.4
Indium	4.23	-.3	Rubidium	24.09	-.4
Lanthanum	.53	-.5	Yttrium	.35	-.4
Lithium	20.48	3.7	Stibium	.2	-.4
Lutetium	0	-.6	Tellurium	.03	-.4
Magnesium	39324.53	-.3	Europium	.05	-.4
Manganum	1100	-.1	Erbium	.03	-.4
Molybdenum	4.35	-.3	Bismuthum	0	-.4
Natrium	12278.88	.2	Cobaltum	1.05	-.5
Neodymium	.34	-.5	Zincum	370.93	-.5
Niccolum	5.49	-.3	Zirconium	0	-.5
Niobium	.03	-.9	Caesium	.1	-.5
Plumbum	7.12	-.3	Lanthanum	.53	-.5
Praseodymium	.09	-.6	Cerium	1.68	-.5
Rhenium	20.54	1.6	Neodymium	.34	-.5
Rubidium	24.09	-.4	Samarium	.08	-.5
Samarium	.08	-.5	Gadolinium	.1	-.5
Scandium	.59	-.6	Terbium	.01	-.5
Selenium	0	-1.3	Dysprosium	.05	-.5
Silicium	0	-.4	Holmium	.01	-.5
Stibium	.2	-.4	Ytterbium	.02	-.5
Strontium	1360	.7	Uranium	.03	-.5
Tantalum	0	-.7	Borium	279.77	-.6
Tellurium	.03	-.4	Aluminium	1230	-.6
Terbium	.01	-.5	Scandium	.59	-.6
Thallium	.12	-.1	Cadmium	.3	-.6
Thorium	.07	-.6	Praseodymium	.09	-.6
Thulium	0	-.6	Thulium	0	-.6
Titanium	100.25	-1.6	Lutetium	0	-.6
Tungstenium	0	-.6	Tungstenium	0	-.6
Uranium	.03	-.5	Thorium	.07	-.6
Vanadium	2.03	-.3	Tantalum	0	-.7
Ytterbium	.02	-.5	Cuprum	16.87	-.9
Yttrium	.35	-.4	Niobium	.03	-.9
Zincum	370.93	-.5	Selenium	0	-1.3
Zirconium	0	-.5	Titanium	100.25	-1.6

Sanguisorba officinalis

Name	Value	Deviation	Name	Value	Deviation
Aluminium	491	-.7	Borium	442.98	.3
Argentum	0	-.2	Niccolum	12.91	.3
Arsenicum	2.68	-.3	Tellurium	.04	.1
Barium	162.12	.1	Barium	162.12	.1
Beryllium	.01	-.7	Rubidium	174.39	0
Bismuthum	0	-.4	Caesium	.43	0
Borium	442.98	.3	Thallium	.13	0
Cadmium	.55	-.5	Magnesium	48677.9	-.1
Caesium	.43	0	Titanium	232.52	-.1
Calcium	246101.69	-.3	Manganum	1410	-.1
Cerium	1.18	-.7	Ferrum	1965.99	-.2
Chromium	0	-.5	Molybdenum	8.01	-.2
Cobaltum	1.59	-.3	Argentum	0	-.2
Cuprum	31.11	-.7	Natrium	5107.05	-.3
Dysprosium	.02	-.6	Calcium	246101.69	-.3
Erbium	.01	-.6	Cobaltum	1.59	-.3
Europium	.04	-.5	Germanium	0	-.3
Ferrum	1965.99	-.2	Arsenicum	2.68	-.3
Gadolinium	.06	-.6	Strontium	802	-.3
Gallium	.74	-.4	Silicium	0	-.4
Germanium	0	-.3	Scandium	.79	-.4
Hafnium	0	-.6	Vanadium	0	-.4
Holmium	0	-.7	Zincum	415.76	-.4
Indium	2.53	-.4	Gallium	.74	-.4
Lanthanum	.3	-.6	Selenium	1.52	-.4
Lithium	1.28	-.7	Indium	2.53	-.4
Lutetium	0	-.6	Bismuthum	0	-.4
Magnesium	48677.9	-.1	Chromium	0	-.5
Manganum	1410	-.1	Zirconium	0	-.5
Molybdenum	8.01	-.2	Cadmium	.55	-.5
Natrium	5107.05	-.3	Europium	.04	-.5
Neodymium	.17	-.6	Terbium	.01	-.5
Niccolum	12.91	.3	Yttrium	.15	-.6
Niobium	.02	-1	Lanthanum	.3	-.6
Plumbum	3.72	-.6	Neodymium	.17	-.6
Praseodymium	.05	-.7	Samarium	.04	-.6
Rhenium	1.09	-.7	Gadolinium	.06	-.6
Rubidium	174.39	0	Dysprosium	.02	-.6
Samarium	.04	-.6	Erbium	.01	-.6
Scandium	.79	-.4	Thulium	0	-.6
Selenium	1.52	-.4	Ytterbium	.01	-.6
Silicium	0	-.4	Lutetium	0	-.6
Stibium	.09	-.7	Hafnium	0	-.6
Strontium	802	-.3	Tungstenium	0	-.6
Tantalum	0	-.7	Plumbum	3.72	-.6
Tellurium	.04	.1	Uranium	.01	-.6
Terbium	.01	-.5	Lithium	1.28	-.7
Thallium	.13	0	Beryllium	.01	-.7
Thorium	.03	-.7	Aluminium	491	-.7
Thulium	0	-.6	Cuprum	31.11	-.7
Titanium	232.52	-.1	Stibium	.09	-.7
Tungstenium	0	-.6	Cerium	1.18	-.7
Uranium	.01	-.6	Praseodymium	.05	-.7
Vanadium	0	-.4	Holmium	0	-.7
Ytterbium	.01	-.6	Tantalum	0	-.7
Yttrium	.15	-.6	Rhenium	1.09	-.7
Zincum	415.76	-.4	Thorium	.03	-.7
Zirconium	0	-.5	Niobium	.02	-1

Scrophularia nodosa

Name	Value	Deviation	Name	Value	Deviation
Aluminium	2560	-.3	Selenium	5.99	2.4
Argentum	0	-.2	Titanium	343.81	1.2
Arsenicum	2.96	-.2	Stibium	.71	1.1
Barium	106.64	-.3	Rhenium	14.56	.9
Beryllium	.11	-.1	Plumbum	19.06	.8
Bismuthum	0	-.4	Rubidium	444.89	.7
Borium	448.57	.3	Indium	13.26	.7
Cadmium	1.32	-.1	Magnesium	78691.06	.6
Caesium	.23	-.3	Tellurium	.05	.5
Calcium	338261.86	.4	Calcium	338261.86	.4
Cerium	2.61	-.2	Borium	448.57	.3
Chromium	0	-.5	Strontium	1140	.3
Cobaltum	1.62	-.3	Lithium	5.29	.2
Cuprum	75.68	.1	Cuprum	75.68	.1
Dysprosium	.09	-.3	Thulium	.01	.1
Erbium	.05	-.3	Lutetium	.01	.1
Europium	.05	-.4	Gallium	1.09	0
Ferrum	3384.43	-.1	Beryllium	.11	-.1
Gadolinium	.19	-.3	Scandium	1.08	-.1
Gallium	1.09	0	Manganum	1170	-.1
Germanium	0	-.3	Ferrum	3384.43	-.1
Hafnium	.01	-.3	Niobium	.13	-.1
Holmium	.02	-.2	Cadmium	1.32	-.1
Indium	13.26	.7	Vanadium	2.51	-.2
Lanthanum	1.08	-.2	Niccolum	7.29	-.2
Lithium	5.29	.2	Zincum	540.38	-.2
Lutetium	.01	.1	Arsenicum	2.96	-.2
Magnesium	78691.06	.6	Molybdenum	11.46	-.2
Manganum	1170	-.1	Argentum	0	-.2
Molybdenum	11.46	-.2	Lanthanum	1.08	-.2
Natrium	2776.81	-.5	Cerium	2.61	-.2
Neodymium	.75	-.3	Holmium	.02	-.2
Niccolum	7.29	-.2	Uranium	.09	-.2
Niobium	.13	-.1	Aluminium	2560	-.3
Plumbum	19.06	.8	Cobaltum	1.62	-.3
Praseodymium	.21	-.3	Germanium	0	-.3
Rhenium	14.56	.9	Yttrium	.56	-.3
Rubidium	444.89	.7	Caesium	.23	-.3
Samarium	.17	-.3	Barium	106.64	-.3
Scandium	1.08	-.1	Praseodymium	.21	-.3
Selenium	5.99	2.4	Neodymium	.75	-.3
Silicium	0	-.4	Samarium	.17	-.3
Stibium	.71	1.1	Gadolinium	.19	-.3
Strontium	1140	.3	Dysprosium	.09	-.3
Tantalum	0	-.7	Erbium	.05	-.3
Tellurium	.05	.5	Ytterbium	.04	-.3
Terbium	.02	-.4	Hafnium	.01	-.3
Thallium	.06	-.3	Thallium	.06	-.3
Thorium	.2	-.3	Thorium	.2	-.3
Thulium	.01	.1	Silicium	0	-.4
Titanium	343.81	1.2	Europium	.05	-.4
Tungstenium	0	-.6	Terbium	.02	-.4
Uranium	.09	-.2	Bismuthum	0	-.4
Vanadium	2.51	-.2	Natrium	2776.81	-.5
Ytterbium	.04	-.3	Chromium	0	-.5
Yttrium	.56	-.3	Zirconium	0	-.5
Zincum	540.38	-.2	Tungstenium	0	-.6
Zirconium	0	-.5	Tantalum	0	-.7

Scutellaria lateriflora

Name	Value	Deviation	Name	Value	Deviation
Aluminium	5960	.5	Magnesium	150000	2.1
Argentum	0	-.2	Barium	357.75	1.7
Arsenicum	3.05	-.2	Silicium	3449.87	1
Barium	357.75	1.7	Scandium	2.26	1
Beryllium	.06	-.4	Titanium	326.97	1
Bismuthum	0	-.4	Germanium	.27	1
Borium	399.91	.1	Caesium	1.08	1
Cadmium	.99	-.2	Calcium	416481.65	.9
Caesium	1.08	1	Cuprum	114.7	.8
Calcium	416481.65	.9	Gallium	1.52	.6
Cerium	2.46	-.3	Tantalum	.01	.6
Chromium	0	-.5	Aluminium	5960	.5
Cobaltum	1.35	-.4	Strontium	1230	.5
Cuprum	114.7	.8	Selenium	2.73	.4
Dysprosium	.17	0	Europium	.12	.4
Erbium	.07	-.1	Rubidium	236.23	.2
Europium	.12	.4	Borium	399.91	.1
Ferrum	3179.78	-.1	Niccolum	10.1	.1
Gadolinium	.3	-.1	Tellurium	.04	.1
Gallium	1.52	.6	Thulium	.01	.1
Germanium	.27	1	Lutetium	.01	.1
Hafnium	.02	0	Manganum	3000	0
Holmium	.03	0	Terbium	.04	0
Indium	5.27	-.1	Dysprosium	.17	0
Lanthanum	.96	-.3	Holmium	.03	0
Lithium	2.63	-.4	Hafnium	.02	0
Lutetium	.01	.1	Natrium	8022.94	-.1
Magnesium	150000	2.1	Ferrum	3179.78	-.1
Manganum	3000	0	Niobium	.13	-.1
Molybdenum	6.89	-.3	Indium	5.27	-.1
Natrium	8022.94	-.1	Samarium	.24	-.1
Neodymium	.91	-.2	Gadolinium	.3	-.1
Niccolum	10.1	.1	Erbium	.07	-.1
Niobium	.13	-.1	Plumbum	9.14	-.1
Plumbum	9.14	-.1	Thorium	.28	-.1
Praseodymium	.23	-.2	Uranium	.12	-.1
Rhenium	2.67	-.5	Vanadium	4.42	-.2
Rubidium	236.23	.2	Arsenicum	3.05	-.2
Samarium	.24	-.1	Yttrium	.69	-.2
Scandium	2.26	1	Argentum	0	-.2
Selenium	2.73	.4	Cadmium	.99	-.2
Silicium	3449.87	1	Stibium	.27	-.2
Stibium	.27	-.2	Praseodymium	.23	-.2
Strontium	1230	.5	Neodymium	.91	-.2
Tantalum	.01	.6	Molybdenum	6.89	-.3
Tellurium	.04	.1	Lanthanum	.96	-.3
Terbium	.04	0	Cerium	2.46	-.3
Thallium	.05	-.4	Ytterbium	.04	-.3
Thorium	.28	-.1	Lithium	2.63	-.4
Thulium	.01	.1	Beryllium	.06	-.4
Titanium	326.97	1	Cobaltum	1.35	-.4
Tungstenium	0	-.6	Thallium	.05	-.4
Uranium	.12	-.1	Bismuthum	0	-.4
Vanadium	4.42	-.2	Chromium	0	-.5
Ytterbium	.04	-.3	Zincum	313.52	-.5
Yttrium	.69	-.2	Zirconium	0	-.5
Zincum	313.52	-.5	Rhenium	2.67	-.5
Zirconium	0	-.5	Tungstenium	0	-.6

Solidago virgaurea

Name	Value	Deviation
Aluminium	573	-.7
Argentum	0	-.2
Arsenicum	1.31	-.7
Barium	28.84	-.9
Beryllium	.04	-.5
Bismuthum	0	-.4
Borium	401.95	.1
Cadmium	1.18	-.2
Caesium	.12	-.5
Calcium	187131.92	-.6
Cerium	1.34	-.6
Chromium	0	-.5
Cobaltum	.7	-.6
Cuprum	40.62	-.5
Dysprosium	.05	-.5
Erbium	.02	-.5
Europium	.02	-.7
Ferrum	1696.64	-.2
Gadolinium	.09	-.5
Gallium	.62	-.5
Germanium	0	-.3
Hafnium	0	-.6
Holmium	.01	-.5
Indium	1.7	-.5
Lanthanum	.25	-.7
Lithium	.87	-.8
Lutetium	0	-.6
Magnesium	31801.01	-.4
Manganum	861.33	-.1
Molybdenum	7.62	-.2
Natrium	2057.43	-.5
Neodymium	.24	-.6
Niccolum	0	-.8
Niobium	.03	-.9
Plumbum	3.78	-.6
Praseodymium	.06	-.6
Rhenium	4.74	-.2
Rubidium	51.55	-.3
Samarium	.08	-.5
Scandium	.56	-.6
Selenium	2.39	.2
Silicium	0	-.4
Stibium	.11	-.7
Strontium	455	-.9
Tantalum	0	-.7
Tellurium	.02	-.8
Terbium	.01	-.5
Thallium	.11	-.1
Thorium	.06	-.6
Thulium	0	-.6
Titanium	176.05	-.7
Tungstenium	0	-.6
Uranium	.02	-.6
Vanadium	0	-.4
Ytterbium	.01	-.6
Yttrium	.18	-.6
Zincum	289.36	-.6
Zirconium	0	-.5

Name	Value	Deviation
Selenium	2.39	.2
Borium	401.95	.1
Manganum	861.33	-.1
Thallium	.11	-.1
Ferrum	1696.64	-.2
Molybdenum	7.62	-.2
Argentum	0	-.2
Cadmium	1.18	-.2
Rhenium	4.74	-.2
Germanium	0	-.3
Rubidium	51.55	-.3
Magnesium	31801.01	-.4
Silicium	0	-.4
Vanadium	0	-.4
Bismuthum	0	-.4
Beryllium	.04	-.5
Natrium	2057.43	-.5
Chromium	0	-.5
Cuprum	40.62	-.5
Gallium	.62	-.5
Zirconium	0	-.5
Indium	1.7	-.5
Caesium	.12	-.5
Samarium	.08	-.5
Gadolinium	.09	-.5
Terbium	.01	-.5
Dysprosium	.05	-.5
Holmium	.01	-.5
Erbium	.02	-.5
Calcium	187131.92	-.6
Scandium	.56	-.6
Cobaltum	.7	-.6
Zincum	289.36	-.6
Yttrium	.18	-.6
Cerium	1.34	-.6
Praseodymium	.06	-.6
Neodymium	.24	-.6
Thulium	0	-.6
Ytterbium	.01	-.6
Lutetium	0	-.6
Hafnium	0	-.6
Tungstenium	0	-.6
Plumbum	3.78	-.6
Thorium	.06	-.6
Uranium	.02	-.6
Aluminium	573	-.7
Titanium	176.05	-.7
Arsenicum	1.31	-.7
Stibium	.11	-.7
Lanthanum	.25	-.7
Europium	.02	-.7
Tantalum	0	-.7
Lithium	.87	-.8
Niccolum	0	-.8
Tellurium	.02	-.8
Strontium	455	-.9
Niobium	.03	-.9
Barium	28.84	-.9

Spilanthes oleracea

Name	Value	Deviation	Name	Value	Deviation
Aluminium	1290	-.5	Borium	701.6	1.7
Argentum	0	-.2	Rhenium	8.33	.2
Arsenicum	2.59	-.4	Calcium	272156.7	-.1
Barium	20.91	-1	Thallium	.12	-.1
Beryllium	.04	-.5	Vanadium	2.48	-.2
Bismuthum	0	-.4	Manganum	203.54	-.2
Borium	701.6	1.7	Ferrum	2201.65	-.2
Cadmium	.35	-.6	Selenium	1.84	-.2
Caesium	.09	-.6	Argentum	0	-.2
Calcium	272156.7	-.1	Chromium	1.37	-.3
Cerium	1.7	-.5	Germanium	0	-.3
Chromium	1.37	-.3	Molybdenum	2.46	-.3
Cobaltum	.97	-.5	Hafnium	.01	-.3
Cuprum	12.97	-1	Magnesium	32249.76	-.4
Dysprosium	.06	-.4	Silicium	0	-.4
Erbium	.03	-.4	Scandium	.79	-.4
Europium	.03	-.6	Arsenicum	2.59	-.4
Ferrum	2201.65	-.2	Rubidium	35.06	-.4
Gadolinium	.14	-.4	Strontium	728	-.4
Gallium	.51	-.7	Yttrium	.36	-.4
Germanium	0	-.3	Samarium	.12	-.4
Hafnium	.01	-.3	Gadolinium	.14	-.4
Holmium	.01	-.5	Dysprosium	.06	-.4
Indium	1.23	-.6	Erbium	.03	-.4
Lanthanum	.52	-.5	Bismuthum	0	-.4
Lithium	1.26	-.7	Uranium	.06	-.4
Lutetium	0	-.6	Beryllium	.04	-.5
Magnesium	32249.76	-.4	Aluminium	1290	-.5
Manganum	203.54	-.2	Cobaltum	.97	-.5
Molybdenum	2.46	-.3	Zirconium	0	-.5
Natrium	716.59	-.6	Lanthanum	.52	-.5
Neodymium	.43	-.5	Cerium	1.7	-.5
Niccolum	0	-.8	Praseodymium	.12	-.5
Niobium	.05	-.7	Neodymium	.43	-.5
Plumbum	4.2	-.6	Terbium	.01	-.5
Praseodymium	.12	-.5	Holmium	.01	-.5
Rhenium	8.33	.2	Ytterbium	.02	-.5
Rubidium	35.06	-.4	Thorium	.11	-.5
Samarium	.12	-.4	Natrium	716.59	-.6
Scandium	.79	-.4	Zincum	268.66	-.6
Selenium	1.84	-.2	Cadmium	.35	-.6
Silicium	0	-.4	Indium	1.23	-.6
Stibium	.04	-.9	Caesium	.09	-.6
Strontium	728	-.4	Europium	.03	-.6
Tantalum	0	-.7	Thulium	0	-.6
Tellurium	.02	-.8	Lutetium	0	-.6
Terbium	.01	-.5	Tungstenium	0	-.6
Thallium	.12	-.1	Plumbum	4.2	-.6
Thorium	.11	-.5	Lithium	1.26	-.7
Thulium	0	-.6	Gallium	.51	-.7
Titanium	129.55	-1.3	Niobium	.05	-.7
Tungstenium	0	-.6	Tantalum	0	-.7
Uranium	.06	-.4	Niccolum	0	-.8
Vanadium	2.48	-.2	Tellurium	.02	-.8
Ytterbium	.02	-.5	Stibium	.04	-.9
Yttrium	.36	-.4	Cuprum	12.97	-1
Zincum	268.66	-.6	Barium	20.91	-1
Zirconium	0	-.5	Titanium	129.55	-1.3

Stachys officinalis

Name	Value	Deviation	Name	Value	Deviation
Aluminium	947	-.6	Rubidium	230.84	.1
Argentum	0	-.2	Tellurium	.04	.1
Arsenicum	1.7	-.6	Borium	376.41	0
Barium	96.15	-.4	Selenium	1.96	-.1
Beryllium	.02	-.6	Caesium	.4	-.1
Bismuthum	0	-.4	Rhenium	5.85	-.1
Borium	376.41	0	Manganum	613.43	-.2
Cadmium	.07	-.7	Ferrum	1200.39	-.2
Caesium	.4	-.1	Argentum	0	-.2
Calcium	143061.13	-.9	Germanium	0	-.3
Cerium	1.32	-.6	Molybdenum	2.49	-.3
Chromium	0	-.5	Hafnium	.01	-.3
Cobaltum	.76	-.6	Silicium	0	-.4
Cuprum	44.41	-.4	Vanadium	0	-.4
Dysprosium	.03	-.6	Cuprum	44.41	-.4
Erbium	.01	-.6	Barium	96.15	-.4
Europium	.03	-.6	Thallium	.04	-.4
Ferrum	1200.39	-.2	Bismuthum	0	-.4
Gadolinium	.07	-.6	Magnesium	28067.73	-.5
Gallium	.48	-.7	Chromium	0	-.5
Germanium	0	-.3	Zirconium	0	-.5
Hafnium	.01	-.3	Indium	1.97	-.5
Holmium	0	-.7	Terbium	.01	-.5
Indium	1.97	-.5	Beryllium	.02	-.6
Lanthanum	.25	-.7	Natrium	803.34	-.6
Lithium	0	-1	Aluminium	947	-.6
Lutetium	0	-.6	Cobaltum	.76	-.6
Magnesium	28067.73	-.5	Arsenicum	1.7	-.6
Manganum	613.43	-.2	Yttrium	.14	-.6
Molybdenum	2.49	-.3	Cerium	1.32	-.6
Natrium	803.34	-.6	Neodymium	.19	-.6
Neodymium	.19	-.6	Samarium	.05	-.6
Niccolum	0	-.8	Europium	.03	-.6
Niobium	.02	-1	Gadolinium	.07	-.6
Plumbum	2.75	-.7	Dysprosium	.03	-.6
Praseodymium	.05	-.7	Erbium	.01	-.6
Rhenium	5.85	-.1	Thulium	0	-.6
Rubidium	230.84	.1	Ytterbium	.01	-.6
Samarium	.05	-.6	Lutetium	0	-.6
Scandium	0	-1.2	Tungstenium	0	-.6
Selenium	1.96	-.1	Uranium	.02	-.6
Silicium	0	-.4	Zincum	210.35	-.7
Stibium	.1	-.7	Gallium	.48	-.7
Strontium	343	-1.1	Cadmium	.07	-.7
Tantalum	0	-.7	Stibium	.1	-.7
Tellurium	.04	.1	Lanthanum	.25	-.7
Terbium	.01	-.5	Praseodymium	.05	-.7
Thallium	.04	-.4	Holmium	0	-.7
Thorium	.05	-.7	Tantalum	0	-.7
Thulium	0	-.6	Plumbum	2.75	-.7
Titanium	173.81	-.8	Thorium	.05	-.7
Tungstenium	0	-.6	Titanium	173.81	-.8
Uranium	.02	-.6	Niccolum	0	-.8
Vanadium	0	-.4	Calcium	143061.13	-.9
Ytterbium	.01	-.6	Lithium	0	-1
Yttrium	.14	-.6	Niobium	.02	-1
Zincum	210.35	-.7	Strontium	343	-1.1
Zirconium	0	-.5	Scandium	0	-1.2

Stachys officinalis R

Name	Value	Deviation	Name	Value	Deviation
Aluminium	561	-.7	Rubidium	255.48	.2
Argentum	0	-.2	Borium	396.98	.1
Arsenicum	1.34	-.7	Selenium	2.18	0
Barium	99.89	-.4	Caesium	.43	0
Beryllium	.02	-.6	Manganum	647.68	-.2
Bismuthum	0	-.4	Ferrum	1173.14	-.2
Borium	396.98	.1	Argentum	0	-.2
Cadmium	.06	-.7	Germanium	0	-.3
Caesium	.43	0	Molybdenum	2.83	-.3
Calcium	153671.02	-.9	Hafnium	.01	-.3
Cerium	1.18	-.7	Rhenium	4.51	-.3
Chromium	0	-.5	Silicium	0	-.4
Cobaltum	.75	-.6	Vanadium	0	-.4
Cuprum	46.87	-.4	Cuprum	46.87	-.4
Dysprosium	.02	-.6	Indium	2.49	-.4
Erbium	.01	-.6	Tellurium	.03	-.4
Europium	.02	-.7	Barium	99.89	-.4
Ferrum	1173.14	-.2	Thallium	.03	-.4
Gadolinium	.05	-.6	Bismuthum	0	-.4
Gallium	.46	-.7	Magnesium	29853.32	-.5
Germanium	0	-.3	Chromium	0	-.5
Hafnium	.01	-.3	Zirconium	0	-.5
Holmium	0	-.7	Stibium	.16	-.5
Indium	2.49	-.4	Beryllium	.02	-.6
Lanthanum	.18	-.7	Natrium	886.6	-.6
Lithium	.78	-.8	Cobaltum	.75	-.6
Lutetium	0	-.6	Yttrium	.12	-.6
Magnesium	29853.32	-.5	Gadolinium	.05	-.6
Manganum	647.68	-.2	Dysprosium	.02	-.6
Molybdenum	2.83	-.3	Erbium	.01	-.6
Natrium	886.6	-.6	Thulium	0	-.6
Neodymium	.13	-.7	Ytterbium	.01	-.6
Niccolum	0	-.8	Lutetium	0	-.6
Niobium	.02	-1	Tungstenium	0	-.6
Plumbum	2.56	-.7	Uranium	.01	-.6
Praseodymium	.03	-.7	Aluminium	561	-.7
Rhenium	4.51	-.3	Titanium	180.72	-.7
Rubidium	255.48	.2	Zincum	217.5	-.7
Samarium	.03	-.7	Gallium	.46	-.7
Scandium	0	-1.2	Arsenicum	1.34	-.7
Selenium	2.18	0	Cadmium	.06	-.7
Silicium	0	-.4	Lanthanum	.18	-.7
Stibium	.16	-.5	Cerium	1.18	-.7
Strontium	369	-1	Praseodymium	.03	-.7
Tantalum	0	-.7	Neodymium	.13	-.7
Tellurium	.03	-.4	Samarium	.03	-.7
Terbium	0	-.7	Europium	.02	-.7
Thallium	.03	-.4	Terbium	0	-.7
Thorium	.03	-.7	Holmium	0	-.7
Thulium	0	-.6	Tantalum	0	-.7
Titanium	180.72	-.7	Plumbum	2.56	-.7
Tungstenium	0	-.6	Thorium	.03	-.7
Uranium	.01	-.6	Lithium	.78	-.8
Vanadium	0	-.4	Niccolum	0	-.8
Ytterbium	.01	-.6	Calcium	153671.02	-.9
Yttrium	.12	-.6	Strontium	369	-1
Zincum	217.5	-.7	Niobium	.02	-1
Zirconium	0	-.5	Scandium	0	-1.2

Stachys sylvatica

Name	Value	Deviation	Name	Value	Deviation
Aluminium	2740	-.2	Selenium	4.33	1.4
Argentum	0	-.2	Germanium	.19	.6
Arsenicum	0	-1.1	Magnesium	59248.84	.2
Barium	145.72	0	Titanium	255.85	.2
Beryllium	.09	-.2	Thulium	.01	.1
Bismuthum	0	-.4	Lutetium	.01	.1
Borium	337.23	-.2	Scandium	1.18	0
Cadmium	.19	-.6	Molybdenum	21.92	0
Caesium	.28	-.2	Indium	6.97	0
Calcium	237175.73	-.3	Stibium	.33	0
Cerium	2.76	-.2	Barium	145.72	0
Chromium	0	-.5	Thallium	.14	0
Cobaltum	1.28	-.4	Manganum	722.55	-.1
Cuprum	43.44	-.4	Niobium	.12	-.1
Dysprosium	.09	-.3	Lanthanum	1.22	-.1
Erbium	.05	-.3	Rhenium	6.31	-.1
Europium	.06	-.3	Plumbum	9.53	-.1
Ferrum	2771.01	-.2	Beryllium	.09	-.2
Gadolinium	.22	-.2	Borium	337.23	-.2
Gallium	.77	-.3	Aluminium	2740	-.2
Germanium	.19	.6	Vanadium	2.93	-.2
Hafnium	.01	-.3	Ferrum	2771.01	-.2
Holmium	.02	-.2	Niccolum	6.84	-.2
Indium	6.97	0	Argentum	0	-.2
Lanthanum	1.22	-.1	Caesium	.28	-.2
Lithium	2.64	-.4	Cerium	2.76	-.2
Lutetium	.01	.1	Gadolinium	.22	-.2
Magnesium	59248.84	.2	Holmium	.02	-.2
Manganum	722.55	-.1	Thorium	.22	-.2
Molybdenum	21.92	0	Uranium	.09	-.2
Natrium	5270.58	-.3	Natrium	5270.58	-.3
Neodymium	.74	-.3	Calcium	237175.73	-.3
Niccolum	6.84	-.2	Gallium	.77	-.3
Niobium	.12	-.1	Yttrium	.53	-.3
Plumbum	9.53	-.1	Praseodymium	.22	-.3
Praseodymium	.22	-.3	Neodymium	.74	-.3
Rhenium	6.31	-.1	Samarium	.17	-.3
Rubidium	42.48	-.4	Europium	.06	-.3
Samarium	.17	-.3	Dysprosium	.09	-.3
Scandium	1.18	0	Erbium	.05	-.3
Selenium	4.33	1.4	Ytterbium	.04	-.3
Silicium	0	-.4	Hafnium	.01	-.3
Stibium	.33	0	Lithium	2.64	-.4
Strontium	677	-.5	Silicium	0	-.4
Tantalum	0	-.7	Cobaltum	1.28	-.4
Tellurium	.02	-.8	Cuprum	43.44	-.4
Terbium	.02	-.4	Rubidium	42.48	-.4
Thallium	.14	0	Terbium	.02	-.4
Thorium	.22	-.2	Bismuthum	0	-.4
Thulium	.01	.1	Chromium	0	-.5
Titanium	255.85	.2	Strontium	677	-.5
Tungstenium	0	-.6	Zirconium	0	-.5
Uranium	.09	-.2	Cadmium	.19	-.6
Vanadium	2.93	-.2	Tungstenium	0	-.6
Ytterbium	.04	-.3	Zincum	226.39	-.7
Yttrium	.53	-.3	Tantalum	0	-.7
Zincum	226.39	-.7	Tellurium	.02	-.8
Zirconium	0	-.5	Arsenicum	0	-1.1

Syringa vulgaris

Name	Value	Deviation
Aluminium	2140	-.4
Argentum	0	-.2
Arsenicum	4.51	.2
Barium	33.17	-.9
Beryllium	.09	-.2
Bismuthum	0	-.4
Borium	647.33	1.4
Cadmium	1.04	-.2
Caesium	.72	.5
Calcium	125209.02	-1.1
Cerium	2.39	-.3
Chromium	4.89	.3
Cobaltum	13.94	4.7
Cuprum	100.76	.6
Dysprosium	.07	-.4
Erbium	.04	-.3
Europium	.04	-.5
Ferrum	2614.65	-.2
Gadolinium	.18	-.3
Gallium	1.67	.8
Germanium	0	-.3
Hafnium	.04	.5
Holmium	.01	-.5
Indium	79.71	7.8
Lanthanum	1.02	-.2
Lithium	1.9	-.5
Lutetium	0	-.6
Magnesium	38464.79	-.3
Manganum	6470	.2
Molybdenum	1.73	-.4
Natrium	7732.65	-.1
Neodymium	.64	-.3
Niccolum	46.24	3.4
Niobium	.2	.5
Plumbum	11.03	0
Praseodymium	.2	-.3
Rhenium	4.15	-.3
Rubidium	862.05	1.8
Samarium	.13	-.4
Scandium	.71	-.5
Selenium	0	-1.3
Silicium	0	-.4
Stibium	1.05	2.2
Strontium	294	-1.2
Tantalum	.01	.6
Tellurium	.03	-.4
Terbium	.02	-.4
Thallium	.01	-.5
Thorium	.2	-.3
Thulium	0	-.6
Titanium	547.71	3.6
Tungstenium	0	-.6
Uranium	.06	-.4
Vanadium	4.4	-.2
Ytterbium	.03	-.4
Yttrium	.46	-.4
Zincum	2790	3.1
Zirconium	1.17	.7

Name	Value	Deviation
Indium	79.71	7.8
Cobaltum	13.94	4.7
Titanium	547.71	3.6
Niccolum	46.24	3.4
Zincum	2790	3.1
Stibium	1.05	2.2
Rubidium	862.05	1.8
Borium	647.33	1.4
Gallium	1.67	.8
Zirconium	1.17	.7
Cuprum	100.76	.6
Tantalum	.01	.6
Niobium	.2	.5
Caesium	.72	.5
Hafnium	.04	.5
Chromium	4.89	.3
Manganum	6470	.2
Arsenicum	4.51	.2
Plumbum	11.03	0
Natrium	7732.65	-.1
Beryllium	.09	-.2
Vanadium	4.4	-.2
Ferrum	2614.65	-.2
Argentum	0	-.2
Cadmium	1.04	-.2
Lanthanum	1.02	-.2
Magnesium	38464.79	-.3
Germanium	0	-.3
Cerium	2.39	-.3
Praseodymium	.2	-.3
Neodymium	.64	-.3
Gadolinium	.18	-.3
Erbium	.04	-.3
Rhenium	4.15	-.3
Thorium	.2	-.3
Aluminium	2140	-.4
Silicium	0	-.4
Yttrium	.46	-.4
Molybdenum	1.73	-.4
Tellurium	.03	-.4
Samarium	.13	-.4
Terbium	.02	-.4
Dysprosium	.07	-.4
Ytterbium	.03	-.4
Bismuthum	0	-.4
Uranium	.06	-.4
Lithium	1.9	-.5
Scandium	.71	-.5
Europium	.04	-.5
Holmium	.01	-.5
Thallium	.01	-.5
Thulium	0	-.6
Lutetium	0	-.6
Tungstenium	0	-.6
Barium	33.17	-.9
Calcium	125209.02	-1.1
Strontium	294	-1.2
Selenium	0	-1.3

Taraxacum officinale

Name	Value	Deviation	Name	Value	Deviation
Aluminium	9540	1.2	Lanthanum	5.48	2.2
Argentum	0	-.2	Thorium	1.22	2.2
Arsenicum	4.76	.3	Praseodymium	1.2	2.1
Barium	61.49	-.7	Thulium	.04	2.1
Beryllium	.31	.9	Cerium	9.87	2
Bismuthum	.05	1.5	Holmium	.11	2
Borium	176.58	-1.1	Neodymium	4.23	1.9
Cadmium	1.16	-.2	Samarium	1.02	1.9
Caesium	.62	.3	Gadolinium	1.18	1.9
Calcium	58742.8	-1.5	Silicium	5334.08	1.8
Cerium	9.87	2	Scandium	3.06	1.8
Chromium	11.51	1.4	Terbium	.13	1.8
Cobaltum	2.41	.1	Yttrium	3.44	1.7
Cuprum	45.95	-.4	Dysprosium	.57	1.7
Dysprosium	.57	1.7	Erbium	.3	1.7
Erbium	.3	1.7	Ytterbium	.22	1.6
Europium	.2	1.3	Bismuthum	.05	1.5
Ferrum	9135.73	.1	Chromium	11.51	1.4
Gadolinium	1.18	1.9	Lutetium	.03	1.4
Gallium	1.91	1	Europium	.2	1.3
Germanium	.22	.7	Aluminium	9540	1.2
Hafnium	.03	.3	Natrium	25299.36	1.1
Holmium	.11	2	Gallium	1.91	1
Indium	9.54	.3	Uranium	.34	1
Lanthanum	5.48	2.2	Beryllium	.31	.9
Lithium	7.47	.7	Niobium	.24	.9
Lutetium	.03	1.4	Plumbum	19.95	.8
Magnesium	9722.49	-.9	Lithium	7.47	.7
Manganum	389.59	-.2	Germanium	.22	.7
Molybdenum	4.36	-.3	Tantalum	.01	.6
Natrium	25299.36	1.1	Tellurium	.05	.5
Neodymium	4.23	1.9	Vanadium	15.52	.3
Niccolum	9.36	0	Arsenicum	4.76	.3
Niobium	.24	.9	Indium	9.54	.3
Plumbum	19.95	.8	Caesium	.62	.3
Praseodymium	1.2	2.1	Hafnium	.03	.3
Rhenium	.28	-.8	Zirconium	.64	.2
Rubidium	24.39	-.4	Titanium	250.9	.1
Samarium	1.02	1.9	Ferrum	9135.73	.1
Scandium	3.06	1.8	Cobaltum	2.41	.1
Selenium	1.75	-.2	Niccolum	9.36	0
Silicium	5334.08	1.8	Manganum	389.59	-.2
Stibium	.22	-.3	Selenium	1.75	-.2
Strontium	228	-1.3	Argentum	0	-.2
Tantalum	.01	.6	Cadmium	1.16	-.2
Tellurium	.05	.5	Molybdenum	4.36	-.3
Terbium	.13	1.8	Stibium	.22	-.3
Thallium	.04	-.4	Cuprum	45.95	-.4
Thorium	1.22	2.2	Rubidium	24.39	-.4
Thulium	.04	2.1	Thallium	.04	-.4
Titanium	250.9	.1	Tungstenium	0	-.6
Tungstenium	0	-.6	Barium	61.49	-.7
Uranium	.34	1	Zincum	141.56	-.8
Vanadium	15.52	.3	Rhenium	.28	-.8
Ytterbium	.22	1.6	Magnesium	9722.49	-.9
Yttrium	3.44	1.7	Borium	176.58	-1.1
Zincum	141.56	-.8	Strontium	228	-1.3
Zirconium	.64	.2	Calcium	58742.8	-1.5

Teucrium chamaedrys

Name	Value	Deviation	Name	Value	Deviation
Aluminium	2440	-.3	Molybdenum	121.82	1.8
Argentum	0	-.2	Borium	605.26	1.2
Arsenicum	6.72	.9	Rhenium	16.4	1.1
Barium	192.23	.4	Arsenicum	6.72	.9
Beryllium	.08	-.3	Calcium	345834	.4
Bismuthum	0	-.4	Barium	192.23	.4
Borium	605.26	1.2	Strontium	1120	.3
Cadmium	.5	-.5	Zirconium	.61	.2
Caesium	.1	-.5	Titanium	252.8	.1
Calcium	345834	.4	Thulium	.01	.1
Cerium	2.05	-.4	Lutetium	.01	.1
Chromium	1.73	-.2	Hafnium	.02	0
Cobaltum	1.45	-.3	Cuprum	61.35	-.1
Cuprum	61.35	-.1	Natrium	6596.94	-.2
Dysprosium	.09	-.3	Vanadium	3.35	-.2
Erbium	.05	-.3	Chromium	1.73	-.2
Europium	.06	-.3	Manganum	261.29	-.2
Ferrum	2808.92	-.2	Ferrum	2808.92	-.2
Gadolinium	.17	-.4	Argentum	0	-.2
Gallium	.7	-.4	Stibium	.26	-.2
Germanium	0	-.3	Holmium	.02	-.2
Hafnium	.02	0	Thorium	.22	-.2
Holmium	.02	-.2	Beryllium	.08	-.3
Indium	3.89	-.3	Aluminium	2440	-.3
Lanthanum	.7	-.4	Cobaltum	1.45	-.3
Lithium	2.03	-.5	Germanium	0	-.3
Lutetium	.01	.1	Yttrium	.5	-.3
Magnesium	31648.13	-.4	Indium	3.89	-.3
Manganum	261.29	-.2	Samarium	.15	-.3
Molybdenum	121.82	1.8	Europium	.06	-.3
Natrium	6596.94	-.2	Dysprosium	.09	-.3
Neodymium	.58	-.4	Erbium	.05	-.3
Niccolum	5.25	-.4	Ytterbium	.04	-.3
Niobium	.08	-.5	Uranium	.07	-.3
Plumbum	4.86	-.5	Magnesium	31648.13	-.4
Praseodymium	.16	-.4	Silicium	0	-.4
Rhenium	16.4	1.1	Scandium	.76	-.4
Rubidium	35.67	-.4	Niccolum	5.25	-.4
Samarium	.15	-.3	Gallium	.7	-.4
Scandium	.76	-.4	Rubidium	35.67	-.4
Selenium	0	-1.3	Lanthanum	.7	-.4
Silicium	0	-.4	Cerium	2.05	-.4
Stibium	.26	-.2	Praseodymium	.16	-.4
Strontium	1120	.3	Neodymium	.58	-.4
Tantalum	0	-.7	Gadolinium	.17	-.4
Tellurium	.02	-.8	Terbium	.02	-.4
Terbium	.02	-.4	Thallium	.05	-.4
Thallium	.05	-.4	Bismuthum	0	-.4
Thorium	.22	-.2	Lithium	2.03	-.5
Thulium	.01	.1	Zincum	352.73	-.5
Titanium	252.8	.1	Niobium	.08	-.5
Tungstenium	0	-.6	Cadmium	.5	-.5
Uranium	.07	-.3	Caesium	.1	-.5
Vanadium	3.35	-.2	Plumbum	4.86	-.5
Ytterbium	.04	-.3	Tungstenium	0	-.6
Yttrium	.5	-.3	Tantalum	0	-.7
Zincum	352.73	-.5	Tellurium	.02	-.8
Zirconium	.61	.2	Selenium	0	-1.3

Thuja occidentalis

Name	Value	Deviation	Name	Value	Deviation
Aluminium	1830	-.4	Strontium	2640	3
Argentum	0	-.2	Calcium	688350.03	2.7
Arsenicum	3.03	-.2	Niccolum	28.81	1.8
Barium	183.42	.3	Tellurium	.08	1.8
Beryllium	.09	-.2	Stibium	.7	1.1
Bismuthum	0	-.4	Cobaltum	4.14	.7
Borium	455.3	.4	Cadmium	2.77	.6
Cadmium	2.77	.6	Borium	455.3	.4
Caesium	.09	-.6	Barium	183.42	.3
Calcium	688350.03	2.7	Ferrum	11262.42	.1
Cerium	1.83	-.5	Indium	7.92	.1
Chromium	2.85	0	Chromium	2.85	0
Cobaltum	4.14	.7	Manganum	2920	0
Cuprum	50.06	-.3	Hafnium	.02	0
Dysprosium	.05	-.5	Zincum	593.62	-.1
Erbium	.02	-.5	Plumbum	9.18	-.1
Europium	.06	-.3	Beryllium	.09	-.2
Ferrum	11262.42	.1	Vanadium	3.87	-.2
Gadolinium	.13	-.4	Gallium	.87	-.2
Gallium	.87	-.2	Arsenicum	3.03	-.2
Germanium	0	-.3	Argentum	0	-.2
Hafnium	.02	0	Magnesium	35989.89	-.3
Holmium	.01	-.5	Scandium	.93	-.3
Indium	7.92	.1	Cuprum	50.06	-.3
Lanthanum	.68	-.4	Germanium	0	-.3
Lithium	2.16	-.5	Molybdenum	3.73	-.3
Lutetium	0	-.6	Europium	.06	-.3
Magnesium	35989.89	-.3	Rhenium	3.85	-.3
Manganum	2920	0	Thallium	.07	-.3
Molybdenum	3.73	-.3	Natrium	4245.57	-.4
Natrium	4245.57	-.4	Aluminium	1830	-.4
Neodymium	.44	-.5	Silicium	0	-.4
Niccolum	28.81	1.8	Selenium	1.41	-.4
Niobium	.07	-.6	Rubidium	14.76	-.4
Plumbum	9.18	-.1	Lanthanum	.68	-.4
Praseodymium	.13	-.5	Gadolinium	.13	-.4
Rhenium	3.85	-.3	Bismuthum	0	-.4
Rubidium	14.76	-.4	Uranium	.05	-.4
Samarium	.09	-.5	Lithium	2.16	-.5
Scandium	.93	-.3	Yttrium	.32	-.5
Selenium	1.41	-.4	Zirconium	0	-.5
Silicium	0	-.4	Cerium	1.83	-.5
Stibium	.7	1.1	Praseodymium	.13	-.5
Strontium	2640	3	Neodymium	.44	-.5
Tantalum	0	-.7	Samarium	.09	-.5
Tellurium	.08	1.8	Terbium	.01	-.5
Terbium	.01	-.5	Dysprosium	.05	-.5
Thallium	.07	-.3	Holmium	.01	-.5
Thorium	.12	-.5	Erbium	.02	-.5
Thulium	0	-.6	Ytterbium	.02	-.5
Titanium	176.84	-.7	Thorium	.12	-.5
Tungstenium	0	-.6	Niobium	.07	-.6
Uranium	.05	-.4	Caesium	.09	-.6
Vanadium	3.87	-.2	Thulium	0	-.6
Ytterbium	.02	-.5	Lutetium	0	-.6
Yttrium	.32	-.5	Tungstenium	0	-.6
Zincum	593.62	-.1	Titanium	176.84	-.7
Zirconium	0	-.5	Tantalum	0	-.7

Tormentilla erecta - Potentilla erecta

Name	Value	Deviation	Name	Value	Deviation
Aluminium	4630	.2	Cadmium	10.04	4.1
Argentum	0	-.2	Thallium	1.14	4.1
Arsenicum	15.07	3.5	Plumbum	54.99	4
Barium	541.14	3.1	Arsenicum	15.07	3.5
Beryllium	.22	.4	Barium	541.14	3.1
Bismuthum	0	-.4	Zincum	2380	2.5
Borium	345.91	-.2	Europium	.26	2
Cadmium	10.04	4.1	Germanium	.43	1.7
Caesium	.37	-.1	Niccolum	21.65	1.2
Calcium	130715.93	-1	Titanium	338.07	1.1
Cerium	6.03	.8	Chromium	9.63	1.1
Chromium	9.63	1.1	Yttrium	2.37	1
Cobaltum	3.86	.6	Lanthanum	3.2	1
Cuprum	109.05	.7	Terbium	.09	1
Dysprosium	.37	.9	Holmium	.07	1
Erbium	.19	.9	Thorium	.75	1
Europium	.26	2	Neodymium	2.51	.9
Ferrum	7076.18	0	Samarium	.63	.9
Gadolinium	.74	.9	Gadolinium	.74	.9
Gallium	1.36	.4	Dysprosium	.37	.9
Germanium	.43	1.7	Erbium	.19	.9
Hafnium	.03	.3	Cerium	6.03	.8
Holmium	.07	1	Praseodymium	.68	.8
Indium	6.11	-.1	Ytterbium	.14	.8
Lanthanum	3.2	1	Lutetium	.02	.8
Lithium	4.75	.1	Cuprum	109.05	.7
Lutetium	.02	.8	Thulium	.02	.7
Magnesium	31359.45	-.4	Cobaltum	3.86	.6
Manganum	1130	-.1	Beryllium	.22	.4
Molybdenum	9.59	-.2	Gallium	1.36	.4
Natrium	8485.86	-.1	Vanadium	15.5	.3
Neodymium	2.51	.9	Hafnium	.03	.3
Niccolum	21.65	1.2	Uranium	.2	.3
Niobium	.12	-.1	Aluminium	4630	.2
Plumbum	54.99	4	Lithium	4.75	.1
Praseodymium	.68	.8	Ferrum	7076.18	0
Rhenium	.54	-.7	Natrium	8485.86	-.1
Rubidium	28.37	-.4	Manganum	1130	-.1
Samarium	.63	.9	Niobium	.12	-.1
Scandium	1.01	-.2	Indium	6.11	-.1
Selenium	0	-1.3	Stibium	.3	-.1
Silicium	0	-.4	Caesium	.37	-.1
Stibium	.3	-.1	Borium	345.91	-.2
Strontium	807	-.2	Scandium	1.01	-.2
Tantalum	0	-.7	Strontium	807	-.2
Tellurium	.02	-.8	Molybdenum	9.59	-.2
Terbium	.09	1	Argentum	0	-.2
Thallium	1.14	4.1	Magnesium	31359.45	-.4
Thorium	.75	1	Silicium	0	-.4
Thulium	.02	.7	Rubidium	28.37	-.4
Titanium	338.07	1.1	Bismuthum	0	-.4
Tungstenium	0	-.6	Zirconium	0	-.5
Uranium	.2	.3	Tungstenium	0	-.6
Vanadium	15.5	.3	Tantalum	0	-.7
Ytterbium	.14	.8	Rhenium	.54	-.7
Yttrium	2.37	1	Tellurium	.02	-.8
Zincum	2380	2.5	Calcium	130715.93	-1
Zirconium	0	-.5	Selenium	0	-1.3

Tussilago farfara

Name	Value	Deviation	Name	Value	Deviation
Aluminium	2050	-.4	Selenium	8.59	4
Argentum	0	-.2	Tellurium	.07	1.4
Arsenicum	0	-1.1	Rubidium	405.02	.6
Barium	19.93	-1	Calcium	365022.25	.5
Beryllium	.04	-.5	Caesium	.73	.5
Bismuthum	0	-.4	Zirconium	.64	.2
Borium	404.68	.1	Borium	404.68	.1
Cadmium	.13	-.7	Magnesium	56882.81	.1
Caesium	.73	.5	Hafnium	.02	0
Calcium	365022.25	.5	Cobaltum	1.92	-.1
Cerium	1.39	-.6	Scandium	1.05	-.2
Chromium	0	-.5	Manganum	324.73	-.2
Cobaltum	1.92	-.1	Ferrum	2179.15	-.2
Cuprum	32.16	-.7	Strontium	820	-.2
Dysprosium	.05	-.5	Molybdenum	11.82	-.2
Erbium	.02	-.5	Argentum	0	-.2
Europium	.02	-.7	Stibium	.26	-.2
Ferrum	2179.15	-.2	Germanium	0	-.3
Gadolinium	.09	-.5	Rhenium	4.45	-.3
Gallium	.57	-.6	Aluminium	2050	-.4
Germanium	0	-.3	Silicium	0	-.4
Hafnium	.02	0	Vanadium	0	-.4
Holmium	.01	-.5	Indium	3.13	-.4
Indium	3.13	-.4	Bismuthum	0	-.4
Lanthanum	.3	-.6	Beryllium	.04	-.5
Lithium	1.61	-.6	Natrium	1641.55	-.5
Lutetium	0	-.6	Chromium	0	-.5
Magnesium	56882.81	.1	Samarium	.07	-.5
Manganum	324.73	-.2	Gadolinium	.09	-.5
Molybdenum	11.82	-.2	Terbium	.01	-.5
Natrium	1641.55	-.5	Dysprosium	.05	-.5
Neodymium	.28	-.6	Holmium	.01	-.5
Niccolum	0	-.8	Erbium	.02	-.5
Niobium	.05	-.7	Thallium	.02	-.5
Plumbum	3.66	-.6	Thorium	.1	-.5
Praseodymium	.08	-.6	Uranium	.04	-.5
Rhenium	4.45	-.3	Lithium	1.61	-.6
Rubidium	405.02	.6	Gallium	.57	-.6
Samarium	.07	-.5	Yttrium	.18	-.6
Scandium	1.05	-.2	Lanthanum	.3	-.6
Selenium	8.59	4	Cerium	1.39	-.6
Silicium	0	-.4	Praseodymium	.08	-.6
Stibium	.26	-.2	Neodymium	.28	-.6
Strontium	820	-.2	Thulium	0	-.6
Tantalum	0	-.7	Ytterbium	.01	-.6
Tellurium	.07	1.4	Lutetium	0	-.6
Terbium	.01	-.5	Tungstenium	0	-.6
Thallium	.02	-.5	Plumbum	3.66	-.6
Thorium	.1	-.5	Cuprum	32.16	-.7
Thulium	0	-.6	Niobium	.05	-.7
Titanium	126.15	-1.3	Cadmium	.13	-.7
Tungstenium	0	-.6	Europium	.02	-.7
Uranium	.04	-.5	Tantalum	0	-.7
Vanadium	0	-.4	Niccolum	0	-.8
Ytterbium	.01	-.6	Zincum	158.96	-.8
Yttrium	.18	-.6	Barium	19.93	-1
Zincum	158.96	-.8	Arsenicum	0	-1.1
Zirconium	.64	.2	Titanium	126.15	-1.3

Tussilago petasites

Name	Value	Deviation	Name	Value	Deviation
Aluminium	1660	-.5	Selenium	9.71	4.7
Argentum	0	-.2	Rubidium	928.3	2
Arsenicum	0	-1.1	Tellurium	.07	1.4
Barium	21.79	-1	Cuprum	116.17	.9
Beryllium	.04	-.5	Caesium	.7	.4
Bismuthum	0	-.4	Borium	390.2	0
Borium	390.2	0	Natrium	8993.25	0
Cadmium	.12	-.7	Scandium	1.25	0
Caesium	.7	.4	Titanium	241.29	0
Calcium	215175.49	-.5	Stibium	.29	-.1
Cerium	1.96	-.4	Manganum	335.83	-.2
Chromium	0	-.5	Ferrum	2210.65	-.2
Cobaltum	1.16	-.4	Niccolum	7.52	-.2
Cuprum	116.17	.9	Argentum	0	-.2
Dysprosium	.07	-.4	Germanium	0	-.3
Erbium	.04	-.3	Indium	3.66	-.3
Europium	.03	-.6	Erbium	.04	-.3
Ferrum	2210.65	-.2	Plumbum	7.53	-.3
Gadolinium	.15	-.4	Uranium	.08	-.3
Gallium	.61	-.5	Lithium	2.43	-.4
Germanium	0	-.3	Magnesium	31420.58	-.4
Hafnium	0	-.6	Silicium	0	-.4
Holmium	.01	-.5	Vanadium	0	-.4
Indium	3.66	-.3	Cobaltum	1.16	-.4
Lanthanum	.7	-.4	Yttrium	.41	-.4
Lithium	2.43	-.4	Molybdenum	1.83	-.4
Lutetium	0	-.6	Lanthanum	.7	-.4
Magnesium	31420.58	-.4	Cerium	1.96	-.4
Manganum	335.83	-.2	Praseodymium	.15	-.4
Molybdenum	1.83	-.4	Neodymium	.57	-.4
Natrium	8993.25	0	Samarium	.13	-.4
Neodymium	.57	-.4	Gadolinium	.15	-.4
Niccolum	7.52	-.2	Terbium	.02	-.4
Niobium	.07	-.6	Dysprosium	.07	-.4
Plumbum	7.53	-.3	Ytterbium	.03	-.4
Praseodymium	.15	-.4	Bismuthum	0	-.4
Rhenium	1.87	-.6	Thorium	.15	-.4
Rubidium	928.3	2	Beryllium	.04	-.5
Samarium	.13	-.4	Aluminium	1660	-.5
Scandium	1.25	0	Calcium	215175.49	-.5
Selenium	9.71	4.7	Chromium	0	-.5
Silicium	0	-.4	Gallium	.61	-.5
Stibium	.29	-.1	Zirconium	0	-.5
Strontium	479	-.8	Holmium	.01	-.5
Tantalum	0	-.7	Thallium	.01	-.5
Tellurium	.07	1.4	Zincum	259.38	-.6
Terbium	.02	-.4	Niobium	.07	-.6
Thallium	.01	-.5	Europium	.03	-.6
Thorium	.15	-.4	Thulium	0	-.6
Thulium	0	-.6	Lutetium	0	-.6
Titanium	241.29	0	Hafnium	0	-.6
Tungstenium	0	-.6	Tungstenium	0	-.6
Uranium	.08	-.3	Rhenium	1.87	-.6
Vanadium	0	-.4	Cadmium	.12	-.7
Ytterbium	.03	-.4	Tantalum	0	-.7
Yttrium	.41	-.4	Strontium	479	-.8
Zincum	259.38	-.6	Barium	21.79	-1
Zirconium	0	-.5	Arsenicum	0	-1.1

Ulmus campestris

Name	Value	Deviation	Name	Value	Deviation
Aluminium	3230	-.1	Strontium	2960	3.6
Argentum	0	-.2	Calcium	647363.69	2.4
Arsenicum	4.4	.2	Scandium	3.1	1.9
Barium	307.08	1.3	Cuprum	166.05	1.8
Beryllium	.09	-.2	Borium	691.91	1.6
Bismuthum	0	-.4	Silicium	4194.29	1.3
Borium	691.91	1.6	Barium	307.08	1.3
Cadmium	.37	-.5	Plumbum	21.64	1
Caesium	.33	-.2	Tellurium	.05	.5
Calcium	647363.69	2.4	Selenium	2.75	.4
Cerium	2.42	-.3	Arsenicum	4.4	.2
Chromium	3.03	0	Indium	8.54	.2
Cobaltum	2.28	0	Europium	.1	.2
Cuprum	166.05	1.8	Lithium	4.51	.1
Dysprosium	.12	-.2	Ferrum	10654.24	.1
Erbium	.05	-.3	Thulium	.01	.1
Europium	.1	.2	Chromium	3.03	0
Ferrum	10654.24	.1	Cobaltum	2.28	0
Gadolinium	.23	-.2	Molybdenum	20.28	0
Gallium	.79	-.3	Hafnium	.02	0
Germanium	0	-.3	Natrium	7364.25	-.1
Hafnium	.02	0	Aluminium	3230	-.1
Holmium	.02	-.2	Vanadium	5.46	-.1
Indium	8.54	.2	Beryllium	.09	-.2
Lanthanum	1.08	-.2	Manganum	231.23	-.2
Lithium	4.51	.1	Argentum	0	-.2
Lutetium	0	-.6	Caesium	.33	-.2
Magnesium	31884.62	-.4	Lanthanum	1.08	-.2
Manganum	231.23	-.2	Samarium	.19	-.2
Molybdenum	20.28	0	Gadolinium	.23	-.2
Natrium	7364.25	-.1	Terbium	.03	-.2
Neodymium	.77	-.3	Dysprosium	.12	-.2
Niccolum	6.31	-.3	Holmium	.02	-.2
Niobium	.07	-.6	Titanium	216.85	-.3
Plumbum	21.64	1	Niccolum	6.31	-.3
Praseodymium	.22	-.3	Zincum	504.45	-.3
Rhenium	.62	-.7	Gallium	.79	-.3
Rubidium	31.35	-.4	Germanium	0	-.3
Samarium	.19	-.2	Yttrium	.53	-.3
Scandium	3.1	1.9	Stibium	.23	-.3
Selenium	2.75	.4	Cerium	2.42	-.3
Silicium	4194.29	1.3	Praseodymium	.22	-.3
Stibium	.23	-.3	Neodymium	.77	-.3
Strontium	2960	3.6	Erbium	.05	-.3
Tantalum	0	-.7	Uranium	.07	-.3
Tellurium	.05	.5	Magnesium	31884.62	-.4
Terbium	.03	-.2	Rubidium	31.35	-.4
Thallium	.01	-.5	Ytterbium	.03	-.4
Thorium	.17	-.4	Bismuthum	0	-.4
Thulium	.01	.1	Thorium	.17	-.4
Titanium	216.85	-.3	Zirconium	0	-.5
Tungstenium	0	-.6	Cadmium	.37	-.5
Uranium	.07	-.3	Thallium	.01	-.5
Vanadium	5.46	-.1	Niobium	.07	-.6
Ytterbium	.03	-.4	Lutetium	0	-.6
Yttrium	.53	-.3	Tungstenium	0	-.6
Zincum	504.45	-.3	Tantalum	0	-.7
Zirconium	0	-.5	Rhenium	.62	-.7

Urtica dioica

Name	Value	Deviation	Name	Value	Deviation
Aluminium	9320	1.2	Scandium	3.07	1.9
Argentum	0	-.2	Thorium	.85	1.3
Arsenicum	2.81	-.3	Aluminium	9320	1.2
Barium	87.85	-.5	Silicium	3554.03	1.1
Beryllium	.26	.6	Uranium	.35	1
Bismuthum	0	-.4	Neodymium	2.53	.9
Borium	358.51	-.1	Erbium	.19	.9
Cadmium	.42	-.5	Chromium	7.62	.8
Caesium	.77	.5	Selenium	3.42	.8
Calcium	387320.91	.7	Lanthanum	2.93	.8
Cerium	6.11	.8	Cerium	6.11	.8
Chromium	7.62	.8	Praseodymium	.65	.8
Cobaltum	2.43	.1	Samarium	.6	.8
Cuprum	37.15	-.6	Gadolinium	.69	.8
Dysprosium	.35	.8	Terbium	.08	.8
Erbium	.19	.9	Dysprosium	.35	.8
Europium	.15	.7	Holmium	.06	.8
Ferrum	11048.72	.1	Ytterbium	.14	.8
Gadolinium	.69	.8	Lutetium	.02	.8
Gallium	1.53	.6	Calcium	387320.91	.7
Germanium	.11	.2	Europium	.15	.7
Hafnium	.04	.5	Thulium	.02	.7
Holmium	.06	.8	Beryllium	.26	.6
Indium	9.42	.3	Gallium	1.53	.6
Lanthanum	2.93	.8	Yttrium	1.85	.6
Lithium	6.54	.5	Lithium	6.54	.5
Lutetium	.02	.8	Vanadium	18.83	.5
Magnesium	28428.42	-.5	Zirconium	.92	.5
Manganum	488.27	-.2	Caesium	.77	.5
Molybdenum	36.87	.3	Hafnium	.04	.5
Natrium	4639.03	-.3	Strontium	1180	.4
Neodymium	2.53	.9	Niobium	.17	.3
Niccolum	7.72	-.1	Molybdenum	36.87	.3
Niobium	.17	.3	Indium	9.42	.3
Plumbum	12.19	.1	Germanium	.11	.2
Praseodymium	.65	.8	Titanium	249.11	.1
Rhenium	3.57	-.4	Ferrum	11048.72	.1
Rubidium	34.98	-.4	Cobaltum	2.43	.1
Samarium	.6	.8	Plumbum	12.19	.1
Scandium	3.07	1.9	Borium	358.51	-.1
Selenium	3.42	.8	Niccolum	7.72	-.1
Silicium	3554.03	1.1	Manganum	488.27	-.2
Stibium	.09	-.7	Argentum	0	-.2
Strontium	1180	.4	Natrium	4639.03	-.3
Tantalum	0	-.7	Arsenicum	2.81	-.3
Tellurium	.03	-.4	Thallium	.07	-.3
Terbium	.08	.8	Rubidium	34.98	-.4
Thallium	.07	-.3	Tellurium	.03	-.4
Thorium	.85	1.3	Rhenium	3.57	-.4
Thulium	.02	.7	Bismuthum	0	-.4
Titanium	249.11	.1	Magnesium	28428.42	-.5
Tungstenium	0	-.6	Cadmium	.42	-.5
Uranium	.35	1	Barium	87.85	-.5
Vanadium	18.83	.5	Cuprum	37.15	-.6
Ytterbium	.14	.8	Tungstenium	0	-.6
Yttrium	1.85	.6	Zincum	172.97	-.7
Zincum	172.97	-.7	Stibium	.09	-.7
Zirconium	.92	.5	Tantalum	0	-.7

Urtica urens 1

Name	Value	Deviation	Name	Value	Deviation
Aluminium	1160	-.6	Rhenium	22.47	1.8
Argentum	0	-.2	Calcium	484108.89	1.3
Arsenicum	2.79	-.3	Strontium	1660	1.3
Barium	143.29	0	Barium	143.29	0
Beryllium	.05	-.5	Thallium	.15	0
Bismuthum	0	-.4	Ferrum	3268.67	-.1
Borium	273.52	-.6	Manganum	348.08	-.2
Cadmium	.27	-.6	Molybdenum	12.2	-.2
Caesium	.06	-.6	Argentum	0	-.2
Calcium	484108.89	1.3	Scandium	.87	-.3
Cerium	1.7	-.5	Vanadium	1.49	-.3
Chromium	0	-.5	Germanium	0	-.3
Cobaltum	1.24	-.4	Arsenicum	2.79	-.3
Cuprum	20.66	-.9	Selenium	1.65	-.3
Dysprosium	.06	-.4	Hafnium	.01	-.3
Erbium	.03	-.4	Silicium	0	-.4
Europium	.05	-.4	Cobaltum	1.24	-.4
Ferrum	3268.67	-.1	Rubidium	10.54	-.4
Gadolinium	.12	-.5	Europium	.05	-.4
Gallium	.39	-.8	Dysprosium	.06	-.4
Germanium	0	-.3	Erbium	.03	-.4
Hafnium	.01	-.3	Bismuthum	0	-.4
Holmium	.01	-.5	Beryllium	.05	-.5
Indium	1.06	-.6	Chromium	0	-.5
Lanthanum	.49	-.5	Yttrium	.33	-.5
Lutetium	0	-.6	Zirconium	0	-.5
Magnesium	23126.44	-.6	Lanthanum	.49	-.5
Manganum	348.08	-.2	Cerium	1.7	-.5
Molybdenum	12.2	-.2	Praseodymium	.11	-.5
Natrium	940.04	-.6	Neodymium	.4	-.5
Neodymium	.4	-.5	Samarium	.1	-.5
Niccolum	0	-.8	Gadolinium	.12	-.5
Niobium	.04	-.8	Terbium	.01	-.5
Plumbum	2.71	-.7	Holmium	.01	-.5
Praseodymium	.11	-.5	Ytterbium	.02	-.5
Rhenium	22.47	1.8	Thorium	.13	-.5
Rubidium	10.54	-.4	Uranium	.04	-.5
Samarium	.1	-.5	Borium	273.52	-.6
Scandium	.87	-.3	Natrium	940.04	-.6
Selenium	1.65	-.3	Magnesium	23126.44	-.6
Silicium	0	-.4	Aluminium	1160	-.6
Stibium	.03	-.9	Cadmium	.27	-.6
Strontium	1660	1.3	Indium	1.06	-.6
Tantalum	0	-.7	Caesium	.06	-.6
Tellurium	.01	-1.3	Thulium	0	-.6
Terbium	.01	-.5	Lutetium	0	-.6
Thallium	.15	0	Tungstenium	0	-.6
Thorium	.13	-.5	Zincum	197.64	-.7
Thulium	0	-.6	Tantalum	0	-.7
Titanium	124.64	-1.3	Plumbum	2.71	-.7
Tungstenium	0	-.6	Niccolum	0	-.8
Uranium	.04	-.5	Gallium	.39	-.8
Vanadium	1.49	-.3	Niobium	.04	-.8
Ytterbium	.02	-.5	Cuprum	20.66	-.9
Yttrium	.33	-.5	Stibium	.03	-.9
Zincum	197.64	-.7	Titanium	124.64	-1.3
Zirconium	0	-.5	Tellurium	.01	-1.3
Lithium	1.74	-.6	Lithium	1.74	-.6

Urtica urens 2

Name	Value	Deviation	Name	Value	Deviation
Aluminium	1810	-.4	Calcium	537165.59	1.7
Argentum	0	-.2	Strontium	1610	1.2
Arsenicum	2.64	-.3	Silicium	2733.05	.7
Barium	101.88	-.4	Scandium	1.84	.6
Beryllium	.08	-.3	Rhenium	9.48	.3
Bismuthum	0	-.4	Titanium	244.09	0
Borium	297.67	-.5	Ferrum	3641.89	-.1
Cadmium	.14	-.7	Selenium	2.01	-.1
Caesium	.15	-.5	Molybdenum	14.1	-.1
Calcium	537165.59	1.7	Chromium	1.69	-.2
Cerium	1.91	-.5	Manganum	68.47	-.2
Chromium	1.69	-.2	Argentum	0	-.2
Cobaltum	1.32	-.4	Beryllium	.08	-.3
Cuprum	30.92	-.7	Vanadium	2.15	-.3
Dysprosium	.06	-.4	Germanium	0	-.3
Erbium	.03	-.4	Arsenicum	2.64	-.3
Europium	.04	-.5	Hafnium	.01	-.3
Ferrum	3641.89	-.1	Thallium	.06	-.3
Gadolinium	.13	-.4	Aluminium	1810	-.4
Gallium	.54	-.6	Cobaltum	1.32	-.4
Germanium	0	-.3	Rubidium	16.99	-.4
Hafnium	.01	-.3	Yttrium	.34	-.4
Holmium	.01	-.5	Barium	101.88	-.4
Indium	2.08	-.5	Gadolinium	.13	-.4
Lanthanum	.56	-.5	Dysprosium	.06	-.4
Lithium	2.15	-.5	Erbium	.03	-.4
Lutetium	0	-.6	Bismuthum	0	-.4
Magnesium	21589.59	-.6	Thorium	.14	-.4
Manganum	68.47	-.2	Uranium	.06	-.4
Molybdenum	14.1	-.1	Lithium	2.15	-.5
Natrium	2666.76	-.5	Borium	297.67	-.5
Neodymium	.45	-.5	Natrium	2666.76	-.5
Niccolum	0	-.8	Zirconium	0	-.5
Niobium	.05	-.7	Indium	2.08	-.5
Plumbum	3.37	-.6	Caesium	.15	-.5
Praseodymium	.12	-.5	Lanthanum	.56	-.5
Rhenium	9.48	.3	Cerium	1.91	-.5
Rubidium	16.99	-.4	Praseodymium	.12	-.5
Samarium	.1	-.5	Neodymium	.45	-.5
Scandium	1.84	.6	Samarium	.1	-.5
Selenium	2.01	-.1	Europium	.04	-.5
Silicium	2733.05	.7	Terbium	.01	-.5
Stibium	.05	-.8	Holmium	.01	-.5
Strontium	1610	1.2	Ytterbium	.02	-.5
Tantalum	0	-.7	Magnesium	21589.59	-.6
Tellurium	.01	-1.3	Gallium	.54	-.6
Terbium	.01	-.5	Thulium	0	-.6
Thallium	.06	-.3	Lutetium	0	-.6
Thorium	.14	-.4	Tungstenium	0	-.6
Thulium	0	-.6	Plumbum	3.37	-.6
Titanium	244.09	0	Cuprum	30.92	-.7
Tungstenium	0	-.6	Niobium	.05	-.7
Uranium	.06	-.4	Cadmium	.14	-.7
Vanadium	2.15	-.3	Tantalum	0	-.7
Ytterbium	.02	-.5	Niccolum	0	-.8
Yttrium	.34	-.4	Zincum	137.64	-.8
Zincum	137.64	-.8	Stibium	.05	-.8
Zirconium	0	-.5	Tellurium	.01	-1.3

Verbena officinalis

Name	Value	Deviation	Name	Value	Deviation
Aluminium	779	-.6	Ferrum	300106.38	9.6
Argentum	0	-.2	Germanium	1.95	9
Arsenicum	11.46	2.4	Chromium	33.42	5.1
Barium	96.93	-.4	Niccolum	43.64	3.2
Beryllium	.03	-.6	Cuprum	208.63	2.5
Bismuthum	0	-.4	Arsenicum	11.46	2.4
Borium	176.02	-1.1	Cobaltum	5.7	1.4
Cadmium	.21	-.6	Gallium	1.61	.7
Caesium	.04	-.6	Stibium	.43	.3
Calcium	204269.86	-.5	Zirconium	.52	.1
Cerium	1.33	-.6	Manganum	1150	-.1
Chromium	33.42	5.1	Molybdenum	10.46	-.2
Cobaltum	5.7	1.4	Argentum	0	-.2
Cuprum	208.63	2.5	Thallium	.08	-.2
Dysprosium	.04	-.5	Vanadium	1.55	-.3
Erbium	.02	-.5	Hafnium	.01	-.3
Europium	.03	-.6	Silicium	0	-.4
Ferrum	300106.38	9.6	Rubidium	22.99	-.4
Gadolinium	.08	-.6	Barium	96.93	-.4
Gallium	1.61	.7	Bismuthum	0	-.4
Germanium	1.95	9	Natrium	2198.2	-.5
Hafnium	.01	-.3	Calcium	204269.86	-.5
Holmium	.01	-.5	Yttrium	.23	-.5
Indium	0	-.7	Terbium	.01	-.5
Lanthanum	.28	-.7	Dysprosium	.04	-.5
Lithium	.97	-.7	Holmium	.01	-.5
Lutetium	0	-.6	Erbium	.02	-.5
Magnesium	16164.3	-.8	Ytterbium	.02	-.5
Manganum	1150	-.1	Plumbum	4.74	-.5
Molybdenum	10.46	-.2	Uranium	.03	-.5
Natrium	2198.2	-.5	Beryllium	.03	-.6
Neodymium	.24	-.6	Aluminium	779	-.6
Niccolum	43.64	3.2	Scandium	.64	-.6
Niobium	.05	-.7	Selenium	1.21	-.6
Plumbum	4.74	-.5	Cadmium	.21	-.6
Praseodymium	.06	-.6	Caesium	.04	-.6
Rhenium	1.8	-.6	Cerium	1.33	-.6
Rubidium	22.99	-.4	Praseodymium	.06	-.6
Samarium	.06	-.6	Neodymium	.24	-.6
Scandium	.64	-.6	Samarium	.06	-.6
Selenium	1.21	-.6	Europium	.03	-.6
Silicium	0	-.4	Gadolinium	.08	-.6
Stibium	.43	.3	Thulium	0	-.6
Strontium	570	-.7	Lutetium	0	-.6
Tantalum	0	-.7	Tungstenium	0	-.6
Tellurium	.02	-.8	Rhenium	1.8	-.6
Terbium	.01	-.5	Thorium	.07	-.6
Thallium	.08	-.2	Lithium	.97	-.7
Thorium	.07	-.6	Zincum	235.23	-.7
Thulium	0	-.6	Strontium	570	-.7
Titanium	169.61	-.8	Niobium	.05	-.7
Tungstenium	0	-.6	Indium	0	-.7
Uranium	.03	-.5	Lanthanum	.28	-.7
Vanadium	1.55	-.3	Tantalum	0	-.7
Ytterbium	.02	-.5	Magnesium	16164.3	-.8
Yttrium	.23	-.5	Titanium	169.61	-.8
Zincum	235.23	-.7	Tellurium	.02	-.8
Zirconium	.52	.1	Borium	176.02	-1.1

Vinca minor

Name	Value	Deviation	Name	Value	Deviation
Aluminium	5830	.4	Borium	894.74	2.7
Argentum	0	-.2	Bismuthum	.07	2.3
Arsenicum	2.49	-.4	Indium	17.52	1.2
Barium	288.92	1.1	Silicium	3629.71	1.1
Beryllium	.1	-.2	Calcium	447404.96	1.1
Bismuthum	.07	2.3	Barium	288.92	1.1
Borium	894.74	2.7	Strontium	1510	1
Cadmium	2.03	.3	Germanium	.24	.8
Caesium	.22	-.3	Stibium	.48	.5
Calcium	447404.96	1.1	Tellurium	.05	.5
Cerium	3.42	0	Aluminium	5830	.4
Chromium	0	-.5	Scandium	1.66	.4
Cobaltum	1.62	-.3	Cadmium	2.03	.3
Cuprum	57.99	-.2	Thallium	.21	.3
Dysprosium	.17	0	Uranium	.21	.3
Erbium	.07	-.1	Selenium	2.47	.2
Europium	.1	.2	Europium	.1	.2
Ferrum	4279.79	-.1	Plumbum	12.68	.2
Gadolinium	.33	0	Thulium	.01	.1
Gallium	.99	-.1	Lutetium	.01	.1
Germanium	.24	.8	Natrium	9357.24	0
Hafnium	.01	-.3	Titanium	238.66	0
Holmium	.03	0	Cerium	3.42	0
Indium	17.52	1.2	Neodymium	1.21	0
Lanthanum	1.35	-.1	Samarium	.27	0
Lithium	2.99	-.3	Gadolinium	.33	0
Lutetium	.01	.1	Terbium	.04	0
Magnesium	45427.12	-.1	Dysprosium	.17	0
Manganum	655.33	-.2	Holmium	.03	0
Molybdenum	9.91	-.2	Thorium	.31	0
Natrium	9357.24	0	Magnesium	45427.12	-.1
Neodymium	1.21	0	Ferrum	4279.79	-.1
Niccolum	6.91	-.2	Gallium	.99	-.1
Niobium	.13	-.1	Niobium	.13	-.1
Plumbum	12.68	.2	Lanthanum	1.35	-.1
Praseodymium	.31	-.1	Praseodymium	.31	-.1
Rhenium	4.58	-.3	Erbium	.07	-.1
Rubidium	26.65	-.4	Ytterbium	.05	-.1
Samarium	.27	0	Beryllium	.1	-.2
Scandium	1.66	.4	Vanadium	4.18	-.2
Selenium	2.47	.2	Manganum	655.33	-.2
Silicium	3629.71	1.1	Niccolum	6.91	-.2
Stibium	.48	.5	Cuprum	57.99	-.2
Strontium	1510	1	Zincum	552.48	-.2
Tantalum	0	-.7	Yttrium	.71	-.2
Tellurium	.05	.5	Molybdenum	9.91	-.2
Terbium	.04	0	Argentum	0	-.2
Thallium	.21	.3	Lithium	2.99	-.3
Thorium	.31	0	Cobaltum	1.62	-.3
Thulium	.01	.1	Caesium	.22	-.3
Titanium	238.66	0	Hafnium	.01	-.3
Tungstenium	0	-.6	Rhenium	4.58	-.3
Uranium	.21	.3	Arsenicum	2.49	-.4
Vanadium	4.18	-.2	Rubidium	26.65	-.4
Ytterbium	.05	-.1	Chromium	0	-.5
Yttrium	.71	-.2	Zirconium	0	-.5
Zincum	552.48	-.2	Tungstenium	0	-.6
Zirconium	0	-.5	Tantalum	0	-.7

Viola tricolor

Name	Value	Deviation	Name	Value	Deviation
Aluminium	2130	-.4	Barium	403.49	2
Argentum	0	-.2	Rhenium	13.47	.8
Arsenicum	2.75	-.3	Chromium	4.98	.3
Barium	403.49	2	Molybdenum	37.89	.3
Beryllium	.05	-.5	Cadmium	2.14	.3
Bismuthum	0	-.4	Germanium	.11	.2
Borium	140.15	-1.3	Zirconium	.53	.1
Cadmium	2.14	.3	Europium	.09	.1
Caesium	.21	-.4	Thulium	.01	.1
Calcium	106920.98	-1.2	Lutetium	.01	.1
Cerium	2.41	-.3	Hafnium	.02	0
Chromium	4.98	.3	Manganum	295.58	-.2
Cobaltum	.63	-.7	Ferrum	1329.13	-.2
Cuprum	29.44	-.7	Argentum	0	-.2
Dysprosium	.08	-.4	Holmium	.02	-.2
Erbium	.04	-.3	Thallium	.09	-.2
Europium	.09	.1	Vanadium	2.21	-.3
Ferrum	1329.13	-.2	Arsenicum	2.75	-.3
Gadolinium	.18	-.3	Cerium	2.41	-.3
Gallium	.49	-.7	Neodymium	.69	-.3
Germanium	.11	.2	Samarium	.15	-.3
Hafnium	.02	0	Gadolinium	.18	-.3
Holmium	.02	-.2	Erbium	.04	-.3
Indium	1.97	-.5	Ytterbium	.04	-.3
Lanthanum	.72	-.4	Thorium	.2	-.3
Lithium	1.86	-.5	Uranium	.08	-.3
Lutetium	.01	.1	Aluminium	2130	-.4
Magnesium	22176.67	-.6	Silicium	0	-.4
Manganum	295.58	-.2	Zincum	395.59	-.4
Molybdenum	37.89	.3	Rubidium	23.83	-.4
Natrium	1032.53	-.6	Yttrium	.46	-.4
Neodymium	.69	-.3	Niobium	.09	-.4
Niccolum	0	-.8	Caesium	.21	-.4
Niobium	.09	-.4	Lanthanum	.72	-.4
Plumbum	3.95	-.6	Praseodymium	.17	-.4
Praseodymium	.17	-.4	Terbium	.02	-.4
Rhenium	13.47	.8	Dysprosium	.08	-.4
Rubidium	23.83	-.4	Bismuthum	0	-.4
Samarium	.15	-.3	Lithium	1.86	-.5
Scandium	.68	-.5	Beryllium	.05	-.5
Selenium	1.24	-.5	Scandium	.68	-.5
Silicium	0	-.4	Selenium	1.24	-.5
Stibium	.17	-.5	Indium	1.97	-.5
Strontium	426	-.9	Stibium	.17	-.5
Tantalum	0	-.7	Natrium	1032.53	-.6
Tellurium	0	-1.7	Magnesium	22176.67	-.6
Terbium	.02	-.4	Tungstenium	0	-.6
Thallium	.09	-.2	Plumbum	3.95	-.6
Thorium	.2	-.3	Titanium	181.07	-.7
Thulium	.01	.1	Cobaltum	.63	-.7
Titanium	181.07	-.7	Cuprum	29.44	-.7
Tungstenium	0	-.6	Gallium	.49	-.7
Uranium	.08	-.3	Tantalum	0	-.7
Vanadium	2.21	-.3	Niccolum	0	-.8
Ytterbium	.04	-.3	Strontium	426	-.9
Yttrium	.46	-.4	Calcium	106920.98	-1.2
Zincum	395.59	-.4	Borium	140.15	-1.3
Zirconium	.53	.1	Tellurium	0	-1.7

Zizia aurea - Thaspium aureum

Name	Value	Deviation	Name	Value	Deviation
Aluminium	3580	0	Silicium	6423.81	2.3
Argentum	0	-.2	Scandium	2.7	1.5
Arsenicum	2.91	-.3	Strontium	1270	.6
Barium	197.94	.4	Tantalum	.01	.6
Beryllium	.09	-.2	Calcium	343240.22	.4
Bismuthum	0	-.4	Barium	197.94	.4
Borium	378.38	0	Germanium	.13	.3
Cadmium	.27	-.6	Natrium	12695.19	.2
Caesium	.46	0	Chromium	3.68	.1
Calcium	343240.22	.4	Zirconium	.54	.1
Cerium	3.3	0	Thulium	.01	.1
Chromium	3.68	.1	Ytterbium	.07	.1
Cobaltum	1.34	-.4	Lutetium	.01	.1
Cuprum	36.26	-.6	Borium	378.38	0
Dysprosium	.16	0	Aluminium	3580	0
Erbium	.08	0	Vanadium	7.63	0
Europium	.08	0	Caesium	.46	0
Ferrum	3378.4	-.1	Cerium	3.3	0
Gadolinium	.32	0	Neodymium	1.19	0
Gallium	.67	-.5	Samarium	.27	0
Germanium	.13	.3	Europium	.08	0
Hafnium	.02	0	Gadolinium	.32	0
Holmium	.03	0	Dysprosium	.16	0
Indium	5.26	-.1	Holmium	.03	0
Lanthanum	1.34	-.1	Erbium	.08	0
Lithium	3.51	-.2	Hafnium	.02	0
Lutetium	.01	.1	Ferrum	3378.4	-.1
Magnesium	21295.63	-.6	Rubidium	139.31	-.1
Manganum	660.71	-.2	Yttrium	.9	-.1
Molybdenum	3.19	-.3	Indium	5.26	-.1
Natrium	12695.19	.2	Lanthanum	1.34	-.1
Neodymium	1.19	0	Praseodymium	.3	-.1
Niccolum	6.67	-.2	Thorium	.28	-.1
Niobium	.1	-.3	Uranium	.13	-.1
Plumbum	7.69	-.3	Lithium	3.51	-.2
Praseodymium	.3	-.1	Beryllium	.09	-.2
Rhenium	3.34	-.4	Manganum	660.71	-.2
Rubidium	139.31	-.1	Niccolum	6.67	-.2
Samarium	.27	0	Argentum	0	-.2
Scandium	2.7	1.5	Terbium	.03	-.2
Selenium	1.49	-.4	Arsenicum	2.91	-.3
Silicium	6423.81	2.3	Niobium	.1	-.3
Stibium	.12	-.6	Molybdenum	3.19	-.3
Strontium	1270	.6	Plumbum	7.69	-.3
Tantalum	.01	.6	Cobaltum	1.34	-.4
Tellurium	0	-1.7	Selenium	1.49	-.4
Terbium	.03	-.2	Rhenium	3.34	-.4
Thallium	.04	-.4	Thallium	.04	-.4
Thorium	.28	-.1	Bismuthum	0	-.4
Thulium	.01	.1	Zincum	367.43	-.5
Titanium	177.96	-.7	Gallium	.67	-.5
Tungstenium	0	-.6	Magnesium	21295.63	-.6
Uranium	.13	-.1	Cuprum	36.26	-.6
Vanadium	7.63	0	Cadmium	.27	-.6
Ytterbium	.07	.1	Stibium	.12	-.6
Yttrium	.9	-.1	Tungstenium	0	-.6
Zincum	367.43	-.5	Titanium	177.96	-.7
Zirconium	.54	.1	Tellurium	0	-1.7

Part 2

Elements

Element	Mean	Mean SD
Aluminium	3805.89	4682.5
Argentum	.01	0
Arsenicum	3.76	3.3
Barium	146.6	125.6
Beryllium	.14	.2
Bismuthum	.01	0
Borium	384.13	188.3
Cadmium	1.5	2.1
Caesium	.43	.6
Calcium	284409.21	150332.3
Cerium	3.38	3.3
Chromium	3.07	6
Cobaltum	2.28	2.5
Cuprum	68.31	55.4
Dysprosium	.16	.2
Erbium	.08	.1
Europium	.08	.1
Ferrum	7603.47	30415.7
Gadolinium	.33	.4
Gallium	1.05	.8
Germanium	.06	.2
Hafnium	.02	0
Holmium	.03	0
Indium	6.59	9.4
Kalium	0	0
Lanthanum	1.47	1.8
Lithium	4.22	4.4
Lutetium	.01	0
Magnesium	51772.05	47130.7
Manganum	1450.04	15464
Molybdenum	21.52	55.7
Natrium	9353.91	14172.8
Neodymium	1.17	1.6
Niccolum	9.17	10.8
Niobium	.14	.1
Plumbum	10.51	11.2
Praseodymium	.33	.4
Rhenium	6.79	8.5
Rubidium	174.92	376.3
Samarium	.28	.4
Scandium	1.21	1
Selenium	2.12	1.6
Silicium	924.63	2426.2
Stibium	.33	.3
Strontium	946.17	558.7
Tantalum	.01	0
Tellurium	.04	0
Terbium	.04	.1
Thallium	.14	.2
Thorium	.32	.4
Thulium	.01	0
Titanium	239.99	86.4
Tungstenium	.18	.3
Uranium	.14	.2
Vanadium	8.02	22.8
Ytterbium	.06	.1
Yttrium	.97	1.4
Zincum	686.41	685.9
Zirconium	.47	1

Element	Blank	Coal	Reference	Deviation
Aluminium	-2	2340	1572.96	308.96
Argentum	-0.2	-.2	-.2	0
Arsenicum	-1	-1	-1	0
Barium	-3	414.73	328.23	57.2
Beryllium	-0.005	.4	.44	.1
Bismuthum	-0.05	.95	.54	.39
Borium	-5	60.41	69.87	16.49
Cadmium	-0.01	.04	.05	.01
Caesium	-0.001	.05	.05	.01
Calcium	-0.1	.28	.34	.06
Cerium	-0.01	4.84	5.11	.56
Chromium	-1	2.43	2.31	.62
Cobaltum	-0.01	.67	.69	.08
Cuprum	-0.2	3.45	3.57	.4
Dysprosium	-0.001	.24	.28	.03
Erbium	-0.001	.15	.17	.02
Europium	-0.001	.13	.14	.06
Ferrum	-0.01	.11	.12	.02
Gadolinium	-0.01	.41	.44	.07
Gallium	-0.1	.89	.99	.19
Germanium	-0.1	.16	.15	.11
Hafnium	-0.01	.13	.12	.05
Holmium	-0.001	.05	.05	.01
Indium	-1	6.11	7.17	1.34
Kalium	-0.01	-.01	-.01	0
Lanthanum	-0.002	1.91	2.07	.25
Lithium	-0.5	-.5	.61	.37
Lutetium	-0.001	.02	.02	0
Magnesium	-0.01	.03	.04	.01
Manganum	-0.1	10.7	12.37	2.26
Molybdenum	-0.1	.27	.3	.09
Natrium	-0.01	.06	.06	.01
Neodymium	-0.002	1.51	1.74	.19
Niccolum	-5	-5	-5	0
Niobium	-0.005	.56	.57	.14
Plumbum	-0.1	1.59	1.79	.16
Praseodymium	-0.002	.45	.47	.05
Rhenium	-0.1	.5	.49	.18
Rubidium	-0.01	.37	.45	.14
Samarium	-0.001	.34	.38	.05
Scandium	-0.5	.49	.71	.18
Selenium	-1	-1	-1	0
Silicium	-0.2	-.2	-.2	0
Stibium	-0.02	.05	.06	.03
Strontium	-0.1	160.4	129.95	41.61
Tantalum	-0.001	.03	.03	.01
Tellurium	-0.01	.03	.04	.02
Terbium	-0.001	.05	.06	.01
Thallium	-0.001	.07	.08	.01
Thorium	-0.001	.63	.65	.13
Thulium	-0.001	.02	.02	0
Titanium	-1	130	139.56	15.49
Tungstenium	-0.5	-.5	-.5	0
Uranium		.26	.29	.04
Vanadium	-1	3.38	3.65	.97
Ytterbium	-0.001	.14	.15	.02
Yttrium	-0.001	1.98	2	.22
Zincum	-1	2.92	4	1.03
Zirconium		0	0	0

Aluminium

Name	Value	Deviation	Name	Value	Deviation
Aconitum napellus	5790	.4	Marrubium vulgare	2190	-.3
Adonis vernalis	6110	.5	Melilotus officinalis	1110	-.6
Agrimonia eupatoria	794	-.6	Melissa officinalis	4540	.2
Alchemilla vulgaris	5830	.4	Mentha arvensis	1960	-.4
Anethum graveolens	934	-.6	Mercurialis perennis	5340	.3
Angelica archangelica	5780	.4	Milium solis	1500	-.5
Anthemis nobilis	4450	.1	Ocimum canum	7600	.8
Aquilegia vulgaris	1030	-.6	Ocimum canum R	6430	.6
Belladonna	3050	-.2	Oenanthe crocata	5550	.4
Bellis perennis	12100	1.8	Ononis spinosa	3020	-.2
Bryonia dioica	7280	.7	Osmundo regalis	35500	6.8
Cardamine pratensis	2540	-.3	Osmundo regalis 2	427	-.7
Cardamine pratensis R	1590	-.5	Paeonia officinalis	3410	-.1
Cardiospermum halicaca	2650	-.2	Petroselinum crispum	6770	.6
Castanea vesca	4260	.1	Plantago major	3410	-.1
Centaurium erythraea	9940	1.3	Polygonum aviculare	3130	-.1
Chamomilla	3660	0	Primula veris	4820	.2
Chelidonium majus	20800	3.6	Prunus padus	1910	-.4
Chenopodium anthelmint	488	-.7	Psoralea bituminosa	6330	.5
Cicuta virosa	7050	.7	Rhus toxicodendron	1320	-.5
Cimicifuga racemosa	2660	-.2	Rumex acetosa	16900	2.8
Cinnamodendron cortiso	1610	-.5	Ruta graveolens	1310	-.5
Clematis erecta	2410	-.3	Salicaria purpurea	2050	-.4
Collinsonia canadensis	5600	.4	Salix purpurea	1580	-.5
Conium maculatum	486	-.7	Salvia sclarea	1230	-.6
Cupressus lawsoniana	3920	0	Sanguisorba officinali	491	-.7
Echinacea angustifolia	1960	-.4	Scrophularia nodosa	2560	-.3
Echinacea purpurea	402	-.7	Scutellaria lateriflor	5960	.5
Escholtzia californica	3150	-.1	Solidago virgaurea	573	-.7
Faba vulgaris	999	-.6	Spilanthes oleracea	1290	-.5
Fagopyrum esculentum	804	-.6	Stachys officinalis	947	-.6
Fraxinus excelsior	2020	-.4	Stachys officinalis R	561	-.7
Galium aparine	1950	-.4	Stachys sylvatica	2740	-.2
Glechoma hederacea	1910	-.4	Syringa vulgaris	2140	-.4
Gnaphalium leontopodiu	8770	1.1	Taraxacum officinale	9540	1.2
Hedera helix	1420	-.5	Teucrium chamaedrys	2440	-.3
Hydrophyllum virginicu	779	-.6	Thuja occidentalis	1830	-.4
Hyoscyamus niger	2430	-.3	Tormentilla erecta	4630	.2
Hyssopus officinalis	1670	-.5	Tussilago farfara	2050	-.4
Lappa arctium	4410	.1	Tussilago petasites	1660	-.5
Lapsana communis	4330	.1	Ulmus campestris	3230	-.1
Laurocerasus	1650	-.5	Urtica dioica	9320	1.2
Leonurus cardiaca	750	-.7	Urtica urens	1160	-.6
Lespedeza sieboldii	965	-.6	Urtica urens 2	1810	-.4
Linum usitatissimum	506	-.7	Verbena officinalis	779	-.6
Lycopus europaeus	1160	-.6	Vinca minor	5830	.4
Malva sylvestris	2890	-.2	Viola tricolor	2130	-.4
Mandragora officinalis	1040	-.6	Zizia aurea	3580	0

Aluminium

Name	Value	Deviation	Name	Value	Deviation
Osmundo regalis	35500	6.8	Scrophularia nodosa	2560	-.3
Chelidonium majus	20800	3.6	Teucrium chamaedrys	2440	-.3
Rumex acetosa	16900	2.8	Echinacea angustifolia	1960	-.4
Bellis perennis	12100	1.8	Fraxinus excelsior	2020	-.4
Centaurium erythraea	9940	1.3	Galium aparine	1950	-.4
Taraxacum officinale	9540	1.2	Glechoma hederacea	1910	-.4
Urtica dioica	9320	1.2	Mentha arvensis	1960	-.4
Gnaphalium leontopodiu	8770	1.1	Prunus padus	1910	-.4
Ocimum canum	7600	.8	Salicaria purpurea	2050	-.4
Bryonia dioica	7280	.7	Syringa vulgaris	2140	-.4
Cicuta virosa	7050	.7	Thuja occidentalis	1830	-.4
Ocimum canum R	6430	.6	Tussilago farfara	2050	-.4
Petroselinum crispum	6770	.6	Urtica urens 2	1810	-.4
Adonis vernalis	6110	.5	Viola tricolor	2130	-.4
Psoralea bituminosa	6330	.5	Cardamine pratensis R	1590	-.5
Scutellaria lateriflor	5960	.5	Cinnamodendron cortiso	1610	-.5
Aconitum napellus	5790	.4	Hedera helix	1420	-.5
Alchemilla vulgaris	5830	.4	Hyssopus officinalis	1670	-.5
Angelica archangelica	5780	.4	Laurocerasus	1650	-.5
Collinsonia canadensis	5600	.4	Milium solis	1500	-.5
Oenanthe crocata	5550	.4	Rhus toxicodendron	1320	-.5
Vinca minor	5830	.4	Ruta graveolens	1310	-.5
Mercurialis perennis	5340	.3	Salix purpurea	1580	-.5
Melissa officinalis	4540	.2	Spilanthes oleracea	1290	-.5
Primula veris	4820	.2	Tussilago petasites	1660	-.5
Tormentilla erecta	4630	.2	Agrimonia eupatoria	794	-.6
Anthemis nobilis	4450	.1	Anethum graveolens	934	-.6
Castanea vesca	4260	.1	Aquilegia vulgaris	1030	-.6
Lappa arctium	4410	.1	Faba vulgaris	999	-.6
Lapsana communis	4330	.1	Fagopyrum esculentum	804	-.6
Chamomilla	3660	0	Hydrophyllum virginicu	779	-.6
Cupressus lawsoniana	3920	0	Lespedeza sieboldii	965	-.6
Zizia aurea	3580	0	Lycopus europaeus	1160	-.6
Escholtzia californica	3150	-.1	Mandragora officinalis	1040	-.6
Paeonia officinalis	3410	-.1	Melilotus officinalis	1110	-.6
Plantago major	3410	-.1	Salvia sclarea	1230	-.6
Polygonum aviculare	3130	-.1	Stachys officinalis	947	-.6
Ulmus campestris	3230	-.1	Urtica urens	1160	-.6
Belladonna	3050	-.2	Verbena officinalis	779	-.6
Cardiospermum halicaca	2650	-.2	Chenopodium anthelmint	488	-.7
Cimicifuga racemosa	2660	-.2	Conium maculatum	486	-.7
Malva sylvestris	2890	-.2	Echinacea purpurea	402	-.7
Ononis spinosa	3020	-.2	Leonurus cardiaca	750	-.7
Stachys sylvatica	2740	-.2	Linum usitatissimum	506	-.7
Cardamine pratensis	2540	-.3	Osmundo regalis 2	427	-.7
Clematis erecta	2410	-.3	Sanguisorba officinali	491	-.7
Hyoscyamus niger	2430	-.3	Solidago virgaurea	573	-.7
Marrubium vulgare	2190	-.3	Stachys officinalis R	561	-.7

Argentum

Name	Value	Deviation	Name	Value	Deviation
Aconitum napellus	0	-.2	Marrubium vulgare	0	-.2
Adonis vernalis	0	-.2	Melilotus officinalis	0	-.2
Agrimonia eupatoria	0	-.2	Melissa officinalis	0	-.2
Alchemilla vulgaris	0	-.2	Mentha arvensis	0	-.2
Anethum graveolens	0	-.2	Mercurialis perennis	0	-.2
Angelica archangelica	0	-.2	Milium solis	0	-.2
Anthemis nobilis	0	-.2	Ocimum canum	0	-.2
Aquilegia vulgaris	0	-.2	Ocimum canum R	0	-.2
Belladonna	0	-.2	Oenanthe crocata	0	-.2
Bellis perennis	0	-.2	Ononis spinosa	0	-.2
Bryonia dioica	.2	4	Osmundo regalis	.22	4.4
Cardamine pratensis	0	-.2	Osmundo regalis 2	0	-.2
Cardamine pratensis R	0	-.2	Paeonia officinalis	0	-.2
Cardiospermum halicaca	0	-.2	Petroselinum crispum	0	-.2
Castanea vesca	0	-.2	Plantago major	0	-.2
Centaurium erythraea	0	-.2	Polygonum aviculare	0	-.2
Chamomilla	0	-.2	Primula veris	0	-.2
Chelidonium majus	0	-.2	Prunus padus	0	-.2
Chenopodium anthelmint	0	-.2	Psoralea bituminosa	0	-.2
Cicuta virosa	0	-.2	Rhus toxicodendron	0	-.2
Cimicifuga racemosa	0	-.2	Rumex acetosa	0	-.2
Cinnamodendron cortiso	0	-.2	Ruta graveolens	0	-.2
Clematis erecta	0	-.2	Salicaria purpurea	0	-.2
Collinsonia canadensis	0	-.2	Salix purpurea	0	-.2
Conium maculatum	.2	4	Salvia sclarea	0	-.2
Cupressus lawsoniana	.2	4	Sanguisorba officinali	0	-.2
Echinacea angustifolia	0	-.2	Scrophularia nodosa	0	-.2
Echinacea purpurea	0	-.2	Scutellaria lateriflor	0	-.2
Escholtzia californica	0	-.2	Solidago virgaurea	0	-.2
Faba vulgaris	0	-.2	Spilanthes oleracea	0	-.2
Fagopyrum esculentum	0	-.2	Stachys officinalis	0	-.2
Fraxinus excelsior	0	-.2	Stachys officinalis R	0	-.2
Galium aparine	0	-.2	Stachys sylvatica	0	-.2
Glechoma hederacea	0	-.2	Syringa vulgaris	0	-.2
Gnaphalium leontopodiu	0	-.2	Taraxacum officinale	0	-.2
Hedera helix	.23	4.7	Teucrium chamaedrys	0	-.2
Hydrophyllum virginicu	0	-.2	Thuja occidentalis	0	-.2
Hyoscyamus niger	0	-.2	Tormentilla erecta	0	-.2
Hyssopus officinalis	0	-.2	Tussilago farfara	0	-.2
Lappa arctium	0	-.2	Tussilago petasites	0	-.2
Lapsana communis	0	-.2	Ulmus campestris	0	-.2
Laurocerasus	0	-.2	Urtica dioica	0	-.2
Leonurus cardiaca	0	-.2	Urtica urens	0	-.2
Lespedeza sieboldii	0	-.2	Urtica urens 2	0	-.2
Linum usitatissimum	0	-.2	Verbena officinalis	0	-.2
Lycopus europaeus	0	-.2	Vinca minor	0	-.2
Malva sylvestris	0	-.2	Viola tricolor	0	-.2
Mandragora officinalis	0	-.2	Zizia aurea	0	-.2

Argentum

Name	Value	Deviation	Name	Value	Deviation
Hedera helix	.23	4.7	Mandragora officinalis	0	-.2
Osmundo regalis	.22	4.4	Marrubium vulgare	0	-.2
Bryonia dioica	.2	4	Melilotus officinalis	0	-.2
Conium maculatum	.2	4	Melissa officinalis	0	-.2
Cupressus lawsoniana	.2	4	Mentha arvensis	0	-.2
Aconitum napellus	0	-.2	Mercurialis perennis	0	-.2
Adonis vernalis	0	-.2	Milium solis	0	-.2
Agrimonia eupatoria	0	-.2	Ocimum canum	0	-.2
Alchemilla vulgaris	0	-.2	Ocimum canum R	0	-.2
Anethum graveolens	0	-.2	Oenanthe crocata	0	-.2
Angelica archangelica	0	-.2	Ononis spinosa	0	-.2
Anthemis nobilis	0	-.2	Osmundo regalis 2	0	-.2
Aquilegia vulgaris	0	-.2	Paeonia officinalis	0	-.2
Belladonna	0	-.2	Petroselinum crispum	0	-.2
Bellis perennis	0	-.2	Plantago major	0	-.2
Cardamine pratensis	0	-.2	Polygonum aviculare	0	-.2
Cardamine pratensis R	0	-.2	Primula veris	0	-.2
Cardiospermum halicaca	0	-.2	Prunus padus	0	-.2
Castanea vesca	0	-.2	Psoralea bituminosa	0	-.2
Centaurium erythraea	0	-.2	Rhus toxicodendron	0	-.2
Chamomilla	0	-.2	Rumex acetosa	0	-.2
Chelidonium majus	0	-.2	Ruta graveolens	0	-.2
Chenopodium anthelmint	0	-.2	Salicaria purpurea	0	-.2
Cicuta virosa	0	-.2	Salix purpurea	0	-.2
Cimicifuga racemosa	0	-.2	Salvia sclarea	0	-.2
Cinnamodendron cortiso	0	-.2	Sanguisorba officinali	0	-.2
Clematis erecta	0	-.2	Scrophularia nodosa	0	-.2
Collinsonia canadensis	0	-.2	Scutellaria lateriflor	0	-.2
Echinacea angustifolia	0	-.2	Solidago virgaurea	0	-.2
Echinacea purpurea	0	-.2	Spilanthes oleracea	0	-.2
Escholtzia californica	0	-.2	Stachys officinalis	0	-.2
Faba vulgaris	0	-.2	Stachys officinalis R	0	-.2
Fagopyrum esculentum	0	-.2	Stachys sylvatica	0	-.2
Fraxinus excelsior	0	-.2	Syringa vulgaris	0	-.2
Galium aparine	0	-.2	Taraxacum officinale	0	-.2
Glechoma hederacea	0	-.2	Teucrium chamaedrys	0	-.2
Gnaphalium leontopodiu	0	-.2	Thuja occidentalis	0	-.2
Hydrophyllum virginicu	0	-.2	Tormentilla erecta	0	-.2
Hyoscyamus niger	0	-.2	Tussilago farfara	0	-.2
Hyssopus officinalis	0	-.2	Tussilago petasites	0	-.2
Lappa arctium	0	-.2	Ulmus campestris	0	-.2
Lapsana communis	0	-.2	Urtica dioica	0	-.2
Laurocerasus	0	-.2	Urtica urens	0	-.2
Leonurus cardiaca	0	-.2	Urtica urens 2	0	-.2
Lespedeza sieboldii	0	-.2	Verbena officinalis	0	-.2
Linum usitatissimum	0	-.2	Vinca minor	0	-.2
Lycopus europaeus	0	-.2	Viola tricolor	0	-.2
Malva sylvestris	0	-.2	Zizia aurea	0	-.2

Arsenicum

Name	Value	Deviation	Name	Value	Deviation
Aconitum napellus	2.95	-.2	Marrubium vulgare	5.44	.5
Adonis vernalis	3.13	-.2	Melilotus officinalis	1.95	-.6
Agrimonia eupatoria	0	-1.1	Melissa officinalis	5.85	.6
Alchemilla vulgaris	3.34	-.1	Mentha arvensis	1.37	-.7
Anethum graveolens	0	-1.1	Mercurialis perennis	3.58	-.1
Angelica archangelica	3.08	-.2	Milium solis	5.93	.7
Anthemis nobilis	1.74	-.6	Ocimum canum	7.35	1.1
Aquilegia vulgaris	0	-1.1	Ocimum canum R	7.89	1.3
Belladonna	1.73	-.6	Oenanthe crocata	5.05	.4
Bellis perennis	6.72	.9	Ononis spinosa	6.38	.8
Bryonia dioica	4.66	.3	Osmundo regalis	11.29	2.3
Cardamine pratensis	1.06	-.8	Osmundo regalis 2	1.61	-.7
Cardamine pratensis R	1.47	-.7	Paeonia officinalis	12.82	2.8
Cardiospermum halicaca	6.66	.9	Petroselinum crispum	2.07	-.5
Castanea vesca	5.38	.5	Plantago major	2.1	-.5
Centaurium erythraea	3.93	.1	Polygonum aviculare	0	-1.1
Chamomilla	2.01	-.5	Primula veris	2.97	-.2
Chelidonium majus	7.56	1.2	Prunus padus	8.93	1.6
Chenopodium anthelmint	0	-1.1	Psoralea bituminosa	3.53	-.1
Cicuta virosa	11.81	2.5	Rhus toxicodendron	2.85	-.3
Cimicifuga racemosa	13.97	3.1	Rumex acetosa	5.38	.5
Cinnamodendron cortiso	0	-1.1	Ruta graveolens	3.04	-.2
Clematis erecta	5.73	.6	Salicaria purpurea	1.31	-.7
Collinsonia canadensis	8.85	1.6	Salix purpurea	2.05	-.5
Conium maculatum	4.15	.1	Salvia sclarea	2.58	-.4
Cupressus lawsoniana	8.16	1.3	Sanguisorba officinali	2.68	-.3
Echinacea angustifolia	1.43	-.7	Scrophularia nodosa	2.96	-.2
Echinacea purpurea	0	-1.1	Scutellaria lateriflor	3.05	-.2
Escholtzia californica	0	-1.1	Solidago virgaurea	1.31	-.7
Faba vulgaris	4.15	.1	Spilanthes oleracea	2.59	-.4
Fagopyrum esculentum	1.6	-.7	Stachys officinalis	1.7	-.6
Fraxinus excelsior	1.77	-.6	Stachys officinalis R	1.34	-.7
Galium aparine	4.22	.1	Stachys sylvatica	0	-1.1
Glechoma hederacea	0	-1.1	Syringa vulgaris	4.51	.2
Gnaphalium leontopodiu	6.02	.7	Taraxacum officinale	4.76	.3
Hedera helix	1.41	-.7	Teucrium chamaedrys	6.72	.9
Hydrophyllum virginicu	9.22	1.7	Thuja occidentalis	3.03	-.2
Hyoscyamus niger	1.15	-.8	Tormentilla erecta	15.07	3.5
Hyssopus officinalis	2.64	-.3	Tussilago farfara	0	-1.1
Lappa arctium	2.16	-.5	Tussilago petasites	0	-1.1
Lapsana communis	1.78	-.6	Ulmus campestris	4.4	.2
Laurocerasus	1.93	-.6	Urtica dioica	2.81	-.3
Leonurus cardiaca	3.65	0	Urtica urens	2.79	-.3
Lespedeza sieboldii	3.9	0	Urtica urens 2	2.64	-.3
Linum usitatissimum	5.85	.6	Verbena officinalis	11.46	2.4
Lycopus europaeus	1.77	-.6	Vinca minor	2.49	-.4
Malva sylvestris	1.09	-.8	Viola tricolor	2.75	-.3
Mandragora officinalis	1.65	-.6	Zizia aurea	2.91	-.3

Arsenicum

Name	Value	Deviation	Name	Value	Deviation
Tormentilla erecta	15.07	3.5	Rhus toxicodendron	2.85	-.3
Cimicifuga racemosa	13.97	3.1	Sanguisorba officinali	2.68	-.3
Paeonia officinalis	12.82	2.8	Urtica dioica	2.81	-.3
Cicuta virosa	11.81	2.5	Urtica urens	2.79	-.3
Verbena officinalis	11.46	2.4	Urtica urens 2	2.64	-.3
Osmundo regalis	11.29	2.3	Viola tricolor	2.75	-.3
Hydrophyllum virginicu	9.22	1.7	Zizia aurea	2.91	-.3
Collinsonia canadensis	8.85	1.6	Salvia sclarea	2.58	-.4
Prunus padus	8.93	1.6	Spilanthes oleracea	2.59	-.4
Cupressus lawsoniana	8.16	1.3	Vinca minor	2.49	-.4
Ocimum canum R	7.89	1.3	Chamomilla	2.01	-.5
Chelidonium majus	7.56	1.2	Lappa arctium	2.16	-.5
Ocimum canum	7.35	1.1	Petroselinum crispum	2.07	-.5
Bellis perennis	6.72	.9	Plantago major	2.1	-.5
Cardiospermum halicaca	6.66	.9	Salix purpurea	2.05	-.5
Teucrium chamaedrys	6.72	.9	Anthemis nobilis	1.74	-.6
Ononis spinosa	6.38	.8	Belladonna	1.73	-.6
Gnaphalium leontopodiu	6.02	.7	Fraxinus excelsior	1.77	-.6
Milium solis	5.93	.7	Lapsana communis	1.78	-.6
Clematis erecta	5.73	.6	Laurocerasus	1.93	-.6
Linum usitatissimum	5.85	.6	Lycopus europaeus	1.77	-.6
Melissa officinalis	5.85	.6	Mandragora officinalis	1.65	-.6
Castanea vesca	5.38	.5	Melilotus officinalis	1.95	-.6
Marrubium vulgare	5.44	.5	Stachys officinalis	1.7	-.6
Rumex acetosa	5.38	.5	Cardamine pratensis R	1.47	-.7
Oenanthe crocata	5.05	.4	Echinacea angustifolia	1.43	-.7
Bryonia dioica	4.66	.3	Fagopyrum esculentum	1.6	-.7
Taraxacum officinale	4.76	.3	Hedera helix	1.41	-.7
Syringa vulgaris	4.51	.2	Mentha arvensis	1.37	-.7
Ulmus campestris	4.4	.2	Osmundo regalis 2	1.61	-.7
Centaurium erythraea	3.93	.1	Salicaria purpurea	1.31	-.7
Conium maculatum	4.15	.1	Solidago virgaurea	1.31	-.7
Faba vulgaris	4.15	.1	Stachys officinalis R	1.34	-.7
Galium aparine	4.22	.1	Cardamine pratensis	1.06	-.8
Leonurus cardiaca	3.65	0	Hyoscyamus niger	1.15	-.8
Lespedeza sieboldii	3.9	0	Malva sylvestris	1.09	-.8
Alchemilla vulgaris	3.34	-.1	Agrimonia eupatoria	0	-1.1
Mercurialis perennis	3.58	-.1	Anethum graveolens	0	-1.1
Psoralea bituminosa	3.53	-.1	Aquilegia vulgaris	0	-1.1
Aconitum napellus	2.95	-.2	Chenopodium anthelmint	0	-1.1
Adonis vernalis	3.13	-.2	Cinnamodendron cortiso	0	-1.1
Angelica archangelica	3.08	-.2	Echinacea purpurea	0	-1.1
Primula veris	2.97	-.2	Escholtzia californica	0	-1.1
Ruta graveolens	3.04	-.2	Glechoma hederacea	0	-1.1
Scrophularia nodosa	2.96	-.2	Polygonum aviculare	0	-1.1
Scutellaria lateriflor	3.05	-.2	Stachys sylvatica	0	-1.1
Thuja occidentalis	3.03	-.2	Tussilago farfara	0	-1.1
Hyssopus officinalis	2.64	-.3	Tussilago petasites	0	-1.1

Barium

Name	Value	Deviation	Name	Value	Deviation
Aconitum napellus	159.87	.1	Marrubium vulgare	63.16	-.7
Adonis vernalis	30.17	-.9	Melilotus officinalis	54.58	-.7
Agrimonia eupatoria	24.59	-1	Melissa officinalis	520.32	3
Alchemilla vulgaris	85.18	-.5	Mentha arvensis	270.26	1
Anethum graveolens	79.99	-.5	Mercurialis perennis	98.96	-.4
Angelica archangelica	56.12	-.7	Milium solis	71.08	-.6
Anthemis nobilis	45.78	-.8	Ocimum canum	199.14	.4
Aquilegia vulgaris	20.05	-1	Ocimum canum R	208.45	.5
Belladonna	87.3	-.5	Oenanthe crocata	24.47	-1
Bellis perennis	49.29	-.8	Ononis spinosa	64.37	-.7
Bryonia dioica	89.72	-.5	Osmundo regalis	292.66	1.2
Cardamine pratensis	147.24	0	Osmundo regalis 2	84.69	-.5
Cardamine pratensis R	140.98	0	Paeonia officinalis	161.39	.1
Cardiospermum halicaca	62.86	-.7	Petroselinum crispum	144.13	0
Castanea vesca	320.35	1.4	Plantago major	120.65	-.2
Centaurium erythraea	312.48	1.3	Polygonum aviculare	32.42	-.9
Chamomilla	25.59	-1	Primula veris	26.63	-1
Chelidonium majus	65.58	-.6	Prunus padus	106.23	-.3
Chenopodium anthelmint	23.08	-1	Psoralea bituminosa	246.1	.8
Cicuta virosa	133.9	-.1	Rhus toxicodendron	398.78	2
Cimicifuga racemosa	122.55	-.2	Rumex acetosa	379.28	1.9
Cinnamodendron cortiso	228.36	.7	Ruta graveolens	109.52	-.3
Clematis erecta	120.56	-.2	Salicaria purpurea	70.94	-.6
Collinsonia canadensis	131.96	-.1	Salix purpurea	183.52	.3
Conium maculatum	50.22	-.8	Salvia sclarea	192.77	.4
Cupressus lawsoniana	140.36	0	Sanguisorba officinali	162.12	.1
Echinacea angustifolia	71.26	-.6	Scrophularia nodosa	106.64	-.3
Echinacea purpurea	62.95	-.7	Scutellaria lateriflor	357.75	1.7
Escholtzia californica	109.91	-.3	Solidago virgaurea	28.84	-.9
Faba vulgaris	26.63	-1	Spilanthes oleracea	20.91	-1
Fagopyrum esculentum	67.43	-.6	Stachys officinalis	96.15	-.4
Fraxinus excelsior	194.45	.4	Stachys officinalis R	99.89	-.4
Galium aparine	317.87	1.4	Stachys sylvatica	145.72	0
Glechoma hederacea	601.57	3.6	Syringa vulgaris	33.17	-.9
Gnaphalium leontopodiu	176.54	.2	Taraxacum officinale	61.49	-.7
Hedera helix	323.8	1.4	Teucrium chamaedrys	192.23	.4
Hydrophyllum virginicu	233.79	.7	Thuja occidentalis	183.42	.3
Hyoscyamus niger	89.14	-.5	Tormentilla erecta	541.14	3.1
Hyssopus officinalis	31.57	-.9	Tussilago farfara	19.93	-1
Lappa arctium	61.3	-.7	Tussilago petasites	21.79	-1
Lapsana communis	75.34	-.6	Ulmus campestris	307.08	1.3
Laurocerasus	94.85	-.4	Urtica dioica	87.85	-.5
Leonurus cardiaca	307.24	1.3	Urtica urens	143.29	0
Lespedeza sieboldii	402.84	2	Urtica urens 2	101.88	-.4
Linum usitatissimum	19.51	-1	Verbena officinalis	96.93	-.4
Lycopus europaeus	154.84	.1	Vinca minor	288.92	1.1
Malva sylvestris	60.39	-.7	Viola tricolor	403.49	2
Mandragora officinalis	13.19	-1.1	Zizia aurea	197.94	.4

Barium

Name	Value	Deviation	Name	Value	Deviation
Glechoma hederacea	601.57	3.6	Laurocerasus	94.85	-.4
Tormentilla erecta	541.14	3.1	Mercurialis perennis	98.96	-.4
Melissa officinalis	520.32	3	Stachys officinalis	96.15	-.4
Lespedeza sieboldii	402.84	2	Stachys officinalis R	99.89	-.4
Rhus toxicodendron	398.78	2	Urtica urens 2	101.88	-.4
Viola tricolor	403.49	2	Verbena officinalis	96.93	-.4
Rumex acetosa	379.28	1.9	Alchemilla vulgaris	85.18	-.5
Scutellaria lateriflor	357.75	1.7	Anethum graveolens	79.99	-.5
Castanea vesca	320.35	1.4	Belladonna	87.3	-.5
Galium aparine	317.87	1.4	Bryonia dioica	89.72	-.5
Hedera helix	323.8	1.4	Hyoscyamus niger	89.14	-.5
Centaurium erythraea	312.48	1.3	Osmundo regalis 2	84.69	-.5
Leonurus cardiaca	307.24	1.3	Urtica dioica	87.85	-.5
Ulmus campestris	307.08	1.3	Chelidonium majus	65.58	-.6
Osmundo regalis	292.66	1.2	Echinacea angustifolia	71.26	-.6
Vinca minor	288.92	1.1	Fagopyrum esculentum	67.43	-.6
Mentha arvensis	270.26	1	Lapsana communis	75.34	-.6
Psoralea bituminosa	246.1	.8	Milium solis	71.08	-.6
Cinnamodendron cortiso	228.36	.7	Salicaria purpurea	70.94	-.6
Hydrophyllum virginicu	233.79	.7	Angelica archangelica	56.12	-.7
Ocimum canum R	208.45	.5	Cardiospermum halicaca	62.86	-.7
Fraxinus excelsior	194.45	.4	Echinacea purpurea	62.95	-.7
Ocimum canum	199.14	.4	Lappa arctium	61.3	-.7
Salvia sclarea	192.77	.4	Malva sylvestris	60.39	-.7
Teucrium chamaedrys	192.23	.4	Marrubium vulgare	63.16	-.7
Zizia aurea	197.94	.4	Melilotus officinalis	54.58	-.7
Salix purpurea	183.52	.3	Ononis spinosa	64.37	-.7
Thuja occidentalis	183.42	.3	Taraxacum officinale	61.49	-.7
Gnaphalium leontopodiu	176.54	.2	Anthemis nobilis	45.78	-.8
Aconitum napellus	159.87	.1	Bellis perennis	49.29	-.8
Lycopus europaeus	154.84	.1	Conium maculatum	50.22	-.8
Paeonia officinalis	161.39	.1	Adonis vernalis	30.17	-.9
Sanguisorba officinali	162.12	.1	Hyssopus officinalis	31.57	-.9
Cardamine pratensis	147.24	0	Polygonum aviculare	32.42	-.9
Cardamine pratensis R	140.98	0	Solidago virgaurea	28.84	-.9
Cupressus lawsoniana	140.36	0	Syringa vulgaris	33.17	-.9
Petroselinum crispum	144.13	0	Agrimonia eupatoria	24.59	-1
Stachys sylvatica	145.72	0	Aquilegia vulgaris	20.05	-1
Urtica urens	143.29	0	Chamomilla	25.59	-1
Cicuta virosa	133.9	-.1	Chenopodium anthelmint	23.08	-1
Collinsonia canadensis	131.96	-.1	Faba vulgaris	26.63	-1
Cimicifuga racemosa	122.55	-.2	Linum usitatissimum	19.51	-1
Clematis erecta	120.56	-.2	Oenanthe crocata	24.47	-1
Plantago major	120.65	-.2	Primula veris	26.63	-1
Escholtzia californica	109.91	-.3	Spilanthes oleracea	20.91	-1
Prunus padus	106.23	-.3	Tussilago farfara	19.93	-1
Ruta graveolens	109.52	-.3	Tussilago petasites	21.79	-1
Scrophularia nodosa	106.64	-.3	Mandragora officinalis	13.19	-1.1

Beryllium

Name	Value	Deviation	Name	Value	Deviation
Aconitum napellus	.2	.3	Marrubium vulgare	.07	-.3
Adonis vernalis	.11	-.1	Melilotus officinalis	.16	.1
Agrimonia eupatoria	.03	-.6	Melissa officinalis	.12	-.1
Alchemilla vulgaris	.18	.2	Mentha arvensis	.06	-.4
Anethum graveolens	.03	-.6	Mercurialis perennis	.2	.3
Angelica archangelica	.16	.1	Milium solis	.07	-.3
Anthemis nobilis	.12	-.1	Ocimum canum	.1	-.2
Aquilegia vulgaris	.03	-.6	Ocimum canum R	.12	-.1
Belladonna	.07	-.3	Oenanthe crocata	.19	.3
Bellis perennis	.46	1.7	Ononis spinosa	.12	-.1
Bryonia dioica	.23	.5	Osmundo regalis	1.56	7.5
Cardamine pratensis	.08	-.3	Osmundo regalis 2	.01	-.7
Cardamine pratensis R	.08	-.3	Paeonia officinalis	.29	.8
Cardiospermum halicaca	.14	0	Petroselinum crispum	.21	.4
Castanea vesca	.25	.6	Plantago major	.14	0
Centaurium erythraea	.29	.8	Polygonum aviculare	.07	-.3
Chamomilla	.1	-.2	Primula veris	.12	-.1
Chelidonium majus	.67	2.8	Prunus padus	.06	-.4
Chenopodium anthelmint	.03	-.6	Psoralea bituminosa	.13	0
Cicuta virosa	.37	1.2	Rhus toxicodendron	.1	-.2
Cimicifuga racemosa	.23	.5	Rumex acetosa	.73	3.1
Cinnamodendron cortiso	.05	-.5	Ruta graveolens	.19	.3
Clematis erecta	.05	-.5	Salicaria purpurea	.06	-.4
Collinsonia canadensis	.27	.7	Salix purpurea	.07	-.3
Conium maculatum	.01	-.7	Salvia sclarea	.23	.5
Cupressus lawsoniana	.13	0	Sanguisorba officinali	.01	-.7
Echinacea angustifolia	.08	-.3	Scrophularia nodosa	.11	-.1
Echinacea purpurea	.05	-.5	Scutellaria lateriflor	.06	-.4
Escholtzia californica	.02	-.6	Solidago virgaurea	.04	-.5
Faba vulgaris	.04	-.5	Spilanthes oleracea	.04	-.5
Fagopyrum esculentum	.1	-.2	Stachys officinalis	.02	-.6
Fraxinus excelsior	.07	-.3	Stachys officinalis R	.02	-.6
Galium aparine	.07	-.3	Stachys sylvatica	.09	-.2
Glechoma hederacea	.06	-.4	Syringa vulgaris	.09	-.2
Gnaphalium leontopodiu	.3	.9	Taraxacum officinale	.31	.9
Hedera helix	.07	-.3	Teucrium chamaedrys	.08	-.3
Hydrophyllum virginicu	.05	-.5	Thuja occidentalis	.09	-.2
Hyoscyamus niger	.06	-.4	Tormentilla erecta	.22	.4
Hyssopus officinalis	.03	-.6	Tussilago farfara	.04	-.5
Lappa arctium	.16	.1	Tussilago petasites	.04	-.5
Lapsana communis	.15	.1	Ulmus campestris	.09	-.2
Laurocerasus	.04	-.5	Urtica dioica	.26	.6
Leonurus cardiaca	.04	-.5	Urtica urens	.05	-.5
Lespedeza sieboldii	.03	-.6	Urtica urens 2	.08	-.3
Linum usitatissimum	.01	-.7	Verbena officinalis	.03	-.6
Lycopus europaeus	.03	-.6	Vinca minor	.1	-.2
Malva sylvestris	.03	-.6	Viola tricolor	.05	-.5
Mandragora officinalis	.06	-.4	Zizia aurea	.09	-.2

Beryllium

Name	Value	Deviation	Name	Value	Deviation
Osmundo regalis	1.56	7.5	Cardamine pratensis	.08	-.3
Rumex acetosa	.73	3.1	Cardamine pratensis R	.08	-.3
Chelidonium majus	.67	2.8	Echinacea angustifolia	.08	-.3
Bellis perennis	.46	1.7	Fraxinus excelsior	.07	-.3
Cicuta virosa	.37	1.2	Galium aparine	.07	-.3
Gnaphalium leontopodiu	.3	.9	Hedera helix	.07	-.3
Taraxacum officinale	.31	.9	Marrubium vulgare	.07	-.3
Centaurium erythraea	.29	.8	Milium solis	.07	-.3
Paeonia officinalis	.29	.8	Polygonum aviculare	.07	-.3
Collinsonia canadensis	.27	.7	Salix purpurea	.07	-.3
Castanea vesca	.25	.6	Teucrium chamaedrys	.08	-.3
Urtica dioica	.26	.6	Urtica urens 2	.08	-.3
Bryonia dioica	.23	.5	Glechoma hederacea	.06	-.4
Cimicifuga racemosa	.23	.5	Hyoscyamus niger	.06	-.4
Salvia sclarea	.23	.5	Mandragora officinalis	.06	-.4
Petroselinum crispum	.21	.4	Mentha arvensis	.06	-.4
Tormentilla erecta	.22	.4	Prunus padus	.06	-.4
Aconitum napellus	.2	.3	Salicaria purpurea	.06	-.4
Mercurialis perennis	.2	.3	Scutellaria lateriflor	.06	-.4
Oenanthe crocata	.19	.3	Cinnamodendron cortiso	.05	-.5
Ruta graveolens	.19	.3	Clematis erecta	.05	-.5
Alchemilla vulgaris	.18	.2	Echinacea purpurea	.05	-.5
Angelica archangelica	.16	.1	Faba vulgaris	.04	-.5
Lappa arctium	.16	.1	Hydrophyllum virginicu	.05	-.5
Lapsana communis	.15	.1	Laurocerasus	.04	-.5
Melilotus officinalis	.16	.1	Leonurus cardiaca	.04	-.5
Cardiospermum halicaca	.14	0	Solidago virgaurea	.04	-.5
Cupressus lawsoniana	.13	0	Spilanthes oleracea	.04	-.5
Plantago major	.14	0	Tussilago farfara	.04	-.5
Psoralea bituminosa	.13	0	Tussilago petasites	.04	-.5
Adonis vernalis	.11	-.1	Urtica urens	.05	-.5
Anthemis nobilis	.12	-.1	Viola tricolor	.05	-.5
Melissa officinalis	.12	-.1	Agrimonia eupatoria	.03	-.6
Ocimum canum R	.12	-.1	Anethum graveolens	.03	-.6
Ononis spinosa	.12	-.1	Aquilegia vulgaris	.03	-.6
Primula veris	.12	-.1	Chenopodium anthelmint	.03	-.6
Scrophularia nodosa	.11	-.1	Escholtzia californica	.02	-.6
Chamomilla	.1	-.2	Hyssopus officinalis	.03	-.6
Fagopyrum esculentum	.1	-.2	Lespedeza sieboldii	.03	-.6
Ocimum canum	.1	-.2	Lycopus europaeus	.03	-.6
Rhus toxicodendron	.1	-.2	Malva sylvestris	.03	-.6
Stachys sylvatica	.09	-.2	Stachys officinalis	.02	-.6
Syringa vulgaris	.09	-.2	Stachys officinalis R	.02	-.6
Thuja occidentalis	.09	-.2	Verbena officinalis	.03	-.6
Ulmus campestris	.09	-.2	Conium maculatum	.01	-.7
Vinca minor	.1	-.2	Linum usitatissimum	.01	-.7
Zizia aurea	.09	-.2	Osmundo regalis 2	.01	-.7
Belladonna	.07	-.3	Sanguisorba officinali	.01	-.7

Bismuthum

Name	Value	Deviation	Name	Value	Deviation
Aconitum napellus	0	-.4	Marrubium vulgare	0	-.4
Adonis vernalis	0	-.4	Melilotus officinalis	0	-.4
Agrimonia eupatoria	0	-.4	Melissa officinalis	.05	1.5
Alchemilla vulgaris	.06	1.9	Mentha arvensis	.07	2.3
Anethum graveolens	0	-.4	Mercurialis perennis	0	-.4
Angelica archangelica	0	-.4	Milium solis	0	-.4
Anthemis nobilis	0	-.4	Ocimum canum	0	-.4
Aquilegia vulgaris	0	-.4	Ocimum canum R	0	-.4
Belladonna	0	-.4	Oenanthe crocata	0	-.4
Bellis perennis	.05	1.5	Ononis spinosa	.08	2.7
Bryonia dioica	0	-.4	Osmundo regalis	.1	3.5
Cardamine pratensis	0	-.4	Osmundo regalis 2	0	-.4
Cardamine pratensis R	0	-.4	Paeonia officinalis	0	-.4
Cardiospermum halicaca	.05	1.5	Petroselinum crispum	0	-.4
Castanea vesca	0	-.4	Plantago major	0	-.4
Centaurium erythraea	0	-.4	Polygonum aviculare	0	-.4
Chamomilla	0	-.4	Primula veris	0	-.4
Chelidonium majus	.07	2.3	Prunus padus	0	-.4
Chenopodium anthelmint	0	-.4	Psoralea bituminosa	.07	2.3
Cicuta virosa	.11	3.8	Rhus toxicodendron	0	-.4
Cimicifuga racemosa	0	-.4	Rumex acetosa	0	-.4
Cinnamodendron cortiso	0	-.4	Ruta graveolens	0	-.4
Clematis erecta	0	-.4	Salicaria purpurea	.06	1.9
Collinsonia canadensis	0	-.4	Salix purpurea	0	-.4
Conium maculatum	0	-.4	Salvia sclarea	0	-.4
Cupressus lawsoniana	0	-.4	Sanguisorba officinali	0	-.4
Echinacea angustifolia	0	-.4	Scrophularia nodosa	0	-.4
Echinacea purpurea	0	-.4	Scutellaria lateriflor	0	-.4
Escholtzia californica	0	-.4	Solidago virgaurea	0	-.4
Faba vulgaris	0	-.4	Spilanthes oleracea	0	-.4
Fagopyrum esculentum	0	-.4	Stachys officinalis	0	-.4
Fraxinus excelsior	0	-.4	Stachys officinalis R	0	-.4
Galium aparine	0	-.4	Stachys sylvatica	0	-.4
Glechoma hederacea	0	-.4	Syringa vulgaris	0	-.4
Gnaphalium leontopodiu	0	-.4	Taraxacum officinale	.05	1.5
Hedera helix	0	-.4	Teucrium chamaedrys	0	-.4
Hydrophyllum virginicu	0	-.4	Thuja occidentalis	0	-.4
Hyoscyamus niger	0	-.4	Tormentilla erecta	0	-.4
Hyssopus officinalis	0	-.4	Tussilago farfara	0	-.4
Lappa arctium	0	-.4	Tussilago petasites	0	-.4
Lapsana communis	0	-.4	Ulmus campestris	0	-.4
Laurocerasus	0	-.4	Urtica dioica	0	-.4
Leonurus cardiaca	0	-.4	Urtica urens	0	-.4
Lespedeza sieboldii	0	-.4	Urtica urens 2	0	-.4
Linum usitatissimum	.08	2.7	Verbena officinalis	0	-.4
Lycopus europaeus	0	-.4	Vinca minor	.07	2.3
Malva sylvestris	0	-.4	Viola tricolor	0	-.4
Mandragora officinalis	.05	1.5	Zizia aurea	0	-.4

Bismuthum

Name	Value	Deviation	Name	Value	Deviation
Cicuta virosa	.11	3.8	Hyssopus officinalis	0	-.4
Osmundo regalis	.1	3.5	Lappa arctium	0	-.4
Linum usitatissimum	.08	2.7	Lapsana communis	0	-.4
Ononis spinosa	.08	2.7	Laurocerasus	0	-.4
Chelidonium majus	.07	2.3	Leonurus cardiaca	0	-.4
Mentha arvensis	.07	2.3	Lespedeza sieboldii	0	-.4
Psoralea bituminosa	.07	2.3	Lycopus europaeus	0	-.4
Vinca minor	.07	2.3	Malva sylvestris	0	-.4
Alchemilla vulgaris	.06	1.9	Marrubium vulgare	0	-.4
Salicaria purpurea	.06	1.9	Melilotus officinalis	0	-.4
Bellis perennis	.05	1.5	Mercurialis perennis	0	-.4
Cardiospermum halicaca	.05	1.5	Milium solis	0	-.4
Mandragora officinalis	.05	1.5	Ocimum canum	0	-.4
Melissa officinalis	.05	1.5	Ocimum canum R	0	-.4
Taraxacum officinale	.05	1.5	Oenanthe crocata	0	-.4
Aconitum napellus	0	-.4	Osmundo regalis 2	0	-.4
Adonis vernalis	0	-.4	Paeonia officinalis	0	-.4
Agrimonia eupatoria	0	-.4	Petroselinum crispum	0	-.4
Anethum graveolens	0	-.4	Plantago major	0	-.4
Angelica archangelica	0	-.4	Polygonum aviculare	0	-.4
Anthemis nobilis	0	-.4	Primula veris	0	-.4
Aquilegia vulgaris	0	-.4	Prunus padus	0	-.4
Belladonna	0	-.4	Rhus toxicodendron	0	-.4
Bryonia dioica	0	-.4	Rumex acetosa	0	-.4
Cardamine pratensis	0	-.4	Ruta graveolens	0	-.4
Cardamine pratensis R	0	-.4	Salix purpurea	0	-.4
Castanea vesca	0	-.4	Salvia sclarea	0	-.4
Centaurium erythraea	0	-.4	Sanguisorba officinali	0	-.4
Chamomilla	0	-.4	Scrophularia nodosa	0	-.4
Chenopodium anthelmint	0	-.4	Scutellaria lateriflor	0	-.4
Cimicifuga racemosa	0	-.4	Solidago virgaurea	0	-.4
Cinnamodendron cortiso	0	-.4	Spilanthes oleracea	0	-.4
Clematis erecta	0	-.4	Stachys officinalis	0	-.4
Collinsonia canadensis	0	-.4	Stachys officinalis R	0	-.4
Conium maculatum	0	-.4	Stachys sylvatica	0	-.4
Cupressus lawsoniana	0	-.4	Syringa vulgaris	0	-.4
Echinacea angustifolia	0	-.4	Teucrium chamaedrys	0	-.4
Echinacea purpurea	0	-.4	Thuja occidentalis	0	-.4
Escholtzia californica	0	-.4	Tormentilla erecta	0	-.4
Faba vulgaris	0	-.4	Tussilago farfara	0	-.4
Fagopyrum esculentum	0	-.4	Tussilago petasites	0	-.4
Fraxinus excelsior	0	-.4	Ulmus campestris	0	-.4
Galium aparine	0	-.4	Urtica dioica	0	-.4
Glechoma hederacea	0	-.4	Urtica urens	0	-.4
Gnaphalium leontopodiu	0	-.4	Urtica urens 2	0	-.4
Hedera helix	0	-.4	Verbena officinalis	0	-.4
Hydrophyllum virginicu	0	-.4	Viola tricolor	0	-.4
Hyoscyamus niger	0	-.4	Zizia aurea	0	-.4

Borium

Name	Value	Deviation	Name	Value	Deviation
Aconitum napellus	266.95	-.6	Marrubium vulgare	429.88	.2
Adonis vernalis	107.92	-1.5	Melilotus officinalis	761.29	2
Agrimonia eupatoria	313.53	-.4	Melissa officinalis	379.83	0
Alchemilla vulgaris	396.98	.1	Mentha arvensis	325.86	-.3
Anethum graveolens	325.7	-.3	Mercurialis perennis	294.96	-.5
Angelica archangelica	279.84	-.6	Milium solis	506.12	.6
Anthemis nobilis	435.15	.3	Ocimum canum	194.41	-1
Aquilegia vulgaris	345.7	-.2	Ocimum canum R	211.84	-.9
Belladonna	285.19	-.5	Oenanthe crocata	215.3	-.9
Bellis perennis	66.9	-1.7	Ononis spinosa	493.5	.6
Bryonia dioica	194.03	-1	Osmundo regalis	322.98	-.3
Cardamine pratensis	367.68	-.1	Osmundo regalis 2	550.83	.9
Cardamine pratensis R	352.03	-.2	Paeonia officinalis	368.98	-.1
Cardiospermum halicaca	170.48	-1.1	Petroselinum crispum	497.24	.6
Castanea vesca	870.09	2.6	Plantago major	257.41	-.7
Centaurium erythraea	505.57	.6	Polygonum aviculare	279.1	-.6
Chamomilla	403.98	.1	Primula veris	416.31	.2
Chelidonium majus	92.2	-1.6	Prunus padus	1170	4.2
Chenopodium anthelmint	365.77	-.1	Psoralea bituminosa	441.84	.3
Cicuta virosa	199.11	-1	Rhus toxicodendron	505.67	.6
Cimicifuga racemosa	301.07	-.4	Rumex acetosa	436.84	.3
Cinnamodendron cortiso	457.92	.4	Ruta graveolens	660.93	1.5
Clematis erecta	459.35	.4	Salicaria purpurea	358.33	-.1
Collinsonia canadensis	195.58	-1	Salix purpurea	417.08	.2
Conium maculatum	374.6	-.1	Salvia sclarea	279.77	-.6
Cupressus lawsoniana	694.06	1.6	Sanguisorba officinali	442.98	.3
Echinacea angustifolia	590.32	1.1	Scrophularia nodosa	448.57	.3
Echinacea purpurea	680.34	1.6	Scutellaria lateriflor	399.91	.1
Escholtzia californica	207.52	-.9	Solidago virgaurea	401.95	.1
Faba vulgaris	225.64	-.8	Spilanthes oleracea	701.6	1.7
Fagopyrum esculentum	173.41	-1.1	Stachys officinalis	376.41	0
Fraxinus excelsior	305.37	-.4	Stachys officinalis R	396.98	.1
Galium aparine	193.8	-1	Stachys sylvatica	337.23	-.2
Glechoma hederacea	202.58	-1	Syringa vulgaris	647.33	1.4
Gnaphalium leontopodiu	295.37	-.5	Taraxacum officinale	176.58	-1.1
Hedera helix	373.76	-.1	Teucrium chamaedrys	605.26	1.2
Hydrophyllum virginicu	348.8	-.2	Thuja occidentalis	455.3	.4
Hyoscyamus niger	156.66	-1.2	Tormentilla erecta	345.91	-.2
Hyssopus officinalis	202.97	-1	Tussilago farfara	404.68	.1
Lappa arctium	256.81	-.7	Tussilago petasites	390.2	0
Lapsana communis	290.18	-.5	Ulmus campestris	691.91	1.6
Laurocerasus	491.92	.6	Urtica dioica	358.51	-.1
Leonurus cardiaca	273.92	-.6	Urtica urens	273.52	-.6
Lespedeza sieboldii	742.51	1.9	Urtica urens 2	297.67	-.5
Linum usitatissimum	589.77	1.1	Verbena officinalis	176.02	-1.1
Lycopus europaeus	424.72	.2	Vinca minor	894.74	2.7
Malva sylvestris	411.07	.1	Viola tricolor	140.15	-1.3
Mandragora officinalis	93.64	-1.5	Zizia aurea	378.38	0

Borium

Name	Value	Deviation	Name	Value	Deviation
Prunus padus	1170	4.2	Paeonia officinalis	368.98	-.1
Vinca minor	894.74	2.7	Salicaria purpurea	358.33	-.1
Castanea vesca	870.09	2.6	Urtica dioica	358.51	-.1
Melilotus officinalis	761.29	2	Aquilegia vulgaris	345.7	-.2
Lespedeza sieboldii	742.51	1.9	Cardamine pratensis R	352.03	-.2
Spilanthes oleracea	701.6	1.7	Hydrophyllum virginicu	348.8	-.2
Cupressus lawsoniana	694.06	1.6	Stachys sylvatica	337.23	-.2
Echinacea purpurea	680.34	1.6	Tormentilla erecta	345.91	-.2
Ulmus campestris	691.91	1.6	Anethum graveolens	325.7	-.3
Ruta graveolens	660.93	1.5	Mentha arvensis	325.86	-.3
Syringa vulgaris	647.33	1.4	Osmundo regalis	322.98	-.3
Teucrium chamaedrys	605.26	1.2	Agrimonia eupatoria	313.53	-.4
Echinacea angustifolia	590.32	1.1	Cimicifuga racemosa	301.07	-.4
Linum usitatissimum	589.77	1.1	Fraxinus excelsior	305.37	-.4
Osmundo regalis 2	550.83	.9	Belladonna	285.19	-.5
Centaurium erythraea	505.57	.6	Gnaphalium leontopodiu	295.37	-.5
Laurocerasus	491.92	.6	Lapsana communis	290.18	-.5
Milium solis	506.12	.6	Mercurialis perennis	294.96	-.5
Ononis spinosa	493.5	.6	Urtica urens 2	297.67	-.5
Petroselinum crispum	497.24	.6	Aconitum napellus	266.95	-.6
Rhus toxicodendron	505.67	.6	Angelica archangelica	279.84	-.6
Cinnamodendron cortiso	457.92	.4	Leonurus cardiaca	273.92	-.6
Clematis erecta	459.35	.4	Polygonum aviculare	279.1	-.6
Thuja occidentalis	455.3	.4	Salvia sclarea	279.77	-.6
Anthemis nobilis	435.15	.3	Urtica urens	273.52	-.6
Psoralea bituminosa	441.84	.3	Lappa arctium	256.81	-.7
Rumex acetosa	436.84	.3	Plantago major	257.41	-.7
Sanguisorba officinali	442.98	.3	Faba vulgaris	225.64	-.8
Scrophularia nodosa	448.57	.3	Escholtzia californica	207.52	-.9
Lycopus europaeus	424.72	.2	Ocimum canum R	211.84	-.9
Marrubium vulgare	429.88	.2	Oenanthe crocata	215.3	-.9
Primula veris	416.31	.2	Bryonia dioica	194.03	-1
Salix purpurea	417.08	.2	Cicuta virosa	199.11	-1
Alchemilla vulgaris	396.98	.1	Collinsonia canadensis	195.58	-1
Chamomilla	403.98	.1	Galium aparine	193.8	-1
Malva sylvestris	411.07	.1	Glechoma hederacea	202.58	-1
Scutellaria lateriflor	399.91	.1	Hyssopus officinalis	202.97	-1
Solidago virgaurea	401.95	.1	Ocimum canum	194.41	-1
Stachys officinalis R	396.98	.1	Cardiospermum halicaca	170.48	-1.1
Tussilago farfara	404.68	.1	Fagopyrum esculentum	173.41	-1.1
Melissa officinalis	379.83	0	Taraxacum officinale	176.58	-1.1
Stachys officinalis	376.41	0	Verbena officinalis	176.02	-1.1
Tussilago petasites	390.2	0	Hyoscyamus niger	156.66	-1.2
Zizia aurea	378.38	0	Viola tricolor	140.15	-1.3
Cardamine pratensis	367.68	-.1	Adonis vernalis	107.92	-1.5
Chenopodium anthelmint	365.77	-.1	Mandragora officinalis	93.64	-1.5
Conium maculatum	374.6	-.1	Chelidonium majus	92.2	-1.6
Hedera helix	373.76	-.1	Bellis perennis	66.9	-1.7

Cadmium

Name	Value	Deviation	Name	Value	Deviation
Aconitum napellus	.39	-.5	Marrubium vulgare	.43	-.5
Adonis vernalis	1.3	-.1	Melilotus officinalis	.83	-.3
Agrimonia eupatoria	.16	-.6	Melissa officinalis	.47	-.5
Alchemilla vulgaris	.56	-.5	Mentha arvensis	.33	-.6
Anethum graveolens	2.11	.3	Mercurialis perennis	1.39	-.1
Angelica archangelica	2.28	.4	Milium solis	.29	-.6
Anthemis nobilis	1.69	.1	Ocimum canum	.17	-.6
Aquilegia vulgaris	.34	-.6	Ocimum canum R	.18	-.6
Belladonna	.86	-.3	Oenanthe crocata	.22	-.6
Bellis perennis	1.35	-.1	Ononis spinosa	.36	-.5
Bryonia dioica	1.01	-.2	Osmundo regalis	7.39	2.8
Cardamine pratensis	5.86	2.1	Osmundo regalis 2	7.27	2.8
Cardamine pratensis R	5.79	2.1	Paeonia officinalis	3.55	1
Cardiospermum halicaca	2.89	.7	Petroselinum crispum	.4	-.5
Castanea vesca	2.53	.5	Plantago major	.87	-.3
Centaurium erythraea	4.28	1.3	Polygonum aviculare	.62	-.4
Chamomilla	1.45	0	Primula veris	.25	-.6
Chelidonium majus	1.5	0	Prunus padus	.44	-.5
Chenopodium anthelmint	.19	-.6	Psoralea bituminosa	3.36	.9
Cicuta virosa	.87	-.3	Rhus toxicodendron	.9	-.3
Cimicifuga racemosa	2.25	.4	Rumex acetosa	3.38	.9
Cinnamodendron cortiso	2.43	.4	Ruta graveolens	2.61	.5
Clematis erecta	.72	-.4	Salicaria purpurea	.78	-.3
Collinsonia canadensis	1.18	-.2	Salix purpurea	13.03	5.5
Conium maculatum	.38	-.5	Salvia sclarea	.3	-.6
Cupressus lawsoniana	2.37	.4	Sanguisorba officinali	.55	-.5
Echinacea angustifolia	.17	-.6	Scrophularia nodosa	1.32	-.1
Echinacea purpurea	.16	-.6	Scutellaria lateriflor	.99	-.2
Escholtzia californica	.56	-.5	Solidago virgaurea	1.18	-.2
Faba vulgaris	.3	-.6	Spilanthes oleracea	.35	-.6
Fagopyrum esculentum	1.34	-.1	Stachys officinalis	.07	-.7
Fraxinus excelsior	.72	-.4	Stachys officinalis R	.06	-.7
Galium aparine	.94	-.3	Stachys sylvatica	.19	-.6
Glechoma hederacea	.33	-.6	Syringa vulgaris	1.04	-.2
Gnaphalium leontopodiu	.5	-.5	Taraxacum officinale	1.16	-.2
Hedera helix	2	.2	Teucrium chamaedrys	.5	-.5
Hydrophyllum virginicu	.27	-.6	Thuja occidentalis	2.77	.6
Hyoscyamus niger	1.39	-.1	Tormentilla erecta	10.04	4.1
Hyssopus officinalis	.31	-.6	Tussilago farfara	.13	-.7
Lappa arctium	1.41	0	Tussilago petasites	.12	-.7
Lapsana communis	3.72	1.1	Ulmus campestris	.37	-.5
Laurocerasus	1.27	-.1	Urtica dioica	.42	-.5
Leonurus cardiaca	.2	-.6	Urtica urens	.27	-.6
Lespedeza sieboldii	1.15	-.2	Urtica urens 2	.14	-.7
Linum usitatissimum	2.75	.6	Verbena officinalis	.21	-.6
Lycopus europaeus	.17	-.6	Vinca minor	2.03	.3
Malva sylvestris	.82	-.3	Viola tricolor	2.14	.3
Mandragora officinalis	1.15	-.2	Zizia aurea	.27	-.6

Cadmium

Name	Value	Deviation	Name	Value	Deviation
Salix purpurea	13.03	5.5	Plantago major	.87	-.3
Tormentilla erecta	10.04	4.1	Rhus toxicodendron	.9	-.3
Osmundo regalis	7.39	2.8	Salicaria purpurea	.78	-.3
Osmundo regalis 2	7.27	2.8	Clematis erecta	.72	-.4
Cardamine pratensis	5.86	2.1	Fraxinus excelsior	.72	-.4
Cardamine pratensis R	5.79	2.1	Polygonum aviculare	.62	-.4
Centaurium erythraea	4.28	1.3	Aconitum napellus	.39	-.5
Lapsana communis	3.72	1.1	Alchemilla vulgaris	.56	-.5
Paeonia officinalis	3.55	1	Conium maculatum	.38	-.5
Psoralea bituminosa	3.36	.9	Escholtzia californica	.56	-.5
Rumex acetosa	3.38	.9	Gnaphalium leontopodiu	.5	-.5
Cardiospermum halicaca	2.89	.7	Marrubium vulgare	.43	-.5
Linum usitatissimum	2.75	.6	Melissa officinalis	.47	-.5
Thuja occidentalis	2.77	.6	Ononis spinosa	.36	-.5
Castanea vesca	2.53	.5	Petroselinum crispum	.4	-.5
Ruta graveolens	2.61	.5	Prunus padus	.44	-.5
Angelica archangelica	2.28	.4	Sanguisorba officinali	.55	-.5
Cimicifuga racemosa	2.25	.4	Teucrium chamaedrys	.5	-.5
Cinnamodendron cortiso	2.43	.4	Ulmus campestris	.37	-.5
Cupressus lawsoniana	2.37	.4	Urtica dioica	.42	-.5
Anethum graveolens	2.11	.3	Agrimonia eupatoria	.16	-.6
Vinca minor	2.03	.3	Aquilegia vulgaris	.34	-.6
Viola tricolor	2.14	.3	Chenopodium anthelmint	.19	-.6
Hedera helix	2	.2	Echinacea angustifolia	.17	-.6
Anthemis nobilis	1.69	.1	Echinacea purpurea	.16	-.6
Chamomilla	1.45	0	Faba vulgaris	.3	-.6
Chelidonium majus	1.5	0	Glechoma hederacea	.33	-.6
Lappa arctium	1.41	0	Hydrophyllum virginicu	.27	-.6
Adonis vernalis	1.3	-.1	Hyssopus officinalis	.31	-.6
Bellis perennis	1.35	-.1	Leonurus cardiaca	.2	-.6
Fagopyrum esculentum	1.34	-.1	Lycopus europaeus	.17	-.6
Hyoscyamus niger	1.39	-.1	Mentha arvensis	.33	-.6
Laurocerasus	1.27	-.1	Milium solis	.29	-.6
Mercurialis perennis	1.39	-.1	Ocimum canum	.17	-.6
Scrophularia nodosa	1.32	-.1	Ocimum canum R	.18	-.6
Bryonia dioica	1.01	-.2	Oenanthe crocata	.22	-.6
Collinsonia canadensis	1.18	-.2	Primula veris	.25	-.6
Lespedeza sieboldii	1.15	-.2	Salvia sclarea	.3	-.6
Mandragora officinalis	1.15	-.2	Spilanthes oleracea	.35	-.6
Scutellaria lateriflor	.99	-.2	Stachys sylvatica	.19	-.6
Solidago virgaurea	1.18	-.2	Urtica urens	.27	-.6
Syringa vulgaris	1.04	-.2	Verbena officinalis	.21	-.6
Taraxacum officinale	1.16	-.2	Zizia aurea	.27	-.6
Belladonna	.86	-.3	Stachys officinalis	.07	-.7
Cicuta virosa	.87	-.3	Stachys officinalis R	.06	-.7
Galium aparine	.94	-.3	Tussilago farfara	.13	-.7
Malva sylvestris	.82	-.3	Tussilago petasites	.12	-.7
Melilotus officinalis	.83	-.3	Urtica urens 2	.14	-.7

Caesium

Name	Value	Deviation	Name	Value	Deviation
Aconitum napellus	.4	-.1	Marrubium vulgare	.1	-.5
Adonis vernalis	.58	.2	Melilotus officinalis	4.77	6.9
Agrimonia eupatoria	.06	-.6	Melissa officinalis	.23	-.3
Alchemilla vulgaris	.52	.1	Mentha arvensis	.26	-.3
Anethum graveolens	.09	-.6	Mercurialis perennis	.61	.3
Angelica archangelica	.36	-.1	Milium solis	.1	-.5
Anthemis nobilis	.34	-.2	Ocimum canum	.21	-.4
Aquilegia vulgaris	.06	-.6	Ocimum canum R	.24	-.3
Belladonna	.23	-.3	Oenanthe crocata	.48	.1
Bellis perennis	1.24	1.3	Ononis spinosa	.35	-.1
Bryonia dioica	.54	.2	Osmundo regalis	2.15	2.7
Cardamine pratensis	1.52	1.7	Osmundo regalis 2	1.82	2.2
Cardamine pratensis R	1.44	1.6	Paeonia officinalis	.36	-.1
Cardiospermum halicaca	.19	-.4	Petroselinum crispum	.54	.2
Castanea vesca	.25	-.3	Plantago major	.23	-.3
Centaurium erythraea	.71	.4	Polygonum aviculare	.24	-.3
Chamomilla	.3	-.2	Primula veris	.25	-.3
Chelidonium majus	1.24	1.3	Prunus padus	.07	-.6
Chenopodium anthelmint	.06	-.6	Psoralea bituminosa	.24	-.3
Cicuta virosa	.7	.4	Rhus toxicodendron	.05	-.6
Cimicifuga racemosa	.17	-.4	Rumex acetosa	2.38	3.1
Cinnamodendron cortiso	.16	-.4	Ruta graveolens	.06	-.6
Clematis erecta	.11	-.5	Salicaria purpurea	.2	-.4
Collinsonia canadensis	.39	-.1	Salix purpurea	.06	-.6
Conium maculatum	.1	-.5	Salvia sclarea	.1	-.5
Cupressus lawsoniana	.24	-.3	Sanguisorba officinali	.43	0
Echinacea angustifolia	.16	-.4	Scrophularia nodosa	.23	-.3
Echinacea purpurea	.11	-.5	Scutellaria lateriflor	1.08	1
Escholtzia californica	.39	-.1	Solidago virgaurea	.12	-.5
Faba vulgaris	.37	-.1	Spilanthes oleracea	.09	-.6
Fagopyrum esculentum	.15	-.5	Stachys officinalis	.4	-.1
Fraxinus excelsior	.08	-.6	Stachys officinalis R	.43	0
Galium aparine	.2	-.4	Stachys sylvatica	.28	-.2
Glechoma hederacea	.22	-.3	Syringa vulgaris	.72	.5
Gnaphalium leontopodiu	.89	.7	Taraxacum officinale	.62	.3
Hedera helix	.52	.1	Teucrium chamaedrys	.1	-.5
Hydrophyllum virginicu	.16	-.4	Thuja occidentalis	.09	-.6
Hyoscyamus niger	.22	-.3	Tormentilla erecta	.37	-.1
Hyssopus officinalis	.05	-.6	Tussilago farfara	.73	.5
Lappa arctium	.37	-.1	Tussilago petasites	.7	.4
Lapsana communis	.37	-.1	Ulmus campestris	.33	-.2
Laurocerasus	.26	-.3	Urtica dioica	.77	.5
Leonurus cardiaca	.11	-.5	Urtica urens	.06	-.6
Lespedeza sieboldii	.1	-.5	Urtica urens 2	.15	-.5
Linum usitatissimum	.15	-.5	Verbena officinalis	.04	-.6
Lycopus europaeus	.06	-.6	Vinca minor	.22	-.3
Malva sylvestris	.04	-.6	Viola tricolor	.21	-.4
Mandragora officinalis	.05	-.6	Zizia aurea	.46	0

Caesium

Name	Value	Deviation	Name	Value	Deviation
Melilotus officinalis	4.77	6.9	Melissa officinalis	.23	-.3
Rumex acetosa	2.38	3.1	Mentha arvensis	.26	-.3
Osmundo regalis	2.15	2.7	Ocimum canum R	.24	-.3
Osmundo regalis 2	1.82	2.2	Plantago major	.23	-.3
Cardamine pratensis	1.52	1.7	Polygonum aviculare	.24	-.3
Cardamine pratensis R	1.44	1.6	Primula veris	.25	-.3
Bellis perennis	1.24	1.3	Psoralea bituminosa	.24	-.3
Chelidonium majus	1.24	1.3	Scrophularia nodosa	.23	-.3
Scutellaria lateriflor	1.08	1	Vinca minor	.22	-.3
Gnaphalium leontopodiu	.89	.7	Cardiospermum halicaca	.19	-.4
Syringa vulgaris	.72	.5	Cimicifuga racemosa	.17	-.4
Tussilago farfara	.73	.5	Cinnamodendron cortiso	.16	-.4
Urtica dioica	.77	.5	Echinacea angustifolia	.16	-.4
Centaurium erythraea	.71	.4	Galium aparine	.2	-.4
Cicuta virosa	.7	.4	Hydrophyllum virginicu	.16	-.4
Tussilago petasites	.7	.4	Ocimum canum	.21	-.4
Mercurialis perennis	.61	.3	Salicaria purpurea	.2	-.4
Taraxacum officinale	.62	.3	Viola tricolor	.21	-.4
Adonis vernalis	.58	.2	Clematis erecta	.11	-.5
Bryonia dioica	.54	.2	Conium maculatum	.1	-.5
Petroselinum crispum	.54	.2	Echinacea purpurea	.11	-.5
Alchemilla vulgaris	.52	.1	Fagopyrum esculentum	.15	-.5
Hedera helix	.52	.1	Leonurus cardiaca	.11	-.5
Oenanthe crocata	.48	.1	Lespedeza sieboldii	.1	-.5
Sanguisorba officinali	.43	0	Linum usitatissimum	.15	-.5
Stachys officinalis R	.43	0	Marrubium vulgare	.1	-.5
Zizia aurea	.46	0	Milium solis	.1	-.5
Aconitum napellus	.4	-.1	Salvia sclarea	.1	-.5
Angelica archangelica	.36	-.1	Solidago virgaurea	.12	-.5
Collinsonia canadensis	.39	-.1	Teucrium chamaedrys	.1	-.5
Escholtzia californica	.39	-.1	Urtica urens 2	.15	-.5
Faba vulgaris	.37	-.1	Agrimonia eupatoria	.06	-.6
Lappa arctium	.37	-.1	Anethum graveolens	.09	-.6
Lapsana communis	.37	-.1	Aquilegia vulgaris	.06	-.6
Ononis spinosa	.35	-.1	Chenopodium anthelmint	.06	-.6
Paeonia officinalis	.36	-.1	Fraxinus excelsior	.08	-.6
Stachys officinalis	.4	-.1	Hyssopus officinalis	.05	-.6
Tormentilla erecta	.37	-.1	Lycopus europaeus	.06	-.6
Anthemis nobilis	.34	-.2	Malva sylvestris	.04	-.6
Chamomilla	.3	-.2	Mandragora officinalis	.05	-.6
Stachys sylvatica	.28	-.2	Prunus padus	.07	-.6
Ulmus campestris	.33	-.2	Rhus toxicodendron	.05	-.6
Belladonna	.23	-.3	Ruta graveolens	.06	-.6
Castanea vesca	.25	-.3	Salix purpurea	.06	-.6
Cupressus lawsoniana	.24	-.3	Spilanthes oleracea	.09	-.6
Glechoma hederacea	.22	-.3	Thuja occidentalis	.09	-.6
Hyoscyamus niger	.22	-.3	Urtica urens	.06	-.6
Laurocerasus	.26	-.3	Verbena officinalis	.04	-.6

Calcium

Name	Value	Deviation	Name	Value	Deviation
Aconitum napellus	353296.29	.5	Marrubium vulgare	307417.38	.2
Adonis vernalis	91976.54	-1.3	Melilotus officinalis	527089.08	1.6
Agrimonia eupatoria	233196.3	-.3	Melissa officinalis	325880.37	.3
Alchemilla vulgaris	231759.65	-.4	Mentha arvensis	387424.19	.7
Anethum graveolens	402386.55	.8	Mercurialis perennis	326193.11	.3
Angelica archang	107612.32	-1.2	Milium solis	158249.57	-.8
Anthemis nobilis	248937.64	-.2	Ocimum canum	413067.18	.9
Aquilegia vulgaris	169161.37	-.8	Ocimum canum R	430962.07	1
Belladonna	164358.34	-.8	Oenanthe crocata	86363.63	-1.3
Bellis perennis	32626.53	-1.7	Ononis spinosa	285531.79	0
Bryonia dioica	124760.08	-1.1	Osmundo regalis	194368.22	-.6
Cardamine prat	336765.42	.3	Osmundo regalis 2	35926.12	-1.7
Cardamine prat R	263806.27	-.1	Paeonia officinalis	586652.39	2
Cardiospermum hal	257371.19	-.2	Petroselinum crispum	241300.35	-.3
Castanea vesca	229370.67	-.4	Plantago major	380050.74	.6
Centaurium eryt	132941.54	-1	Polygonum aviculare	302427.37	.1
Chamomilla	136975.57	-1	Primula veris	145533.2	-.9
Chelidonium majus	45142.56	-1.6	Prunus padus	391988.56	.7
Chenopodium anth	265763.31	-.1	Psoralea bituminosa	436703.3	1
Cicuta virosa	112275.19	-1.1	Rhus toxicodendron	536080.95	1.7
Cimicifuga racemosa	166268.35	-.8	Rumex acetosa	293406.38	.1
Cinnamodendron cort	227850.39	-.4	Ruta graveolens	343347.9	.4
Clematis erecta	308022.94	.2	Salicaria purpurea	337844.41	.4
Collinsonia canadensis	84231.72	-1.3	Salix purpurea	765546.53	3.2
Conium maculatum	270868.31	-.1	Salvia sclarea	328233.78	.3
Cupressus lawsoniana	415111.81	.9	Sanguisorba officin	246101.69	-.3
Echinacea angust	347386.63	.4	Scrophularia nodosa	338261.86	.4
Echinacea purpurea	418447.22	.9	Scutellaria lateriflor	416481.65	.9
Escholtzia californ	104626.29	-1.2	Solidago virgaurea	187131.92	-.6
Faba vulgaris	236671.79	-.3	Spilanthes oleracea	272156.7	-.1
Fagopyrum esculentum	345075.14	.4	Stachys officinalis	143061.13	-.9
Fraxinus excelsior	629499.14	2.3	Stachys officinalis R	153671.02	-.9
Galium aparine	274154.23	-.1	Stachys sylvatica	237175.73	-.3
Glechoma hederacea	182997.94	-.7	Syringa vulgaris	125209.02	-1.1
Gnaphalium leont	229674.87	-.4	Taraxacum officin	58742.8	-1.5
Hedera helix	426983.2	.9	Teucrium chamaed	345834	.4
Hydrophyllum virgin	235811.97	-.3	Thuja occidentalis	688350.03	2.7
Hyoscyamus niger	173088.17	-.7	Tormentilla erecta	130715.93	-1
Hyssopus officinalis	301928.79	.1	Tussilago farfara	365022.25	.5
Lappa arctium	147378.66	-.9	Tussilago petasites	215175.49	-.5
Lapsana communis	239345.74	-.3	Ulmus campestris	647363.69	2.4
Laurocerasus	394922.74	.7	Urtica dioica	387320.91	.7
Leonurus cardiaca	300123.85	.1	Urtica urens	484108.89	1.3
Lespedeza sieboldii	379866.28	.6	Urtica urens 2	537165.59	1.7
Linum usitatissimum	221280.44	-.4	Verbena officinalis	204269.86	-.5
Lycopus europaeus	209251.43	-.5	Vinca minor	447404.96	1.1
Malva sylvestris	416222.55	.9	Viola tricolor	106920.98	-1.2
Mandragora officin	27234.99	-1.7	Zizia aurea	343240.22	.4

Calcium

Name	Value	Deviation	Name	Value	Deviation
Salix purpurea	765546.53	3.2	Galium aparine	274154.23	-.1
Thuja occidentalis	688350.03	2.7	Spilanthes oleracea	272156.7	-.1
Ulmus campestris	647363.69	2.4	Anthemis nobilis	248937.64	-.2
Fraxinus excelsior	629499.14	2.3	Cardiospermum hal	257371.19	-.2
Paeonia officinalis	586652.39	2	Agrimonia eupatoria	233196.3	-.3
Rhus toxicodendron	536080.95	1.7	Faba vulgaris	236671.79	-.3
Urtica urens 2	537165.59	1.7	Hydrophyllum virg	235811.97	-.3
Melilotus officinalis	527089.08	1.6	Lapsana communis	239345.74	-.3
Urtica urens	484108.89	1.3	Petroselinum crispum	241300.35	-.3
Vinca minor	447404.96	1.1	Sanguisorba officin	246101.69	-.3
Ocimum canum R	430962.07	1	Stachys sylvatica	237175.73	-.3
Psoralea bituminosa	436703.3	1	Alchemilla vulgaris	231759.65	-.4
Cupressus lawsoniana	415111.81	.9	Castanea vesca	229370.67	-.4
Echinacea purpurea	418447.22	.9	Cinnamodendron cor	227850.39	-.4
Hedera helix	426983.2	.9	Gnaphalium leont	229674.87	-.4
Malva sylvestris	416222.55	.9	Linum usitatissimum	221280.44	-.4
Ocimum canum	413067.18	.9	Lycopus europaeus	209251.43	-.5
Scutellaria lateriflor	416481.65	.9	Tussilago petasites	215175.49	-.5
Anethum graveolens	402386.55	.8	Verbena officinalis	204269.86	-.5
Laurocerasus	394922.74	.7	Osmundo regalis	194368.22	-.6
Mentha arvensis	387424.19	.7	Solidago virgaurea	187131.92	-.6
Prunus padus	391988.56	.7	Glechoma hederacea	182997.94	-.7
Urtica dioica	387320.91	.7	Hyoscyamus niger	173088.17	-.7
Lespedeza sieboldii	379866.28	.6	Aquilegia vulgaris	169161.37	-.8
Plantago major	380050.74	.6	Belladonna	164358.34	-.8
Aconitum napellus	353296.29	.5	Cimicifuga racemosa	166268.35	-.8
Tussilago farfara	365022.25	.5	Milium solis	158249.57	-.8
Echinacea angust	347386.63	.4	Lappa arctium	147378.66	-.9
Fagopyrum esculen	345075.14	.4	Primula veris	145533.2	-.9
Ruta graveolens	343347.9	.4	Stachys officin	143061.13	-.9
Salicaria purpurea	337844.41	.4	Stachys officin R	153671.02	-.9
Scrophularia nodosa	338261.86	.4	Centaurium erythr	132941.54	-1
Teucrium chamaed	345834	.4	Chamomilla	136975.57	-1
Zizia aurea	343240.22	.4	Tormentilla erecta	130715.93	-1
Cardamine pratensis	336765.42	.3	Bryonia dioica	124760.08	-1.1
Melissa officinalis	325880.37	.3	Cicuta virosa	112275.19	-1.1
Mercurialis perennis	326193.11	.3	Syringa vulgaris	125209.02	-1.1
Salvia sclarea	328233.78	.3	Angelica archangel	107612.32	-1.2
Clematis erecta	308022.94	.2	Escholtzia californ	104626.29	-1.2
Marrubium vulgare	307417.38	.2	Viola tricolor	106920.98	-1.2
Hyssopus officinalis	301928.79	.1	Adonis vernalis	91976.54	-1.3
Leonurus cardiaca	300123.85	.1	Collinsonia canadensis	84231.72	-1.3
Polygonum aviculare	302427.37	.1	Oenanthe crocata	86363.63	-1.3
Rumex acetosa	293406.38	.1	Taraxacum officinale	58742.8	-1.5
Ononis spinosa	285531.79	0	Chelidonium majus	45142.56	-1.6
Cardamine prat R	263806.27	-.1	Bellis perennis	32626.53	-1.7
Chenopodium anthel	265763.31	-.1	Mandragora officin	27234.99	-1.7
Conium maculatum	270868.31	-.1	Osmundo regalis 2	35926.12	-1.7

Cerium

Name	Value	Deviation	Name	Value	Deviation
Aconitum napellus	4.63	.4	Marrubium vulgare	2.3	-.3
Adonis vernalis	5.06	.5	Melilotus officinalis	1.61	-.5
Agrimonia eupatoria	1.46	-.6	Melissa officinalis	3.02	-.1
Alchemilla vulgaris	4.56	.4	Mentha arvensis	2.21	-.4
Anethum graveolens	1.51	-.6	Mercurialis perennis	4.52	.3
Angelica archangelica	3.85	.1	Milium solis	2.02	-.4
Anthemis nobilis	3.39	0	Ocimum canum	3.01	-.1
Aquilegia vulgaris	1.37	-.6	Ocimum canum R	3.01	-.1
Belladonna	2.79	-.2	Oenanthe crocata	4.11	.2
Bellis perennis	13.34	3	Ononis spinosa	3.08	-.1
Bryonia dioica	5.08	.5	Osmundo regalis	23.37	6.1
Cardamine pratensis	3.98	.2	Osmundo regalis 2	.93	-.8
Cardamine pratensis R	3.48	0	Paeonia officinalis	3.37	0
Cardiospermum halicaca	2.9	-.1	Petroselinum crispum	5.09	.5
Castanea vesca	2.12	-.4	Plantago major	4.14	.2
Centaurium erythraea	6.49	1	Polygonum aviculare	2.51	-.3
Chamomilla	2.92	-.1	Primula veris	3.17	-.1
Chelidonium majus	14.56	3.4	Prunus padus	2.05	-.4
Chenopodium anthelmint	1.34	-.6	Psoralea bituminosa	2.56	-.3
Cicuta virosa	10.8	2.3	Rhus toxicodendron	1.78	-.5
Cimicifuga racemosa	4.34	.3	Rumex acetosa	12.4	2.8
Cinnamodendron cortiso	2.05	-.4	Ruta graveolens	1.61	-.5
Clematis erecta	1.76	-.5	Salicaria purpurea	1.69	-.5
Collinsonia canadensis	5.13	.5	Salix purpurea	2.16	-.4
Conium maculatum	1.59	-.5	Salvia sclarea	1.68	-.5
Cupressus lawsoniana	3.82	.1	Sanguisorba officinali	1.18	-.7
Echinacea angustifolia	2.02	-.4	Scrophularia nodosa	2.61	-.2
Echinacea purpurea	1.28	-.6	Scutellaria lateriflor	2.46	-.3
Escholtzia californica	1.67	-.5	Solidago virgaurea	1.34	-.6
Faba vulgaris	1.61	-.5	Spilanthes oleracea	1.7	-.5
Fagopyrum esculentum	1.19	-.7	Stachys officinalis	1.32	-.6
Fraxinus excelsior	2.6	-.2	Stachys officinalis R	1.18	-.7
Galium aparine	2.85	-.2	Stachys sylvatica	2.76	-.2
Glechoma hederacea	2.54	-.3	Syringa vulgaris	2.39	-.3
Gnaphalium leontopodiu	7.22	1.2	Taraxacum officinale	9.87	2
Hedera helix	2.18	-.4	Teucrium chamaedrys	2.05	-.4
Hydrophyllum virginicu	1.54	-.6	Thuja occidentalis	1.83	-.5
Hyoscyamus niger	2.62	-.2	Tormentilla erecta	6.03	.8
Hyssopus officinalis	1.54	-.6	Tussilago farfara	1.39	-.6
Lappa arctium	4.28	.3	Tussilago petasites	1.96	-.4
Lapsana communis	4.68	.4	Ulmus campestris	2.42	-.3
Laurocerasus	1.35	-.6	Urtica dioica	6.11	.8
Leonurus cardiaca	1.38	-.6	Urtica urens	1.7	-.5
Lespedeza sieboldii	1.43	-.6	Urtica urens 2	1.91	-.5
Linum usitatissimum	1.28	-.6	Verbena officinalis	1.33	-.6
Lycopus europaeus	1.48	-.6	Vinca minor	3.42	0
Malva sylvestris	1.77	-.5	Viola tricolor	2.41	-.3
Mandragora officinalis	2	-.4	Zizia aurea	3.3	0

Cerium

Name	Value	Deviation	Name	Value	Deviation
Osmundo regalis	23.37	6.1	Syringa vulgaris	2.39	-.3
Chelidonium majus	14.56	3.4	Ulmus campestris	2.42	-.3
Bellis perennis	13.34	3	Viola tricolor	2.41	-.3
Rumex acetosa	12.4	2.8	Castanea vesca	2.12	-.4
Cicuta virosa	10.8	2.3	Cinnamodendron cortiso	2.05	-.4
Taraxacum officinale	9.87	2	Echinacea angustifolia	2.02	-.4
Gnaphalium leontopodiu	7.22	1.2	Hedera helix	2.18	-.4
Centaurium erythraea	6.49	1	Mandragora officinalis	2	-.4
Tormentilla erecta	6.03	.8	Mentha arvensis	2.21	-.4
Urtica dioica	6.11	.8	Milium solis	2.02	-.4
Adonis vernalis	5.06	.5	Prunus padus	2.05	-.4
Bryonia dioica	5.08	.5	Salix purpurea	2.16	-.4
Collinsonia canadensis	5.13	.5	Teucrium chamaedrys	2.05	-.4
Petroselinum crispum	5.09	.5	Tussilago petasites	1.96	-.4
Aconitum napellus	4.63	.4	Clematis erecta	1.76	-.5
Alchemilla vulgaris	4.56	.4	Conium maculatum	1.59	-.5
Lapsana communis	4.68	.4	Escholtzia californica	1.67	-.5
Cimicifuga racemosa	4.34	.3	Faba vulgaris	1.61	-.5
Lappa arctium	4.28	.3	Malva sylvestris	1.77	-.5
Mercurialis perennis	4.52	.3	Melilotus officinalis	1.61	-.5
Cardamine pratensis	3.98	.2	Rhus toxicodendron	1.78	-.5
Oenanthe crocata	4.11	.2	Ruta graveolens	1.61	-.5
Plantago major	4.14	.2	Salicaria purpurea	1.69	-.5
Angelica archangelica	3.85	.1	Salvia sclarea	1.68	-.5
Cupressus lawsoniana	3.82	.1	Spilanthes oleracea	1.7	-.5
Anthemis nobilis	3.39	0	Thuja occidentalis	1.83	-.5
Cardamine pratensis R	3.48	0	Urtica urens	1.7	-.5
Paeonia officinalis	3.37	0	Urtica urens 2	1.91	-.5
Vinca minor	3.42	0	Agrimonia eupatoria	1.46	-.6
Zizia aurea	3.3	0	Anethum graveolens	1.51	-.6
Cardiospermum halicaca	2.9	-.1	Aquilegia vulgaris	1.37	-.6
Chamomilla	2.92	-.1	Chenopodium anthelmint	1.34	-.6
Melissa officinalis	3.02	-.1	Echinacea purpurea	1.28	-.6
Ocimum canum	3.01	-.1	Hydrophyllum virginicu	1.54	-.6
Ocimum canum R	3.01	-.1	Hyssopus officinalis	1.54	-.6
Ononis spinosa	3.08	-.1	Laurocerasus	1.35	-.6
Primula veris	3.17	-.1	Leonurus cardiaca	1.38	-.6
Belladonna	2.79	-.2	Lespedeza sieboldii	1.43	-.6
Fraxinus excelsior	2.6	-.2	Linum usitatissimum	1.28	-.6
Galium aparine	2.85	-.2	Lycopus europaeus	1.48	-.6
Hyoscyamus niger	2.62	-.2	Solidago virgaurea	1.34	-.6
Scrophularia nodosa	2.61	-.2	Stachys officinalis	1.32	-.6
Stachys sylvatica	2.76	-.2	Tussilago farfara	1.39	-.6
Glechoma hederacea	2.54	-.3	Verbena officinalis	1.33	-.6
Marrubium vulgare	2.3	-.3	Fagopyrum esculentum	1.19	-.7
Polygonum aviculare	2.51	-.3	Sanguisorba officinali	1.18	-.7
Psoralea bituminosa	2.56	-.3	Stachys officinalis R	1.18	-.7
Scutellaria lateriflor	2.46	-.3	Osmundo regalis 2	.93	-.8

Chromium

Name	Value	Deviation	Name	Value	Deviation
Aconitum napellus	5.21	.4	Marrubium vulgare	2.2	-.1
Adonis vernalis	7.45	.7	Melilotus officinalis	1.45	-.3
Agrimonia eupatoria	0	-.5	Melissa officinalis	3.97	.2
Alchemilla vulgaris	3.5	.1	Mentha arvensis	0	-.5
Anethum graveolens	0	-.5	Mercurialis perennis	0	-.5
Angelica archangelica	6.64	.6	Milium solis	0	-.5
Anthemis nobilis	0	-.5	Ocimum canum	0	-.5
Aquilegia vulgaris	0	-.5	Ocimum canum R	0	-.5
Belladonna	0	-.5	Oenanthe crocata	2.67	-.1
Bellis perennis	24.14	3.5	Ononis spinosa	0	-.5
Bryonia dioica	4.44	.2	Osmundo regalis	25.56	3.8
Cardamine pratensis	0	-.5	Osmundo regalis 2	0	-.5
Cardamine pratensis R	0	-.5	Paeonia officinalis	3.64	.1
Cardiospermum halicaca	4.08	.2	Petroselinum crispum	3.41	.1
Castanea vesca	0	-.5	Plantago major	1.85	-.2
Centaurium erythraea	8.61	.9	Polygonum aviculare	0	-.5
Chamomilla	0	-.5	Primula veris	0	-.5
Chelidonium majus	25.65	3.8	Prunus padus	0	-.5
Chenopodium anthelmint	0	-.5	Psoralea bituminosa	0	-.5
Cicuta virosa	4.8	.3	Rhus toxicodendron	2.68	-.1
Cimicifuga racemosa	2.57	-.1	Rumex acetosa	18.24	2.5
Cinnamodendron cortiso	0	-.5	Ruta graveolens	0	-.5
Clematis erecta	0	-.5	Salicaria purpurea	0	-.5
Collinsonia canadensis	8.39	.9	Salix purpurea	3.9	.1
Conium maculatum	4.26	.2	Salvia sclarea	1.2	-.3
Cupressus lawsoniana	5.98	.5	Sanguisorba officinali	0	-.5
Echinacea angustifolia	3.34	0	Scrophularia nodosa	0	-.5
Echinacea purpurea	0	-.5	Scutellaria lateriflor	0	-.5
Escholtzia californica	0	-.5	Solidago virgaurea	0	-.5
Faba vulgaris	0	-.5	Spilanthes oleracea	1.37	-.3
Fagopyrum esculentum	0	-.5	Stachys officinalis	0	-.5
Fraxinus excelsior	3.9	.1	Stachys officinalis R	0	-.5
Galium aparine	3.42	.1	Stachys sylvatica	0	-.5
Glechoma hederacea	0	-.5	Syringa vulgaris	4.89	.3
Gnaphalium leontopodiu	5.01	.3	Taraxacum officinale	11.51	1.4
Hedera helix	0	-.5	Teucrium chamaedrys	1.73	-.2
Hydrophyllum virginicu	0	-.5	Thuja occidentalis	2.85	0
Hyoscyamus niger	0	-.5	Tormentilla erecta	9.63	1.1
Hyssopus officinalis	0	-.5	Tussilago farfara	0	-.5
Lappa arctium	0	-.5	Tussilago petasites	0	-.5
Lapsana communis	2	-.2	Ulmus campestris	3.03	0
Laurocerasus	0	-.5	Urtica dioica	7.62	.8
Leonurus cardiaca	0	-.5	Urtica urens	0	-.5
Lespedeza sieboldii	2.58	-.1	Urtica urens 2	1.69	-.2
Linum usitatissimum	0	-.5	Verbena officinalis	33.42	5.1
Lycopus europaeus	0	-.5	Vinca minor	0	-.5
Malva sylvestris	0	-.5	Viola tricolor	4.98	.3
Mandragora officinalis	1.45	-.3	Zizia aurea	3.68	.1

Chromium

Name	Value	Deviation	Name	Value	Deviation
Verbena officinalis	33.42	5.1	Aquilegia vulgaris	0	-.5
Chelidonium majus	25.65	3.8	Belladonna	0	-.5
Osmundo regalis	25.56	3.8	Cardamine pratensis	0	-.5
Bellis perennis	24.14	3.5	Cardamine pratensis R	0	-.5
Rumex acetosa	18.24	2.5	Castanea vesca	0	-.5
Taraxacum officinale	11.51	1.4	Chamomilla	0	-.5
Tormentilla erecta	9.63	1.1	Chenopodium anthelmint	0	-.5
Centaurium erythraea	8.61	.9	Cinnamodendron cortiso	0	-.5
Collinsonia canadensis	8.39	.9	Clematis erecta	0	-.5
Urtica dioica	7.62	.8	Echinacea purpurea	0	-.5
Adonis vernalis	7.45	.7	Escholtzia californica	0	-.5
Angelica archangelica	6.64	.6	Faba vulgaris	0	-.5
Cupressus lawsoniana	5.98	.5	Fagopyrum esculentum	0	-.5
Aconitum napellus	5.21	.4	Glechoma hederacea	0	-.5
Cicuta virosa	4.8	.3	Hedera helix	0	-.5
Gnaphalium leontopodiu	5.01	.3	Hydrophyllum virginicu	0	-.5
Syringa vulgaris	4.89	.3	Hyoscyamus niger	0	-.5
Viola tricolor	4.98	.3	Hyssopus officinalis	0	-.5
Bryonia dioica	4.44	.2	Lappa arctium	0	-.5
Cardiospermum halicaca	4.08	.2	Laurocerasus	0	-.5
Conium maculatum	4.26	.2	Leonurus cardiaca	0	-.5
Melissa officinalis	3.97	.2	Linum usitatissimum	0	-.5
Alchemilla vulgaris	3.5	.1	Lycopus europaeus	0	-.5
Fraxinus excelsior	3.9	.1	Malva sylvestris	0	-.5
Galium aparine	3.42	.1	Mentha arvensis	0	-.5
Paeonia officinalis	3.64	.1	Mercurialis perennis	0	-.5
Petroselinum crispum	3.41	.1	Milium solis	0	-.5
Salix purpurea	3.9	.1	Ocimum canum	0	-.5
Zizia aurea	3.68	.1	Ocimum canum R	0	-.5
Echinacea angustifolia	3.34	0	Ononis spinosa	0	-.5
Thuja occidentalis	2.85	0	Osmundo regalis 2	0	-.5
Ulmus campestris	3.03	0	Polygonum aviculare	0	-.5
Cimicifuga racemosa	2.57	-.1	Primula veris	0	-.5
Lespedeza sieboldii	2.58	-.1	Prunus padus	0	-.5
Marrubium vulgare	2.2	-.1	Psoralea bituminosa	0	-.5
Oenanthe crocata	2.67	-.1	Ruta graveolens	0	-.5
Rhus toxicodendron	2.68	-.1	Salicaria purpurea	0	-.5
Lapsana communis	2	-.2	Sanguisorba officinali	0	-.5
Plantago major	1.85	-.2	Scrophularia nodosa	0	-.5
Teucrium chamaedrys	1.73	-.2	Scutellaria lateriflor	0	-.5
Urtica urens 2	1.69	-.2	Solidago virgaurea	0	-.5
Mandragora officinalis	1.45	-.3	Stachys officinalis	0	-.5
Melilotus officinalis	1.45	-.3	Stachys officinalis R	0	-.5
Salvia sclarea	1.2	-.3	Stachys sylvatica	0	-.5
Spilanthes oleracea	1.37	-.3	Tussilago farfara	0	-.5
Agrimonia eupatoria	0	-.5	Tussilago petasites	0	-.5
Anethum graveolens	0	-.5	Urtica urens	0	-.5
Anthemis nobilis	0	-.5	Vinca minor	0	-.5

Cobaltum

Name	Value	Deviation	Name	Value	Deviation
Aconitum napellus	1.68	-.2	Marrubium vulgare	1	-.5
Adonis vernalis	.97	-.5	Melilotus officinalis	4.31	.8
Agrimonia eupatoria	.73	-.6	Melissa officinalis	1.29	-.4
Alchemilla vulgaris	1.84	-.2	Mentha arvensis	1.27	-.4
Anethum graveolens	1.05	-.5	Mercurialis perennis	2.54	.1
Angelica archangelica	1.18	-.4	Milium solis	1.04	-.5
Anthemis nobilis	1.19	-.4	Ocimum canum	1.38	-.4
Aquilegia vulgaris	.96	-.5	Ocimum canum R	1.53	-.3
Belladonna	.96	-.5	Oenanthe crocata	1.3	-.4
Bellis perennis	2.96	.3	Ononis spinosa	1.5	-.3
Bryonia dioica	1.93	-.1	Osmundo regalis	16.59	5.7
Cardamine pratensis	1.21	-.4	Osmundo regalis 2	.26	-.8
Cardamine pratensis R	1.12	-.5	Paeonia officinalis	2.53	.1
Cardiospermum halicaca	1.43	-.3	Petroselinum crispum	1.93	-.1
Castanea vesca	5.45	1.3	Plantago major	1.65	-.3
Centaurium erythraea	2.84	.2	Polygonum aviculare	1.26	-.4
Chamomilla	1.43	-.3	Primula veris	1.02	-.5
Chelidonium majus	3.9	.6	Prunus padus	1.46	-.3
Chenopodium anthelmint	.9	-.6	Psoralea bituminosa	4.93	1.1
Cicuta virosa	9.91	3.1	Rhus toxicodendron	1.45	-.3
Cimicifuga racemosa	4.83	1	Rumex acetosa	6.9	1.8
Cinnamodendron cortiso	1.23	-.4	Ruta graveolens	2.97	.3
Clematis erecta	1.03	-.5	Salicaria purpurea	5.21	1.2
Collinsonia canadensis	1.84	-.2	Salix purpurea	5.29	1.2
Conium maculatum	.63	-.7	Salvia sclarea	1.05	-.5
Cupressus lawsoniana	5.83	1.4	Sanguisorba officinali	1.59	-.3
Echinacea angustifolia	.99	-.5	Scrophularia nodosa	1.62	-.3
Echinacea purpurea	1.38	-.4	Scutellaria lateriflor	1.35	-.4
Escholtzia californica	.55	-.7	Solidago virgaurea	.7	-.6
Faba vulgaris	1.27	-.4	Spilanthes oleracea	.97	-.5
Fagopyrum esculentum	.9	-.6	Stachys officinalis	.76	-.6
Fraxinus excelsior	1.96	-.1	Stachys officinalis R	.75	-.6
Galium aparine	.95	-.5	Stachys sylvatica	1.28	-.4
Glechoma hederacea	1.35	-.4	Syringa vulgaris	13.94	4.7
Gnaphalium leontopodiu	2.84	.2	Taraxacum officinale	2.41	.1
Hedera helix	1.59	-.3	Teucrium chamaedrys	1.45	-.3
Hydrophyllum virginicu	.77	-.6	Thuja occidentalis	4.14	.7
Hyoscyamus niger	1.09	-.5	Tormentilla erecta	3.86	.6
Hyssopus officinalis	.9	-.6	Tussilago farfara	1.92	-.1
Lappa arctium	2.04	-.1	Tussilago petasites	1.16	-.4
Lapsana communis	1.69	-.2	Ulmus campestris	2.28	0
Laurocerasus	4.96	1.1	Urtica dioica	2.43	.1
Leonurus cardiaca	.81	-.6	Urtica urens	1.24	-.4
Lespedeza sieboldii	1.15	-.5	Urtica urens 2	1.32	-.4
Linum usitatissimum	.78	-.6	Verbena officinalis	5.7	1.4
Lycopus europaeus	2.65	.1	Vinca minor	1.62	-.3
Malva sylvestris	1.34	-.4	Viola tricolor	.63	-.7
Mandragora officinalis	2.05	-.1	Zizia aurea	1.34	-.4

Cobaltum

Name	Value	Deviation
Osmundo regalis	16.59	5.7
Syringa vulgaris	13.94	4.7
Cicuta virosa	9.91	3.1
Rumex acetosa	6.9	1.8
Cupressus lawsoniana	5.83	1.4
Verbena officinalis	5.7	1.4
Castanea vesca	5.45	1.3
Salicaria purpurea	5.21	1.2
Salix purpurea	5.29	1.2
Laurocerasus	4.96	1.1
Psoralea bituminosa	4.93	1.1
Cimicifuga racemosa	4.83	1
Melilotus officinalis	4.31	.8
Thuja occidentalis	4.14	.7
Chelidonium majus	3.9	.6
Tormentilla erecta	3.86	.6
Bellis perennis	2.96	.3
Ruta graveolens	2.97	.3
Centaurium erythraea	2.84	.2
Gnaphalium leontopodiu	2.84	.2
Lycopus europaeus	2.65	.1
Mercurialis perennis	2.54	.1
Paeonia officinalis	2.53	.1
Taraxacum officinale	2.41	.1
Urtica dioica	2.43	.1
Ulmus campestris	2.28	0
Bryonia dioica	1.93	-.1
Fraxinus excelsior	1.96	-.1
Lappa arctium	2.04	-.1
Mandragora officinalis	2.05	-.1
Petroselinum crispum	1.93	-.1
Tussilago farfara	1.92	-.1
Aconitum napellus	1.68	-.2
Alchemilla vulgaris	1.84	-.2
Collinsonia canadensis	1.84	-.2
Lapsana communis	1.69	-.2
Cardiospermum halicaca	1.43	-.3
Chamomilla	1.43	-.3
Hedera helix	1.59	-.3
Ocimum canum R	1.53	-.3
Ononis spinosa	1.5	-.3
Plantago major	1.65	-.3
Prunus padus	1.46	-.3
Rhus toxicodendron	1.45	-.3
Sanguisorba officinali	1.59	-.3
Scrophularia nodosa	1.62	-.3
Teucrium chamaedrys	1.45	-.3
Vinca minor	1.62	-.3

Name	Value	Deviation
Angelica archangelica	1.18	-.4
Anthemis nobilis	1.19	-.4
Cardamine pratensis	1.21	-.4
Cinnamodendron cortiso	1.23	-.4
Echinacea purpurea	1.38	-.4
Faba vulgaris	1.27	-.4
Glechoma hederacea	1.35	-.4
Malva sylvestris	1.34	-.4
Melissa officinalis	1.29	-.4
Mentha arvensis	1.27	-.4
Ocimum canum	1.38	-.4
Oenanthe crocata	1.3	-.4
Polygonum aviculare	1.26	-.4
Scutellaria lateriflor	1.35	-.4
Stachys sylvatica	1.28	-.4
Tussilago petasites	1.16	-.4
Urtica urens	1.24	-.4
Urtica urens 2	1.32	-.4
Zizia aurea	1.34	-.4
Adonis vernalis	.97	-.5
Anethum graveolens	1.05	-.5
Aquilegia vulgaris	.96	-.5
Belladonna	.96	-.5
Cardamine pratensis R	1.12	-.5
Clematis erecta	1.03	-.5
Echinacea angustifolia	.99	-.5
Galium aparine	.95	-.5
Hyoscyamus niger	1.09	-.5
Lespedeza sieboldii	1.15	-.5
Marrubium vulgare	1	-.5
Milium solis	1.04	-.5
Primula veris	1.02	-.5
Salvia sclarea	1.05	-.5
Spilanthes oleracea	.97	-.5
Agrimonia eupatoria	.73	-.6
Chenopodium anthelmint	.9	-.6
Fagopyrum esculentum	.9	-.6
Hydrophyllum virginicu	.77	-.6
Hyssopus officinalis	.9	-.6
Leonurus cardiaca	.81	-.6
Linum usitatissimum	.78	-.6
Solidago virgaurea	.7	-.6
Stachys officinalis	.76	-.6
Stachys officinalis R	.75	-.6
Conium maculatum	.63	-.7
Escholtzia californica	.55	-.7
Viola tricolor	.63	-.7
Osmundo regalis 2	.26	-.8

Cuprum

Name	Value	Deviation	Name	Value	Deviation
Aconitum napellus	58.69	-.2	Marrubium vulgare	23.21	-.8
Adonis vernalis	81.4	.2	Melilotus officinalis	76.02	.1
Agrimonia eupatoria	90.07	.4	Melissa officinalis	64.06	-.1
Alchemilla vulgaris	69.31	0	Mentha arvensis	52.39	-.3
Anethum graveolens	48.32	-.4	Mercurialis perennis	53.82	-.3
Angelica archangelica	68.97	0	Milium solis	37.34	-.6
Anthemis nobilis	64.73	-.1	Ocimum canum	33.26	-.6
Aquilegia vulgaris	81.88	.2	Ocimum canum R	35.6	-.6
Belladonna	49.09	-.3	Oenanthe crocata	62.54	-.1
Bellis perennis	17.68	-.9	Ononis spinosa	30.21	-.7
Bryonia dioica	54.33	-.3	Osmundo regalis	343.14	5
Cardamine pratensis	63.79	-.1	Osmundo regalis 2	130.15	1.1
Cardamine pratensis R	61.9	-.1	Paeonia officinalis	46.96	-.4
Cardiospermum halicaca	36.53	-.6	Petroselinum crispum	47.56	-.4
Castanea vesca	21.73	-.8	Plantago major	21.12	-.9
Centaurium erythraea	54.37	-.3	Polygonum aviculare	63.58	-.1
Chamomilla	65.93	0	Primula veris	18.54	-.9
Chelidonium majus	27.85	-.7	Prunus padus	109.32	.7
Chenopodium anthelmint	60	-.2	Psoralea bituminosa	39.13	-.5
Cicuta virosa	278.16	3.8	Rhus toxicodendron	63.85	-.1
Cimicifuga racemosa	124.5	1	Rumex acetosa	156.97	1.6
Cinnamodendron cortiso	159.67	1.7	Ruta graveolens	68.53	0
Clematis erecta	59.01	-.2	Salicaria purpurea	48.25	-.4
Collinsonia canadensis	149.68	1.5	Salix purpurea	36.61	-.6
Conium maculatum	74.94	.1	Salvia sclarea	16.87	-.9
Cupressus lawsoniana	129.65	1.1	Sanguisorba officinali	31.11	-.7
Echinacea angustifolia	47.9	-.4	Scrophularia nodosa	75.68	.1
Echinacea purpurea	53.61	-.3	Scutellaria lateriflor	114.7	.8
Escholtzia californica	41.42	-.5	Solidago virgaurea	40.62	-.5
Faba vulgaris	28.4	-.7	Spilanthes oleracea	12.97	-1
Fagopyrum esculentum	5.29	-1.1	Stachys officinalis	44.41	-.4
Fraxinus excelsior	44.86	-.4	Stachys officinalis R	46.87	-.4
Galium aparine	47.2	-.4	Stachys sylvatica	43.44	-.4
Glechoma hederacea	72.2	.1	Syringa vulgaris	100.76	.6
Gnaphalium leontopodiu	35.79	-.6	Taraxacum officinale	45.95	-.4
Hedera helix	74.16	.1	Teucrium chamaedrys	61.35	-.1
Hydrophyllum virginicu	91.01	.4	Thuja occidentalis	50.06	-.3
Hyoscyamus niger	25.91	-.8	Tormentilla erecta	109.05	.7
Hyssopus officinalis	37.46	-.6	Tussilago farfara	32.16	-.7
Lappa arctium	256.69	3.4	Tussilago petasites	116.17	.9
Lapsana communis	64.73	-.1	Ulmus campestris	166.05	1.8
Laurocerasus	53.85	-.3	Urtica dioica	37.15	-.6
Leonurus cardiaca	82.55	.3	Urtica urens	20.66	-.9
Lespedeza sieboldii	100.84	.6	Urtica urens 2	30.92	-.7
Linum usitatissimum	37	-.6	Verbena officinalis	208.63	2.5
Lycopus europaeus	76.08	.1	Vinca minor	57.99	-.2
Malva sylvestris	36.04	-.6	Viola tricolor	29.44	-.7
Mandragora officinalis	1.12	-1.2	Zizia aurea	36.26	-.6

Cuprum

Name	Value	Deviation	Name	Value	Deviation
Osmundo regalis	343.14	5	Echinacea purpurea	53.61	-.3
Cicuta virosa	278.16	3.8	Laurocerasus	53.85	-.3
Lappa arctium	256.69	3.4	Mentha arvensis	52.39	-.3
Verbena officinalis	208.63	2.5	Mercurialis perennis	53.82	-.3
Ulmus campestris	166.05	1.8	Thuja occidentalis	50.06	-.3
Cinnamodendron cortiso	159.67	1.7	Anethum graveolens	48.32	-.4
Rumex acetosa	156.97	1.6	Echinacea angustifolia	47.9	-.4
Collinsonia canadensis	149.68	1.5	Fraxinus excelsior	44.86	-.4
Cupressus lawsoniana	129.65	1.1	Galium aparine	47.2	-.4
Osmundo regalis 2	130.15	1.1	Paeonia officinalis	46.96	-.4
Cimicifuga racemosa	124.5	1	Petroselinum crispum	47.56	-.4
Tussilago petasites	116.17	.9	Salicaria purpurea	48.25	-.4
Scutellaria lateriflor	114.7	.8	Stachys officinalis	44.41	-.4
Prunus padus	109.32	.7	Stachys officinalis R	46.87	-.4
Tormentilla erecta	109.05	.7	Stachys sylvatica	43.44	-.4
Lespedeza sieboldii	100.84	.6	Taraxacum officinale	45.95	-.4
Syringa vulgaris	100.76	.6	Escholtzia californica	41.42	-.5
Agrimonia eupatoria	90.07	.4	Psoralea bituminosa	39.13	-.5
Hydrophyllum virginicu	91.01	.4	Solidago virgaurea	40.62	-.5
Leonurus cardiaca	82.55	.3	Cardiospermum halicaca	36.53	-.6
Adonis vernalis	81.4	.2	Gnaphalium leontopodiu	35.79	-.6
Aquilegia vulgaris	81.88	.2	Hyssopus officinalis	37.46	-.6
Conium maculatum	74.94	.1	Linum usitatissimum	37	-.6
Glechoma hederacea	72.2	.1	Malva sylvestris	36.04	-.6
Hedera helix	74.16	.1	Milium solis	37.34	-.6
Lycopus europaeus	76.08	.1	Ocimum canum	33.26	-.6
Melilotus officinalis	76.02	.1	Ocimum canum R	35.6	-.6
Scrophularia nodosa	75.68	.1	Salix purpurea	36.61	-.6
Alchemilla vulgaris	69.31	0	Urtica dioica	37.15	-.6
Angelica archangelica	68.97	0	Zizia aurea	36.26	-.6
Chamomilla	65.93	0	Chelidonium majus	27.85	-.7
Ruta graveolens	68.53	0	Faba vulgaris	28.4	-.7
Anthemis nobilis	64.73	-.1	Ononis spinosa	30.21	-.7
Cardamine pratensis	63.79	-.1	Sanguisorba officinali	31.11	-.7
Cardamine pratensis R	61.9	-.1	Tussilago farfara	32.16	-.7
Lapsana communis	64.73	-.1	Urtica urens 2	30.92	-.7
Melissa officinalis	64.06	-.1	Viola tricolor	29.44	-.7
Oenanthe crocata	62.54	-.1	Castanea vesca	21.73	-.8
Polygonum aviculare	63.58	-.1	Hyoscyamus niger	25.91	-.8
Rhus toxicodendron	63.85	-.1	Marrubium vulgare	23.21	-.8
Teucrium chamaedrys	61.35	-.1	Bellis perennis	17.68	-.9
Aconitum napellus	58.69	-.2	Plantago major	21.12	-.9
Chenopodium anthelmint	60	-.2	Primula veris	18.54	-.9
Clematis erecta	59.01	-.2	Salvia sclarea	16.87	-.9
Vinca minor	57.99	-.2	Urtica urens	20.66	-.9
Belladonna	49.09	-.3	Spilanthes oleracea	12.97	-1
Bryonia dioica	54.33	-.3	Fagopyrum esculentum	5.29	-1.1
Centaurium erythraea	54.37	-.3	Mandragora officinalis	1.12	-1.2

Dysprosium

Name	Value	Deviation	Name	Value	Deviation
Aconitum napellus	.24	.3	Marrubium vulgare	.11	-.2
Adonis vernalis	.23	.3	Melilotus officinalis	.05	-.5
Agrimonia eupatoria	.03	-.6	Melissa officinalis	.16	0
Alchemilla vulgaris	.2	.2	Mentha arvensis	.08	-.4
Anethum graveolens	.04	-.5	Mercurialis perennis	.22	.2
Angelica archangelica	.24	.3	Milium solis	.07	-.4
Anthemis nobilis	.14	-.1	Ocimum canum	.19	.1
Aquilegia vulgaris	.03	-.6	Ocimum canum R	.17	0
Belladonna	.12	-.2	Oenanthe crocata	.21	.2
Bellis perennis	.78	2.6	Ononis spinosa	.17	0
Bryonia dioica	.27	.4	Osmundo regalis	1.82	7.1
Cardamine pratensis	.07	-.4	Osmundo regalis 2	.01	-.7
Cardamine pratensis R	.06	-.4	Paeonia officinalis	.18	.1
Cardiospermum halicaca	.14	-.1	Petroselinum crispum	.29	.5
Castanea vesca	.11	-.2	Plantago major	.13	-.1
Centaurium erythraea	.31	.6	Polygonum aviculare	.11	-.2
Chamomilla	.14	-.1	Primula veris	.16	0
Chelidonium majus	.83	2.8	Prunus padus	.07	-.4
Chenopodium anthelmint	.02	-.6	Psoralea bituminosa	.16	0
Cicuta virosa	.53	1.6	Rhus toxicodendron	.06	-.4
Cimicifuga racemosa	.24	.3	Rumex acetosa	.85	2.9
Cinnamodendron cortiso	.08	-.4	Ruta graveolens	.06	-.4
Clematis erecta	.07	-.4	Salicaria purpurea	.07	-.4
Collinsonia canadensis	.29	.5	Salix purpurea	.06	-.4
Conium maculatum	.02	-.6	Salvia sclarea	.05	-.5
Cupressus lawsoniana	.15	-.1	Sanguisorba officinali	.02	-.6
Echinacea angustifolia	.08	-.4	Scrophularia nodosa	.09	-.3
Echinacea purpurea	.02	-.6	Scutellaria lateriflor	.17	0
Escholtzia californica	.09	-.3	Solidago virgaurea	.05	-.5
Faba vulgaris	.05	-.5	Spilanthes oleracea	.06	-.4
Fagopyrum esculentum	.03	-.6	Stachys officinalis	.03	-.6
Fraxinus excelsior	.1	-.3	Stachys officinalis R	.02	-.6
Galium aparine	.1	-.3	Stachys sylvatica	.09	-.3
Glechoma hederacea	.08	-.4	Syringa vulgaris	.07	-.4
Gnaphalium leontopodiu	.39	1	Taraxacum officinale	.57	1.7
Hedera helix	.07	-.4	Teucrium chamaedrys	.09	-.3
Hydrophyllum virginicu	.05	-.5	Thuja occidentalis	.05	-.5
Hyoscyamus niger	.1	-.3	Tormentilla erecta	.37	.9
Hyssopus officinalis	.07	-.4	Tussilago farfara	.05	-.5
Lappa arctium	.19	.1	Tussilago petasites	.07	-.4
Lapsana communis	.28	.5	Ulmus campestris	.12	-.2
Laurocerasus	.05	-.5	Urtica dioica	.35	.8
Leonurus cardiaca	.04	-.5	Urtica urens	.06	-.4
Lespedeza sieboldii	.03	-.6	Urtica urens 2	.06	-.4
Linum usitatissimum	.02	-.6	Verbena officinalis	.04	-.5
Lycopus europaeus	.04	-.5	Vinca minor	.17	0
Malva sylvestris	.08	-.4	Viola tricolor	.08	-.4
Mandragora officinalis	.1	-.3	Zizia aurea	.16	0

Dysprosium

Name	Value	Deviation	Name	Value	Deviation
Osmundo regalis	1.82	7.1	Stachys sylvatica	.09	-.3
Rumex acetosa	.85	2.9	Teucrium chamaedrys	.09	-.3
Chelidonium majus	.83	2.8	Cardamine pratensis	.07	-.4
Bellis perennis	.78	2.6	Cardamine pratensis R	.06	-.4
Taraxacum officinale	.57	1.7	Cinnamodendron cortiso	.08	-.4
Cicuta virosa	.53	1.6	Clematis erecta	.07	-.4
Gnaphalium leontopodiu	.39	1	Echinacea angustifolia	.08	-.4
Tormentilla erecta	.37	.9	Glechoma hederacea	.08	-.4
Urtica dioica	.35	.8	Hedera helix	.07	-.4
Centaurium erythraea	.31	.6	Hyssopus officinalis	.07	-.4
Collinsonia canadensis	.29	.5	Malva sylvestris	.08	-.4
Lapsana communis	.28	.5	Mentha arvensis	.08	-.4
Petroselinum crispum	.29	.5	Milium solis	.07	-.4
Bryonia dioica	.27	.4	Prunus padus	.07	-.4
Aconitum napellus	.24	.3	Rhus toxicodendron	.06	-.4
Adonis vernalis	.23	.3	Ruta graveolens	.06	-.4
Angelica archangelica	.24	.3	Salicaria purpurea	.07	-.4
Cimicifuga racemosa	.24	.3	Salix purpurea	.06	-.4
Alchemilla vulgaris	.2	.2	Spilanthes oleracea	.06	-.4
Mercurialis perennis	.22	.2	Syringa vulgaris	.07	-.4
Oenanthe crocata	.21	.2	Tussilago petasites	.07	-.4
Lappa arctium	.19	.1	Urtica urens	.06	-.4
Ocimum canum	.19	.1	Urtica urens 2	.06	-.4
Paeonia officinalis	.18	.1	Viola tricolor	.08	-.4
Melissa officinalis	.16	0	Anethum graveolens	.04	-.5
Ocimum canum R	.17	0	Faba vulgaris	.05	-.5
Ononis spinosa	.17	0	Hydrophyllum virginicu	.05	-.5
Primula veris	.16	0	Laurocerasus	.05	-.5
Psoralea bituminosa	.16	0	Leonurus cardiaca	.04	-.5
Scutellaria lateriflor	.17	0	Lycopus europaeus	.04	-.5
Vinca minor	.17	0	Melilotus officinalis	.05	-.5
Zizia aurea	.16	0	Salvia sclarea	.05	-.5
Anthemis nobilis	.14	-.1	Solidago virgaurea	.05	-.5
Cardiospermum halicaca	.14	-.1	Thuja occidentalis	.05	-.5
Chamomilla	.14	-.1	Tussilago farfara	.05	-.5
Cupressus lawsoniana	.15	-.1	Verbena officinalis	.04	-.5
Plantago major	.13	-.1	Agrimonia eupatoria	.03	-.6
Belladonna	.12	-.2	Aquilegia vulgaris	.03	-.6
Castanea vesca	.11	-.2	Chenopodium anthelmint	.02	-.6
Marrubium vulgare	.11	-.2	Conium maculatum	.02	-.6
Polygonum aviculare	.11	-.2	Echinacea purpurea	.02	-.6
Ulmus campestris	.12	-.2	Fagopyrum esculentum	.03	-.6
Escholtzia californica	.09	-.3	Lespedeza sieboldii	.03	-.6
Fraxinus excelsior	.1	-.3	Linum usitatissimum	.02	-.6
Galium aparine	.1	-.3	Sanguisorba officinali	.02	-.6
Hyoscyamus niger	.1	-.3	Stachys officinalis	.03	-.6
Mandragora officinalis	.1	-.3	Stachys officinalis R	.02	-.6
Scrophularia nodosa	.09	-.3	Osmundo regalis 2	.01	-.7

Erbium

Name	Value	Deviation	Name	Value	Deviation
Aconitum napellus	.12	.3	Marrubium vulgare	.06	-.2
Adonis vernalis	.11	.2	Melilotus officinalis	.03	-.4
Agrimonia eupatoria	.01	-.6	Melissa officinalis	.08	0
Alchemilla vulgaris	.11	.2	Mentha arvensis	.04	-.3
Anethum graveolens	.02	-.5	Mercurialis perennis	.12	.3
Angelica archangelica	.12	.3	Milium solis	.03	-.4
Anthemis nobilis	.07	-.1	Ocimum canum	.08	0
Aquilegia vulgaris	.01	-.6	Ocimum canum R	.08	0
Belladonna	.06	-.2	Oenanthe crocata	.12	.3
Bellis perennis	.38	2.4	Ononis spinosa	.09	.1
Bryonia dioica	.13	.4	Osmundo regalis	.99	7.3
Cardamine pratensis	.03	-.4	Osmundo regalis 2	0	-.7
Cardamine pratensis R	.03	-.4	Paeonia officinalis	.09	.1
Cardiospermum halicaca	.07	-.1	Petroselinum crispum	.14	.5
Castanea vesca	.05	-.3	Plantago major	.07	-.1
Centaurium erythraea	.17	.7	Polygonum aviculare	.06	-.2
Chamomilla	.07	-.1	Primula veris	.08	0
Chelidonium majus	.42	2.7	Prunus padus	.04	-.3
Chenopodium anthelmint	.01	-.6	Psoralea bituminosa	.06	-.2
Cicuta virosa	.28	1.6	Rhus toxicodendron	.03	-.4
Cimicifuga racemosa	.13	.4	Rumex acetosa	.44	2.9
Cinnamodendron cortiso	.04	-.3	Ruta graveolens	.03	-.4
Clematis erecta	.03	-.4	Salicaria purpurea	.03	-.4
Collinsonia canadensis	.14	.5	Salix purpurea	.03	-.4
Conium maculatum	.01	-.6	Salvia sclarea	.03	-.4
Cupressus lawsoniana	.08	0	Sanguisorba officinali	.01	-.6
Echinacea angustifolia	.04	-.3	Scrophularia nodosa	.05	-.3
Echinacea purpurea	.01	-.6	Scutellaria lateriflor	.07	-.1
Escholtzia californica	.03	-.4	Solidago virgaurea	.02	-.5
Faba vulgaris	.03	-.4	Spilanthes oleracea	.03	-.4
Fagopyrum esculentum	.02	-.5	Stachys officinalis	.01	-.6
Fraxinus excelsior	.04	-.3	Stachys officinalis R	.01	-.6
Galium aparine	.05	-.3	Stachys sylvatica	.05	-.3
Glechoma hederacea	.04	-.3	Syringa vulgaris	.04	-.3
Gnaphalium leontopodiu	.21	1	Taraxacum officinale	.3	1.7
Hedera helix	.04	-.3	Teucrium chamaedrys	.05	-.3
Hydrophyllum virginicu	.02	-.5	Thuja occidentalis	.02	-.5
Hyoscyamus niger	.05	-.3	Tormentilla erecta	.19	.9
Hyssopus officinalis	.03	-.4	Tussilago farfara	.02	-.5
Lappa arctium	.1	.1	Tussilago petasites	.04	-.3
Lapsana communis	.15	.5	Ulmus campestris	.05	-.3
Laurocerasus	.02	-.5	Urtica dioica	.19	.9
Leonurus cardiaca	.02	-.5	Urtica urens	.03	-.4
Lespedeza sieboldii	.02	-.5	Urtica urens 2	.03	-.4
Linum usitatissimum	.01	-.6	Verbena officinalis	.02	-.5
Lycopus europaeus	.02	-.5	Vinca minor	.07	-.1
Malva sylvestris	.03	-.4	Viola tricolor	.04	-.3
Mandragora officinalis	.05	-.3	Zizia aurea	.08	0

Erbium

Name	Value	Deviation	Name	Value	Deviation
Osmundo regalis	.99	7.3	Mandragora officinalis	.05	-.3
Rumex acetosa	.44	2.9	Mentha arvensis	.04	-.3
Chelidonium majus	.42	2.7	Prunus padus	.04	-.3
Bellis perennis	.38	2.4	Scrophularia nodosa	.05	-.3
Taraxacum officinale	.3	1.7	Stachys sylvatica	.05	-.3
Cicuta virosa	.28	1.6	Syringa vulgaris	.04	-.3
Gnaphalium leontopodiu	.21	1	Teucrium chamaedrys	.05	-.3
Tormentilla erecta	.19	.9	Tussilago petasites	.04	-.3
Urtica dioica	.19	.9	Ulmus campestris	.05	-.3
Centaurium erythraea	.17	.7	Viola tricolor	.04	-.3
Collinsonia canadensis	.14	.5	Cardamine pratensis	.03	-.4
Lapsana communis	.15	.5	Cardamine pratensis R	.03	-.4
Petroselinum crispum	.14	.5	Clematis erecta	.03	-.4
Bryonia dioica	.13	.4	Escholtzia californica	.03	-.4
Cimicifuga racemosa	.13	.4	Faba vulgaris	.03	-.4
Aconitum napellus	.12	.3	Hyssopus officinalis	.03	-.4
Angelica archangelica	.12	.3	Malva sylvestris	.03	-.4
Mercurialis perennis	.12	.3	Melilotus officinalis	.03	-.4
Oenanthe crocata	.12	.3	Milium solis	.03	-.4
Adonis vernalis	.11	.2	Rhus toxicodendron	.03	-.4
Alchemilla vulgaris	.11	.2	Ruta graveolens	.03	-.4
Lappa arctium	.1	.1	Salicaria purpurea	.03	-.4
Ononis spinosa	.09	.1	Salix purpurea	.03	-.4
Paeonia officinalis	.09	.1	Salvia sclarea	.03	-.4
Cupressus lawsoniana	.08	0	Spilanthes oleracea	.03	-.4
Melissa officinalis	.08	0	Urtica urens	.03	-.4
Ocimum canum	.08	0	Urtica urens 2	.03	-.4
Ocimum canum R	.08	0	Anethum graveolens	.02	-.5
Primula veris	.08	0	Fagopyrum esculentum	.02	-.5
Zizia aurea	.08	0	Hydrophyllum virginicu	.02	-.5
Anthemis nobilis	.07	-.1	Laurocerasus	.02	-.5
Cardiospermum halicaca	.07	-.1	Leonurus cardiaca	.02	-.5
Chamomilla	.07	-.1	Lespedeza sieboldii	.02	-.5
Plantago major	.07	-.1	Lycopus europaeus	.02	-.5
Scutellaria lateriflor	.07	-.1	Solidago virgaurea	.02	-.5
Vinca minor	.07	-.1	Thuja occidentalis	.02	-.5
Belladonna	.06	-.2	Tussilago farfara	.02	-.5
Marrubium vulgare	.06	-.2	Verbena officinalis	.02	-.5
Polygonum aviculare	.06	-.2	Agrimonia eupatoria	.01	-.6
Psoralea bituminosa	.06	-.2	Aquilegia vulgaris	.01	-.6
Castanea vesca	.05	-.3	Chenopodium anthelmint	.01	-.6
Cinnamodendron cortiso	.04	-.3	Conium maculatum	.01	-.6
Echinacea angustifolia	.04	-.3	Echinacea purpurea	.01	-.6
Fraxinus excelsior	.04	-.3	Linum usitatissimum	.01	-.6
Galium aparine	.05	-.3	Sanguisorba officinali	.01	-.6
Glechoma hederacea	.04	-.3	Stachys officinalis	.01	-.6
Hedera helix	.04	-.3	Stachys officinalis R	.01	-.6
Hyoscyamus niger	.05	-.3	Osmundo regalis 2	0	-.7

Europium

Name	Value	Deviation	Name	Value	Deviation
Aconitum napellus	.11	.3	Marrubium vulgare	.05	-.4
Adonis vernalis	.08	0	Melilotus officinalis	.03	-.6
Agrimonia eupatoria	.02	-.7	Melissa officinalis	.14	.6
Alchemilla vulgaris	.09	.1	Mentha arvensis	.08	0
Anethum graveolens	.02	-.7	Mercurialis perennis	.1	.2
Angelica archangelica	.09	.1	Milium solis	.04	-.5
Anthemis nobilis	.06	-.3	Ocimum canum	.1	.2
Aquilegia vulgaris	.02	-.7	Ocimum canum R	.09	.1
Belladonna	.06	-.3	Oenanthe crocata	.08	0
Bellis perennis	.29	2.3	Ononis spinosa	.08	0
Bryonia dioica	.1	.2	Osmundo regalis	.67	6.6
Cardamine pratensis	.05	-.4	Osmundo regalis 2	.02	-.7
Cardamine pratensis R	.04	-.5	Paeonia officinalis	.09	.1
Cardiospermum halicaca	.06	-.3	Petroselinum crispum	.13	.5
Castanea vesca	.08	0	Plantago major	.08	0
Centaurium erythraea	.16	.9	Polygonum aviculare	.05	-.4
Chamomilla	.06	-.3	Primula veris	.07	-.2
Chelidonium majus	.34	2.9	Prunus padus	.05	-.4
Chenopodium anthelmint	.01	-.8	Psoralea bituminosa	.09	.1
Cicuta virosa	.23	1.6	Rhus toxicodendron	.09	.1
Cimicifuga racemosa	.1	.2	Rumex acetosa	.38	3.3
Cinnamodendron cortiso	.06	-.3	Ruta graveolens	.04	-.5
Clematis erecta	.04	-.5	Salicaria purpurea	.04	-.5
Collinsonia canadensis	.12	.4	Salix purpurea	.06	-.3
Conium maculatum	.02	-.7	Salvia sclarea	.05	-.4
Cupressus lawsoniana	.08	0	Sanguisorba officinali	.04	-.5
Echinacea angustifolia	.04	-.5	Scrophularia nodosa	.05	-.4
Echinacea purpurea	.02	-.7	Scutellaria lateriflor	.12	.4
Escholtzia californica	.04	-.5	Solidago virgaurea	.02	-.7
Faba vulgaris	.03	-.6	Spilanthes oleracea	.03	-.6
Fagopyrum esculentum	.02	-.7	Stachys officinalis	.03	-.6
Fraxinus excelsior	.06	-.3	Stachys officinalis R	.02	-.7
Galium aparine	.09	.1	Stachys sylvatica	.06	-.3
Glechoma hederacea	.12	.4	Syringa vulgaris	.04	-.5
Gnaphalium leontopodiu	.16	.9	Taraxacum officinale	.2	1.3
Hedera helix	.08	0	Teucrium chamaedrys	.06	-.3
Hydrophyllum virginicu	.05	-.4	Thuja occidentalis	.06	-.3
Hyoscyamus niger	.06	-.3	Tormentilla erecta	.26	2
Hyssopus officinalis	.03	-.6	Tussilago farfara	.02	-.7
Lappa arctium	.08	0	Tussilago petasites	.03	-.6
Lapsana communis	.11	.3	Ulmus campestris	.1	.2
Laurocerasus	.03	-.6	Urtica dioica	.15	.7
Leonurus cardiaca	.07	-.2	Urtica urens	.05	-.4
Lespedeza sieboldii	.08	0	Urtica urens 2	.04	-.5
Linum usitatissimum	.01	-.8	Verbena officinalis	.03	-.6
Lycopus europaeus	.04	-.5	Vinca minor	.1	.2
Malva sylvestris	.04	-.5	Viola tricolor	.09	.1
Mandragora officinalis	.04	-.5	Zizia aurea	.08	0

Europium

Name	Value	Deviation	Name	Value	Deviation
Osmundo regalis	.67	6.6	Cinnamodendron cortiso	.06	-.3
Rumex acetosa	.38	3.3	Fraxinus excelsior	.06	-.3
Chelidonium majus	.34	2.9	Hyoscyamus niger	.06	-.3
Bellis perennis	.29	2.3	Salix purpurea	.06	-.3
Tormentilla erecta	.26	2	Stachys sylvatica	.06	-.3
Cicuta virosa	.23	1.6	Teucrium chamaedrys	.06	-.3
Taraxacum officinale	.2	1.3	Thuja occidentalis	.06	-.3
Centaurium erythraea	.16	.9	Cardamine pratensis	.05	-.4
Gnaphalium leontopodiu	.16	.9	Hydrophyllum virginicu	.05	-.4
Urtica dioica	.15	.7	Marrubium vulgare	.05	-.4
Melissa officinalis	.14	.6	Polygonum aviculare	.05	-.4
Petroselinum crispum	.13	.5	Prunus padus	.05	-.4
Collinsonia canadensis	.12	.4	Salvia sclarea	.05	-.4
Glechoma hederacea	.12	.4	Scrophularia nodosa	.05	-.4
Scutellaria lateriflor	.12	.4	Urtica urens	.05	-.4
Aconitum napellus	.11	.3	Cardamine pratensis R	.04	-.5
Lapsana communis	.11	.3	Clematis erecta	.04	-.5
Bryonia dioica	.1	.2	Echinacea angustifolia	.04	-.5
Cimicifuga racemosa	.1	.2	Escholtzia californica	.04	-.5
Mercurialis perennis	.1	.2	Lycopus europaeus	.04	-.5
Ocimum canum	.1	.2	Malva sylvestris	.04	-.5
Ulmus campestris	.1	.2	Mandragora officinalis	.04	-.5
Vinca minor	.1	.2	Milium solis	.04	-.5
Alchemilla vulgaris	.09	.1	Ruta graveolens	.04	-.5
Angelica archangelica	.09	.1	Salicaria purpurea	.04	-.5
Galium aparine	.09	.1	Sanguisorba officinali	.04	-.5
Ocimum canum R	.09	.1	Syringa vulgaris	.04	-.5
Paeonia officinalis	.09	.1	Urtica urens 2	.04	-.5
Psoralea bituminosa	.09	.1	Faba vulgaris	.03	-.6
Rhus toxicodendron	.09	.1	Hyssopus officinalis	.03	-.6
Viola tricolor	.09	.1	Laurocerasus	.03	-.6
Adonis vernalis	.08	0	Melilotus officinalis	.03	-.6
Castanea vesca	.08	0	Spilanthes oleracea	.03	-.6
Cupressus lawsoniana	.08	0	Stachys officinalis	.03	-.6
Hedera helix	.08	0	Tussilago petasites	.03	-.6
Lappa arctium	.08	0	Verbena officinalis	.03	-.6
Lespedeza sieboldii	.08	0	Agrimonia eupatoria	.02	-.7
Mentha arvensis	.08	0	Anethum graveolens	.02	-.7
Oenanthe crocata	.08	0	Aquilegia vulgaris	.02	-.7
Ononis spinosa	.08	0	Conium maculatum	.02	-.7
Plantago major	.08	0	Echinacea purpurea	.02	-.7
Zizia aurea	.08	0	Fagopyrum esculentum	.02	-.7
Leonurus cardiaca	.07	-.2	Osmundo regalis 2	.02	-.7
Primula veris	.07	-.2	Solidago virgaurea	.02	-.7
Anthemis nobilis	.06	-.3	Stachys officinalis R	.02	-.7
Belladonna	.06	-.3	Tussilago farfara	.02	-.7
Cardiospermum halicaca	.06	-.3	Chenopodium anthelmint	.01	-.8
Chamomilla	.06	-.3	Linum usitatissimum	.01	-.8

Ferrum

Name	Value	Deviation	Name	Value	Deviation
Aconitum napellus	6437.88	0	Marrubium vulgare	2582.39	-.2
Adonis vernalis	2280.18	-.2	Melilotus officinalis	6626.77	0
Agrimonia eupatoria	1521.37	-.2	Melissa officinalis	3967.03	-.1
Alchemilla vulgaris	7218.18	0	Mentha arvensis	6749.6	0
Anethum graveolens	2968.76	-.2	Mercurialis perennis	7855.42	0
Angelica archangelica	4580.4	-.1	Milium solis	1533.16	-.2
Anthemis nobilis	4953.68	-.1	Ocimum canum	2803.07	-.2
Aquilegia vulgaris	1242.35	-.2	Ocimum canum R	4126.46	-.1
Belladonna	2115.5	-.2	Oenanthe crocata	2315.29	-.2
Bellis perennis	10699.85	.1	Ononis spinosa	6226.23	0
Bryonia dioica	5126.81	-.1	Osmundo regalis	22331.96	.5
Cardamine pratensis	2249.1	-.2	Osmundo regalis 2	612.64	-.2
Cardamine pratensis R	2081.8	-.2	Paeonia officinalis	9069.07	0
Cardiospermum halic	5084.95	-.1	Petroselinum crispum	8787.09	0
Castanea vesca	4750.59	-.1	Plantago major	3010.13	-.2
Centaurium erythraea	7567.77	0	Polygonum aviculare	2767.36	-.2
Chamomilla	2268.16	-.2	Primula veris	2148.48	-.2
Chelidonium majus	15150.76	.2	Prunus padus	2971.75	-.2
Chenopodium anth	1722.05	-.2	Psoralea bituminosa	5709.08	-.1
Cicuta virosa	17096.27	.3	Rhus toxicodendron	7549.41	0
Cimicifuga racemosa	2374.43	-.2	Rumex acetosa	15186.62	.2
Cinnamodendron cortiso	1614.75	-.2	Ruta graveolens	2296.9	-.2
Clematis erecta	2315.4	-.2	Salicaria purpurea	3960.36	-.1
Collinsonia canadensis	3426.24	-.1	Salix purpurea	9877.26	.1
Conium maculatum	1429.06	-.2	Salvia sclarea	2321	-.2
Cupressus lawsoniana	10816.19	.1	Sanguisorba officinali	1965.99	-.2
Echinacea angustifolia	2023.96	-.2	Scrophularia nodosa	3384.43	-.1
Echinacea purpurea	2135.29	-.2	Scutellaria lateriflor	3179.78	-.1
Escholtzia californica	658.71	-.2	Solidago virgaurea	1696.64	-.2
Faba vulgaris	2560.69	-.2	Spilanthes oleracea	2201.65	-.2
Fagopyrum esculentum	1835.34	-.2	Stachys officinalis	1200.39	-.2
Fraxinus excelsior	7759.64	0	Stachys officinalis R	1173.14	-.2
Galium aparine	2282.85	-.2	Stachys sylvatica	2771.01	-.2
Glechoma hederacea	2078.66	-.2	Syringa vulgaris	2614.65	-.2
Gnaphalium leontopodiu	8184.59	0	Taraxacum officinale	9135.73	.1
Hedera helix	3837.81	-.1	Teucrium chamaedrys	2808.92	-.2
Hydrophyllum virginicu	1748.18	-.2	Thuja occidentalis	11262.42	.1
Hyoscyamus niger	1818.91	-.2	Tormentilla erecta	7076.18	0
Hyssopus officinalis	1911.16	-.2	Tussilago farfara	2179.15	-.2
Lappa arctium	4941.56	-.1	Tussilago petasites	2210.65	-.2
Lapsana communis	6719.52	0	Ulmus campestris	10654.24	.1
Laurocerasus	2573.33	-.2	Urtica dioica	11048.72	.1
Leonurus cardiaca	2067.61	-.2	Urtica urens	3268.67	-.1
Lespedeza sieboldii	3347.64	-.1	Urtica urens 2	3641.89	-.1
Linum usitatissimum	1642.83	-.2	Verbena officinalis	300106.38	9.6
Lycopus europaeus	1569.71	-.2	Vinca minor	4279.79	-.1
Malva sylvestris	2319.82	-.2	Viola tricolor	1329.13	-.2
Mandragora officinalis	870.46	-.2	Zizia aurea	3378.4	-.1

Ferrum

Name	Value	Deviation	Name	Value	Deviation
Verbena officinalis	300106.38	9.6	Aquilegia vulgaris	1242.35	-.2
Osmundo regalis	22331.96	.5	Belladonna	2115.5	-.2
Cicuta virosa	17096.27	.3	Cardamine pratensis	2249.1	-.2
Chelidonium majus	15150.76	.2	Cardamine pratensis R	2081.8	-.2
Rumex acetosa	15186.62	.2	Chamomilla	2268.16	-.2
Bellis perennis	10699.85	.1	Chenopodium anthelm	1722.05	-.2
Cupressus lawsoniana	10816.19	.1	Cimicifuga racemosa	2374.43	-.2
Salix purpurea	9877.26	.1	Cinnamodendron cortiso	1614.75	-.2
Taraxacum officinale	9135.73	.1	Clematis erecta	2315.4	-.2
Thuja occidentalis	11262.42	.1	Conium maculatum	1429.06	-.2
Ulmus campestris	10654.24	.1	Echinacea angustifolia	2023.96	-.2
Urtica dioica	11048.72	.1	Echinacea purpurea	2135.29	-.2
Aconitum napellus	6437.88	0	Escholtzia californica	658.71	-.2
Alchemilla vulgaris	7218.18	0	Faba vulgaris	2560.69	-.2
Centaurium erythraea	7567.77	0	Fagopyrum esculentum	1835.34	-.2
Fraxinus excelsior	7759.64	0	Galium aparine	2282.85	-.2
Gnaphalium leontop	8184.59	0	Glechoma hederacea	2078.66	-.2
Lapsana communis	6719.52	0	Hydrophyllum virginicu	1748.18	-.2
Melilotus officinalis	6626.77	0	Hyoscyamus niger	1818.91	-.2
Mentha arvensis	6749.6	0	Hyssopus officinalis	1911.16	-.2
Mercurialis perennis	7855.42	0	Laurocerasus	2573.33	-.2
Ononis spinosa	6226.23	0	Leonurus cardiaca	2067.61	-.2
Paeonia officinalis	9069.07	0	Linum usitatissimum	1642.83	-.2
Petroselinum crispum	8787.09	0	Lycopus europaeus	1569.71	-.2
Rhus toxicodendron	7549.41	0	Malva sylvestris	2319.82	-.2
Tormentilla erecta	7076.18	0	Mandragora officinalis	870.46	-.2
Angelica archangelica	4580.4	-.1	Marrubium vulgare	2582.39	-.2
Anthemis nobilis	4953.68	-.1	Milium solis	1533.16	-.2
Bryonia dioica	5126.81	-.1	Ocimum canum	2803.07	-.2
Cardiospermum halic	5084.95	-.1	Oenanthe crocata	2315.29	-.2
Castanea vesca	4750.59	-.1	Osmundo regalis 2	612.64	-.2
Collinsonia canadensis	3426.24	-.1	Plantago major	3010.13	-.2
Hedera helix	3837.81	-.1	Polygonum aviculare	2767.36	-.2
Lappa arctium	4941.56	-.1	Primula veris	2148.48	-.2
Lespedeza sieboldii	3347.64	-.1	Prunus padus	2971.75	-.2
Melissa officinalis	3967.03	-.1	Ruta graveolens	2296.9	-.2
Ocimum canum R	4126.46	-.1	Salvia sclarea	2321	-.2
Psoralea bituminosa	5709.08	-.1	Sanguisorba officinali	1965.99	-.2
Salicaria purpurea	3960.36	-.1	Solidago virgaurea	1696.64	-.2
Scrophularia nodosa	3384.43	-.1	Spilanthes oleracea	2201.65	-.2
Scutellaria lateriflor	3179.78	-.1	Stachys officinalis	1200.39	-.2
Urtica urens	3268.67	-.1	Stachys officinalis R	1173.14	-.2
Urtica urens 2	3641.89	-.1	Stachys sylvatica	2771.01	-.2
Vinca minor	4279.79	-.1	Syringa vulgaris	2614.65	-.2
Zizia aurea	3378.4	-.1	Teucrium chamaedrys	2808.92	-.2
Adonis vernalis	2280.18	-.2	Tussilago farfara	2179.15	-.2
Agrimonia eupatoria	1521.37	-.2	Tussilago petasites	2210.65	-.2
Anethum graveolens	2968.76	-.2	Viola tricolor	1329.13	-.2

Gadolinium

Name	Value	Deviation	Name	Value	Deviation
Aconitum napellus	.46	.3	Marrubium vulgare	.22	-.2
Adonis vernalis	.43	.2	Melilotus officinalis	.13	-.4
Agrimonia eupatoria	.06	-.6	Melissa officinalis	.32	0
Alchemilla vulgaris	.4	.2	Mentha arvensis	.18	-.3
Anethum graveolens	.08	-.6	Mercurialis perennis	.46	.3
Angelica archangelica	.43	.2	Milium solis	.15	-.4
Anthemis nobilis	.29	-.1	Ocimum canum	.36	.1
Aquilegia vulgaris	.06	-.6	Ocimum canum R	.32	0
Belladonna	.23	-.2	Oenanthe crocata	.42	.2
Bellis perennis	1.53	2.7	Ononis spinosa	.34	0
Bryonia dioica	.52	.4	Osmundo regalis	3.39	7
Cardamine pratensis	.22	-.2	Osmundo regalis 2	.03	-.7
Cardamine pratensis R	.18	-.3	Paeonia officinalis	.35	.1
Cardiospermum halicaca	.27	-.1	Petroselinum crispum	.57	.6
Castanea vesca	.2	-.3	Plantago major	.34	0
Centaurium erythraea	.62	.7	Polygonum aviculare	.22	-.2
Chamomilla	.26	-.2	Primula veris	.35	.1
Chelidonium majus	1.6	2.9	Prunus padus	.17	-.4
Chenopodium anthelmint	.06	-.6	Psoralea bituminosa	.28	-.1
Cicuta virosa	1.04	1.6	Rhus toxicodendron	.13	-.4
Cimicifuga racemosa	.45	.3	Rumex acetosa	1.55	2.8
Cinnamodendron cortiso	.14	-.4	Ruta graveolens	.12	-.5
Clematis erecta	.12	-.5	Salicaria purpurea	.14	-.4
Collinsonia canadensis	.54	.5	Salix purpurea	.15	-.4
Conium maculatum	.06	-.6	Salvia sclarea	.1	-.5
Cupressus lawsoniana	.34	0	Sanguisorba officinali	.06	-.6
Echinacea angustifolia	.16	-.4	Scrophularia nodosa	.19	-.3
Echinacea purpurea	.06	-.6	Scutellaria lateriflor	.3	-.1
Escholtzia californica	.17	-.4	Solidago virgaurea	.09	-.5
Faba vulgaris	.11	-.5	Spilanthes oleracea	.14	-.4
Fagopyrum esculentum	.07	-.6	Stachys officinalis	.07	-.6
Fraxinus excelsior	.21	-.3	Stachys officinalis R	.05	-.6
Galium aparine	.22	-.2	Stachys sylvatica	.22	-.2
Glechoma hederacea	.18	-.3	Syringa vulgaris	.18	-.3
Gnaphalium leontopodiu	.75	1	Taraxacum officinale	1.18	1.9
Hedera helix	.17	-.4	Teucrium chamaedrys	.17	-.4
Hydrophyllum virginicu	.09	-.5	Thuja occidentalis	.13	-.4
Hyoscyamus niger	.21	-.3	Tormentilla erecta	.74	.9
Hyssopus officinalis	.13	-.4	Tussilago farfara	.09	-.5
Lappa arctium	.38	.1	Tussilago petasites	.15	-.4
Lapsana communis	.52	.4	Ulmus campestris	.23	-.2
Laurocerasus	.1	-.5	Urtica dioica	.69	.8
Leonurus cardiaca	.09	-.5	Urtica urens	.12	-.5
Lespedeza sieboldii	.08	-.6	Urtica urens 2	.13	-.4
Linum usitatissimum	.05	-.6	Verbena officinalis	.08	-.6
Lycopus europaeus	.09	-.5	Vinca minor	.33	0
Malva sylvestris	.17	-.4	Viola tricolor	.18	-.3
Mandragora officinalis	.19	-.3	Zizia aurea	.32	0

Gadolinium

Name	Value	Deviation	Name	Value	Deviation
Osmundo regalis	3.39	7	Hyoscyamus niger	.21	-.3
Chelidonium majus	1.6	2.9	Mandragora officinalis	.19	-.3
Rumex acetosa	1.55	2.8	Mentha arvensis	.18	-.3
Bellis perennis	1.53	2.7	Scrophularia nodosa	.19	-.3
Taraxacum officinale	1.18	1.9	Syringa vulgaris	.18	-.3
Cicuta virosa	1.04	1.6	Viola tricolor	.18	-.3
Gnaphalium leontopodiu	.75	1	Cinnamodendron cortiso	.14	-.4
Tormentilla erecta	.74	.9	Echinacea angustifolia	.16	-.4
Urtica dioica	.69	.8	Escholtzia californica	.17	-.4
Centaurium erythraea	.62	.7	Hedera helix	.17	-.4
Petroselinum crispum	.57	.6	Hyssopus officinalis	.13	-.4
Collinsonia canadensis	.54	.5	Malva sylvestris	.17	-.4
Bryonia dioica	.52	.4	Melilotus officinalis	.13	-.4
Lapsana communis	.52	.4	Milium solis	.15	-.4
Aconitum napellus	.46	.3	Prunus padus	.17	-.4
Cimicifuga racemosa	.45	.3	Rhus toxicodendron	.13	-.4
Mercurialis perennis	.46	.3	Salicaria purpurea	.14	-.4
Adonis vernalis	.43	.2	Salix purpurea	.15	-.4
Alchemilla vulgaris	.4	.2	Spilanthes oleracea	.14	-.4
Angelica archangelica	.43	.2	Teucrium chamaedrys	.17	-.4
Oenanthe crocata	.42	.2	Thuja occidentalis	.13	-.4
Lappa arctium	.38	.1	Tussilago petasites	.15	-.4
Ocimum canum	.36	.1	Urtica urens 2	.13	-.4
Paeonia officinalis	.35	.1	Clematis erecta	.12	-.5
Primula veris	.35	.1	Faba vulgaris	.11	-.5
Cupressus lawsoniana	.34	0	Hydrophyllum virginicu	.09	-.5
Melissa officinalis	.32	0	Laurocerasus	.1	-.5
Ocimum canum R	.32	0	Leonurus cardiaca	.09	-.5
Ononis spinosa	.34	0	Lycopus europaeus	.09	-.5
Plantago major	.34	0	Ruta graveolens	.12	-.5
Vinca minor	.33	0	Salvia sclarea	.1	-.5
Zizia aurea	.32	0	Solidago virgaurea	.09	-.5
Anthemis nobilis	.29	-.1	Tussilago farfara	.09	-.5
Cardiospermum halicaca	.27	-.1	Urtica urens	.12	-.5
Psoralea bituminosa	.28	-.1	Agrimonia eupatoria	.06	-.6
Scutellaria lateriflor	.3	-.1	Anethum graveolens	.08	-.6
Belladonna	.23	-.2	Aquilegia vulgaris	.06	-.6
Cardamine pratensis	.22	-.2	Chenopodium anthelmint	.06	-.6
Chamomilla	.26	-.2	Conium maculatum	.06	-.6
Galium aparine	.22	-.2	Echinacea purpurea	.06	-.6
Marrubium vulgare	.22	-.2	Fagopyrum esculentum	.07	-.6
Polygonum aviculare	.22	-.2	Lespedeza sieboldii	.08	-.6
Stachys sylvatica	.22	-.2	Linum usitatissimum	.05	-.6
Ulmus campestris	.23	-.2	Sanguisorba officinali	.06	-.6
Cardamine pratensis R	.18	-.3	Stachys officinalis	.07	-.6
Castanea vesca	.2	-.3	Stachys officinalis R	.05	-.6
Fraxinus excelsior	.21	-.3	Verbena officinalis	.08	-.6
Glechoma hederacea	.18	-.3	Osmundo regalis 2	.03	-.7

Gallium

Name	Value	Deviation	Name	Value	Deviation
Aconitum napellus	.93	-.1	Marrubium vulgare	.44	-.7
Adonis vernalis	1.14	.1	Melilotus officinalis	.43	-.7
Agrimonia eupatoria	.35	-.8	Melissa officinalis	.94	-.1
Alchemilla vulgaris	1.28	.3	Mentha arvensis	.55	-.6
Anethum graveolens	.4	-.8	Mercurialis perennis	1.08	0
Angelica archangelica	1.19	.2	Milium solis	.47	-.7
Anthemis nobilis	.91	-.2	Ocimum canum	1.19	.2
Aquilegia vulgaris	.44	-.7	Ocimum canum R	1.06	0
Belladonna	.6	-.5	Oenanthe crocata	1.14	.1
Bellis perennis	2.34	1.6	Ononis spinosa	.61	-.5
Bryonia dioica	1.23	.2	Osmundo regalis	3.94	3.5
Cardamine pratensis	.81	-.3	Osmundo regalis 2	.46	-.7
Cardamine pratensis R	.67	-.5	Paeonia officinalis	1.01	0
Cardiospermum halicaca	.67	-.5	Petroselinum crispum	1.11	.1
Castanea vesca	4.92	4.7	Plantago major	.61	-.5
Centaurium erythraea	3.45	2.9	Polygonum aviculare	.72	-.4
Chamomilla	1.37	.4	Primula veris	.9	-.2
Chelidonium majus	3.42	2.9	Prunus padus	.74	-.4
Chenopodium anthelmint	.56	-.6	Psoralea bituminosa	1.2	.2
Cicuta virosa	3.37	2.8	Rhus toxicodendron	1.33	.3
Cimicifuga racemosa	1.64	.7	Rumex acetosa	3.17	2.6
Cinnamodendron cortiso	.97	-.1	Ruta graveolens	1.12	.1
Clematis erecta	1	-.1	Salicaria purpurea	1.69	.8
Collinsonia canadensis	1.54	.6	Salix purpurea	1.5	.5
Conium maculatum	.66	-.5	Salvia sclarea	.78	-.3
Cupressus lawsoniana	2.31	1.5	Sanguisorba officinali	.74	-.4
Echinacea angustifolia	.62	-.5	Scrophularia nodosa	1.09	0
Echinacea purpurea	.28	-.9	Scutellaria lateriflor	1.52	.6
Escholtzia californica	.71	-.4	Solidago virgaurea	.62	-.5
Faba vulgaris	.37	-.8	Spilanthes oleracea	.51	-.7
Fagopyrum esculentum	.38	-.8	Stachys officinalis	.48	-.7
Fraxinus excelsior	.45	-.7	Stachys officinalis R	.46	-.7
Galium aparine	.62	-.5	Stachys sylvatica	.77	-.3
Glechoma hederacea	.63	-.5	Syringa vulgaris	1.67	.8
Gnaphalium leontopodiu	1.49	.5	Taraxacum officinale	1.91	1
Hedera helix	.78	-.3	Teucrium chamaedrys	.7	-.4
Hydrophyllum virginicu	.58	-.6	Thuja occidentalis	.87	-.2
Hyoscyamus niger	.6	-.5	Tormentilla erecta	1.36	.4
Hyssopus officinalis	.48	-.7	Tussilago farfara	.57	-.6
Lappa arctium	1.01	0	Tussilago petasites	.61	-.5
Lapsana communis	.99	-.1	Ulmus campestris	.79	-.3
Laurocerasus	.77	-.3	Urtica dioica	1.53	.6
Leonurus cardiaca	.43	-.7	Urtica urens	.39	-.8
Lespedeza sieboldii	.46	-.7	Urtica urens 2	.54	-.6
Linum usitatissimum	.39	-.8	Verbena officinalis	1.61	.7
Lycopus europaeus	.41	-.8	Vinca minor	.99	-.1
Malva sylvestris	.7	-.4	Viola tricolor	.49	-.7
Mandragora officinalis	.33	-.9	Zizia aurea	.67	-.5

Gallium

Name	Value	Deviation	Name	Value	Deviation
Castanea vesca	4.92	4.7	Stachys sylvatica	.77	-.3
Osmundo regalis	3.94	3.5	Ulmus campestris	.79	-.3
Centaurium erythraea	3.45	2.9	Escholtzia californica	.71	-.4
Chelidonium majus	3.42	2.9	Malva sylvestris	.7	-.4
Cicuta virosa	3.37	2.8	Polygonum aviculare	.72	-.4
Rumex acetosa	3.17	2.6	Prunus padus	.74	-.4
Bellis perennis	2.34	1.6	Sanguisorba officinali	.74	-.4
Cupressus lawsoniana	2.31	1.5	Teucrium chamaedrys	.7	-.4
Taraxacum officinale	1.91	1	Belladonna	.6	-.5
Salicaria purpurea	1.69	.8	Cardamine pratensis R	.67	-.5
Syringa vulgaris	1.67	.8	Cardiospermum halicaca	.67	-.5
Cimicifuga racemosa	1.64	.7	Conium maculatum	.66	-.5
Verbena officinalis	1.61	.7	Echinacea angustifolia	.62	-.5
Collinsonia canadensis	1.54	.6	Galium aparine	.62	-.5
Scutellaria lateriflor	1.52	.6	Glechoma hederacea	.63	-.5
Urtica dioica	1.53	.6	Hyoscyamus niger	.6	-.5
Gnaphalium leontopodiu	1.49	.5	Ononis spinosa	.61	-.5
Salix purpurea	1.5	.5	Plantago major	.61	-.5
Chamomilla	1.37	.4	Solidago virgaurea	.62	-.5
Tormentilla erecta	1.36	.4	Tussilago petasites	.61	-.5
Alchemilla vulgaris	1.28	.3	Zizia aurea	.67	-.5
Rhus toxicodendron	1.33	.3	Chenopodium anthelmint	.56	-.6
Angelica archangelica	1.19	.2	Hydrophyllum virginicu	.58	-.6
Bryonia dioica	1.23	.2	Mentha arvensis	.55	-.6
Ocimum canum	1.19	.2	Tussilago farfara	.57	-.6
Psoralea bituminosa	1.2	.2	Urtica urens 2	.54	-.6
Adonis vernalis	1.14	.1	Aquilegia vulgaris	.44	-.7
Oenanthe crocata	1.14	.1	Fraxinus excelsior	.45	-.7
Petroselinum crispum	1.11	.1	Hyssopus officinalis	.48	-.7
Ruta graveolens	1.12	.1	Leonurus cardiaca	.43	-.7
Lappa arctium	1.01	0	Lespedeza sieboldii	.46	-.7
Mercurialis perennis	1.08	0	Marrubium vulgare	.44	-.7
Ocimum canum R	1.06	0	Melilotus officinalis	.43	-.7
Paeonia officinalis	1.01	0	Milium solis	.47	-.7
Scrophularia nodosa	1.09	0	Osmundo regalis 2	.46	-.7
Aconitum napellus	.93	-.1	Spilanthes oleracea	.51	-.7
Cinnamodendron cortiso	.97	-.1	Stachys officinalis	.48	-.7
Clematis erecta	1	-.1	Stachys officinalis R	.46	-.7
Lapsana communis	.99	-.1	Viola tricolor	.49	-.7
Melissa officinalis	.94	-.1	Agrimonia eupatoria	.35	-.8
Vinca minor	.99	-.1	Anethum graveolens	.4	-.8
Anthemis nobilis	.91	-.2	Faba vulgaris	.37	-.8
Primula veris	.9	-.2	Fagopyrum esculentum	.38	-.8
Thuja occidentalis	.87	-.2	Linum usitatissimum	.39	-.8
Cardamine pratensis	.81	-.3	Lycopus europaeus	.41	-.8
Hedera helix	.78	-.3	Urtica urens	.39	-.8
Laurocerasus	.77	-.3	Echinacea purpurea	.28	-.9
Salvia sclarea	.78	-.3	Mandragora officinalis	.33	-.9

Germanium

Name	Value	Deviation	Name	Value	Deviation
Aconitum napellus	0	-.3	Marrubium vulgare	0	-.3
Adonis vernalis	0	-.3	Melilotus officinalis	0	-.3
Agrimonia eupatoria	0	-.3	Melissa officinalis	.13	.3
Alchemilla vulgaris	0	-.3	Mentha arvensis	0	-.3
Anethum graveolens	0	-.3	Mercurialis perennis	0	-.3
Angelica archangelica	0	-.3	Milium solis	0	-.3
Anthemis nobilis	0	-.3	Ocimum canum	0	-.3
Aquilegia vulgaris	0	-.3	Ocimum canum R	0	-.3
Belladonna	0	-.3	Oenanthe crocata	0	-.3
Bellis perennis	0	-.3	Ononis spinosa	0	-.3
Bryonia dioica	.13	.3	Osmundo regalis	.32	1.2
Cardamine pratensis	0	-.3	Osmundo regalis 2	0	-.3
Cardamine pratensis R	0	-.3	Paeonia officinalis	.1	.2
Cardiospermum halicaca	0	-.3	Petroselinum crispum	.11	.2
Castanea vesca	.1	.2	Plantago major	.18	.6
Centaurium erythraea	.14	.4	Polygonum aviculare	0	-.3
Chamomilla	0	-.3	Primula veris	0	-.3
Chelidonium majus	0	-.3	Prunus padus	0	-.3
Chenopodium anthelmint	0	-.3	Psoralea bituminosa	0	-.3
Cicuta virosa	0	-.3	Rhus toxicodendron	0	-.3
Cimicifuga racemosa	0	-.3	Rumex acetosa	.12	.3
Cinnamodendron cortiso	0	-.3	Ruta graveolens	0	-.3
Clematis erecta	0	-.3	Salicaria purpurea	0	-.3
Collinsonia canadensis	.1	.2	Salix purpurea	0	-.3
Conium maculatum	.11	.2	Salvia sclarea	0	-.3
Cupressus lawsoniana	.11	.2	Sanguisorba officinali	0	-.3
Echinacea angustifolia	0	-.3	Scrophularia nodosa	0	-.3
Echinacea purpurea	0	-.3	Scutellaria lateriflor	.27	1
Escholtzia californica	0	-.3	Solidago virgaurea	0	-.3
Faba vulgaris	0	-.3	Spilanthes oleracea	0	-.3
Fagopyrum esculentum	0	-.3	Stachys officinalis	0	-.3
Fraxinus excelsior	0	-.3	Stachys officinalis R	0	-.3
Galium aparine	0	-.3	Stachys sylvatica	.19	.6
Glechoma hederacea	.11	.2	Syringa vulgaris	0	-.3
Gnaphalium leontopodiu	.13	.3	Taraxacum officinale	.22	.7
Hedera helix	.1	.2	Teucrium chamaedrys	0	-.3
Hydrophyllum virginicu	0	-.3	Thuja occidentalis	0	-.3
Hyoscyamus niger	0	-.3	Tormentilla erecta	.43	1.7
Hyssopus officinalis	0	-.3	Tussilago farfara	0	-.3
Lappa arctium	.11	.2	Tussilago petasites	0	-.3
Lapsana communis	0	-.3	Ulmus campestris	0	-.3
Laurocerasus	0	-.3	Urtica dioica	.11	.2
Leonurus cardiaca	0	-.3	Urtica urens	0	-.3
Lespedeza sieboldii	.11	.2	Urtica urens 2	0	-.3
Linum usitatissimum	0	-.3	Verbena officinalis	1.95	9
Lycopus europaeus	0	-.3	Vinca minor	.24	.8
Malva sylvestris	.14	.4	Viola tricolor	.11	.2
Mandragora officinalis	0	-.3	Zizia aurea	.13	.3

Germanium

Name	Value	Deviation	Name	Value	Deviation
Verbena officinalis	1.95	9	Echinacea purpurea	0	-.3
Tormentilla erecta	.43	1.7	Escholtzia californica	0	-.3
Osmundo regalis	.32	1.2	Faba vulgaris	0	-.3
Scutellaria lateriflor	.27	1	Fagopyrum esculentum	0	-.3
Vinca minor	.24	.8	Fraxinus excelsior	0	-.3
Taraxacum officinale	.22	.7	Galium aparine	0	-.3
Plantago major	.18	.6	Hydrophyllum virginicu	0	-.3
Stachys sylvatica	.19	.6	Hyoscyamus niger	0	-.3
Centaurium erythraea	.14	.4	Hyssopus officinalis	0	-.3
Malva sylvestris	.14	.4	Lapsana communis	0	-.3
Bryonia dioica	.13	.3	Laurocerasus	0	-.3
Gnaphalium leontopodiu	.13	.3	Leonurus cardiaca	0	-.3
Melissa officinalis	.13	.3	Linum usitatissimum	0	-.3
Rumex acetosa	.12	.3	Lycopus europaeus	0	-.3
Zizia aurea	.13	.3	Mandragora officinalis	0	-.3
Castanea vesca	.1	.2	Marrubium vulgare	0	-.3
Collinsonia canadensis	.1	.2	Melilotus officinalis	0	-.3
Conium maculatum	.11	.2	Mentha arvensis	0	-.3
Cupressus lawsoniana	.11	.2	Mercurialis perennis	0	-.3
Glechoma hederacea	.11	.2	Milium solis	0	-.3
Hedera helix	.1	.2	Ocimum canum	0	-.3
Lappa arctium	.11	.2	Ocimum canum R	0	-.3
Lespedeza sieboldii	.11	.2	Oenanthe crocata	0	-.3
Paeonia officinalis	.1	.2	Ononis spinosa	0	-.3
Petroselinum crispum	.11	.2	Osmundo regalis 2	0	-.3
Urtica dioica	.11	.2	Polygonum aviculare	0	-.3
Viola tricolor	.11	.2	Primula veris	0	-.3
Aconitum napellus	0	-.3	Prunus padus	0	-.3
Adonis vernalis	0	-.3	Psoralea bituminosa	0	-.3
Agrimonia eupatoria	0	-.3	Rhus toxicodendron	0	-.3
Alchemilla vulgaris	0	-.3	Ruta graveolens	0	-.3
Anethum graveolens	0	-.3	Salicaria purpurea	0	-.3
Angelica archangelica	0	-.3	Salix purpurea	0	-.3
Anthemis nobilis	0	-.3	Salvia sclarea	0	-.3
Aquilegia vulgaris	0	-.3	Sanguisorba officinali	0	-.3
Belladonna	0	-.3	Scrophularia nodosa	0	-.3
Bellis perennis	0	-.3	Solidago virgaurea	0	-.3
Cardamine pratensis	0	-.3	Spilanthes oleracea	0	-.3
Cardamine pratensis R	0	-.3	Stachys officinalis	0	-.3
Cardiospermum halicaca	0	-.3	Stachys officinalis R	0	-.3
Chamomilla	0	-.3	Syringa vulgaris	0	-.3
Chelidonium majus	0	-.3	Teucrium chamaedrys	0	-.3
Chenopodium anthelmint	0	-.3	Thuja occidentalis	0	-.3
Cicuta virosa	0	-.3	Tussilago farfara	0	-.3
Cimicifuga racemosa	0	-.3	Tussilago petasites	0	-.3
Cinnamodendron cortiso	0	-.3	Ulmus campestris	0	-.3
Clematis erecta	0	-.3	Urtica urens	0	-.3
Echinacea angustifolia	0	-.3	Urtica urens 2	0	-.3

Hafnium

Name	Value	Deviation	Name	Value	Deviation
Aconitum napellus	.01	-.3	Marrubium vulgare	.02	0
Adonis vernalis	.02	0	Melilotus officinalis	.01	-.3
Agrimonia eupatoria	0	-.6	Melissa officinalis	.03	.3
Alchemilla vulgaris	.04	.5	Mentha arvensis	.02	0
Anethum graveolens	0	-.6	Mercurialis perennis	.05	.8
Angelica archangelica	.04	.5	Milium solis	.01	-.3
Anthemis nobilis	.01	-.3	Ocimum canum	.03	.3
Aquilegia vulgaris	0	-.6	Ocimum canum R	.01	-.3
Belladonna	0	-.6	Oenanthe crocata	.03	.3
Bellis perennis	.03	.3	Ononis spinosa	.03	.3
Bryonia dioica	.03	.3	Osmundo regalis	.28	7.5
Cardamine pratensis	.01	-.3	Osmundo regalis 2	0	-.6
Cardamine pratensis R	.02	0	Paeonia officinalis	.04	.5
Cardiospermum halicaca	.02	0	Petroselinum crispum	.02	0
Castanea vesca	.02	0	Plantago major	.02	0
Centaurium erythraea	.05	.8	Polygonum aviculare	.03	.3
Chamomilla	0	-.6	Primula veris	.03	.3
Chelidonium majus	.06	1.1	Prunus padus	.02	0
Chenopodium anthelmint	0	-.6	Psoralea bituminosa	.02	0
Cicuta virosa	.11	2.6	Rhus toxicodendron	0	-.6
Cimicifuga racemosa	.02	0	Rumex acetosa	.15	3.7
Cinnamodendron cortiso	0	-.6	Ruta graveolens	0	-.6
Clematis erecta	0	-.6	Salicaria purpurea	.01	-.3
Collinsonia canadensis	.04	.5	Salix purpurea	0	-.6
Conium maculatum	0	-.6	Salvia sclarea	.01	-.3
Cupressus lawsoniana	.06	1.1	Sanguisorba officinali	0	-.6
Echinacea angustifolia	.01	-.3	Scrophularia nodosa	.01	-.3
Echinacea purpurea	0	-.6	Scutellaria lateriflor	.02	0
Escholtzia californica	.01	-.3	Solidago virgaurea	0	-.6
Faba vulgaris	.01	-.3	Spilanthes oleracea	.01	-.3
Fagopyrum esculentum	0	-.6	Stachys officinalis	.01	-.3
Fraxinus excelsior	0	-.6	Stachys officinalis R	.01	-.3
Galium aparine	.02	0	Stachys sylvatica	.01	-.3
Glechoma hederacea	0	-.6	Syringa vulgaris	.04	.5
Gnaphalium leontopodiu	.01	-.3	Taraxacum officinale	.03	.3
Hedera helix	.01	-.3	Teucrium chamaedrys	.02	0
Hydrophyllum virginicu	0	-.6	Thuja occidentalis	.02	0
Hyoscyamus niger	.02	0	Tormentilla erecta	.03	.3
Hyssopus officinalis	0	-.6	Tussilago farfara	.02	0
Lappa arctium	.03	.3	Tussilago petasites	0	-.6
Lapsana communis	.03	.3	Ulmus campestris	.02	0
Laurocerasus	.01	-.3	Urtica dioica	.04	.5
Leonurus cardiaca	.01	-.3	Urtica urens	.01	-.3
Lespedeza sieboldii	0	-.6	Urtica urens 2	.01	-.3
Linum usitatissimum	0	-.6	Verbena officinalis	.01	-.3
Lycopus europaeus	0	-.6	Vinca minor	.01	-.3
Malva sylvestris	.02	0	Viola tricolor	.02	0
Mandragora officinalis	.01	-.3	Zizia aurea	.02	0

Hafnium

Name	Value	Deviation
Osmundo regalis	.28	7.5
Rumex acetosa	.15	3.7
Cicuta virosa	.11	2.6
Chelidonium majus	.06	1.1
Cupressus lawsoniana	.06	1.1
Centaurium erythraea	.05	.8
Mercurialis perennis	.05	.8
Alchemilla vulgaris	.04	.5
Angelica archangelica	.04	.5
Collinsonia canadensis	.04	.5
Paeonia officinalis	.04	.5
Syringa vulgaris	.04	.5
Urtica dioica	.04	.5
Bellis perennis	.03	.3
Bryonia dioica	.03	.3
Lappa arctium	.03	.3
Lapsana communis	.03	.3
Melissa officinalis	.03	.3
Ocimum canum	.03	.3
Oenanthe crocata	.03	.3
Ononis spinosa	.03	.3
Polygonum aviculare	.03	.3
Primula veris	.03	.3
Taraxacum officinale	.03	.3
Tormentilla erecta	.03	.3
Adonis vernalis	.02	0
Cardamine pratensis R	.02	0
Cardiospermum halicaca	.02	0
Castanea vesca	.02	0
Cimicifuga racemosa	.02	0
Galium aparine	.02	0
Hyoscyamus niger	.02	0
Malva sylvestris	.02	0
Marrubium vulgare	.02	0
Mentha arvensis	.02	0
Petroselinum crispum	.02	0
Plantago major	.02	0
Prunus padus	.02	0
Psoralea bituminosa	.02	0
Scutellaria lateriflor	.02	0
Teucrium chamaedrys	.02	0
Thuja occidentalis	.02	0
Tussilago farfara	.02	0
Ulmus campestris	.02	0
Viola tricolor	.02	0
Zizia aurea	.02	0
Aconitum napellus	.01	-.3
Anthemis nobilis	.01	-.3
Cardamine pratensis	.01	-.3
Echinacea angustifolia	.01	-.3
Escholtzia californica	.01	-.3
Faba vulgaris	.01	-.3
Gnaphalium leontopodiu	.01	-.3
Hedera helix	.01	-.3
Laurocerasus	.01	-.3
Leonurus cardiaca	.01	-.3
Mandragora officinalis	.01	-.3
Melilotus officinalis	.01	-.3
Milium solis	.01	-.3
Ocimum canum R	.01	-.3
Salicaria purpurea	.01	-.3
Salvia sclarea	.01	-.3
Scrophularia nodosa	.01	-.3
Spilanthes oleracea	.01	-.3
Stachys officinalis	.01	-.3
Stachys officinalis R	.01	-.3
Stachys sylvatica	.01	-.3
Urtica urens	.01	-.3
Urtica urens 2	.01	-.3
Verbena officinalis	.01	-.3
Vinca minor	.01	-.3
Agrimonia eupatoria	0	-.6
Anethum graveolens	0	-.6
Aquilegia vulgaris	0	-.6
Belladonna	0	-.6
Chamomilla	0	-.6
Chenopodium anthelmint	0	-.6
Cinnamodendron cortiso	0	-.6
Clematis erecta	0	-.6
Conium maculatum	0	-.6
Echinacea purpurea	0	-.6
Fagopyrum esculentum	0	-.6
Fraxinus excelsior	0	-.6
Glechoma hederacea	0	-.6
Hydrophyllum virginicu	0	-.6
Hyssopus officinalis	0	-.6
Lespedeza sieboldii	0	-.6
Linum usitatissimum	0	-.6
Lycopus europaeus	0	-.6
Osmundo regalis 2	0	-.6
Rhus toxicodendron	0	-.6
Ruta graveolens	0	-.6
Salix purpurea	0	-.6
Sanguisorba officinali	0	-.6
Solidago virgaurea	0	-.6
Tussilago petasites	0	-.6

Holmium

Name	Value	Deviation	Name	Value	Deviation
Aconitum napellus	.04	.3	Marrubium vulgare	.02	-.2
Adonis vernalis	.04	.3	Melilotus officinalis	.01	-.5
Agrimonia eupatoria	.01	-.5	Melissa officinalis	.03	0
Alchemilla vulgaris	.04	.3	Mentha arvensis	.02	-.2
Anethum graveolens	.01	-.5	Mercurialis perennis	.04	.3
Angelica archangelica	.04	.3	Milium solis	.01	-.5
Anthemis nobilis	.02	-.2	Ocimum canum	.03	0
Aquilegia vulgaris	0	-.7	Ocimum canum R	.03	0
Belladonna	.02	-.2	Oenanthe crocata	.04	.3
Bellis perennis	.14	2.7	Ononis spinosa	.03	0
Bryonia dioica	.04	.3	Osmundo regalis	.31	6.8
Cardamine pratensis	.01	-.5	Osmundo regalis 2	0	-.7
Cardamine pratensis R	.01	-.5	Paeonia officinalis	.03	0
Cardiospermum halicaca	.03	0	Petroselinum crispum	.05	.5
Castanea vesca	.02	-.2	Plantago major	.02	-.2
Centaurium erythraea	.06	.8	Polygonum aviculare	.02	-.2
Chamomilla	.02	-.2	Primula veris	.03	0
Chelidonium majus	.15	2.9	Prunus padus	.01	-.5
Chenopodium anthelmint	0	-.7	Psoralea bituminosa	.02	-.2
Cicuta virosa	.1	1.7	Rhus toxicodendron	.01	-.5
Cimicifuga racemosa	.04	.3	Rumex acetosa	.15	2.9
Cinnamodendron cortiso	.01	-.5	Ruta graveolens	.01	-.5
Clematis erecta	.01	-.5	Salicaria purpurea	.01	-.5
Collinsonia canadensis	.05	.5	Salix purpurea	.01	-.5
Conium maculatum	0	-.7	Salvia sclarea	.01	-.5
Cupressus lawsoniana	.03	0	Sanguisorba officinali	0	-.7
Echinacea angustifolia	.01	-.5	Scrophularia nodosa	.02	-.2
Echinacea purpurea	0	-.7	Scutellaria lateriflor	.03	0
Escholtzia californica	.01	-.5	Solidago virgaurea	.01	-.5
Faba vulgaris	.01	-.5	Spilanthes oleracea	.01	-.5
Fagopyrum esculentum	.01	-.5	Stachys officinalis	0	-.7
Fraxinus excelsior	.02	-.2	Stachys officinalis R	0	-.7
Galium aparine	.02	-.2	Stachys sylvatica	.02	-.2
Glechoma hederacea	.01	-.5	Syringa vulgaris	.01	-.5
Gnaphalium leontopodiu	.07	1	Taraxacum officinale	.11	2
Hedera helix	.01	-.5	Teucrium chamaedrys	.02	-.2
Hydrophyllum virginicu	.01	-.5	Thuja occidentalis	.01	-.5
Hyoscyamus niger	.02	-.2	Tormentilla erecta	.07	1
Hyssopus officinalis	.01	-.5	Tussilago farfara	.01	-.5
Lappa arctium	.03	0	Tussilago petasites	.01	-.5
Lapsana communis	.05	.5	Ulmus campestris	.02	-.2
Laurocerasus	.01	-.5	Urtica dioica	.06	.8
Leonurus cardiaca	.01	-.5	Urtica urens	.01	-.5
Lespedeza sieboldii	.01	-.5	Urtica urens 2	.01	-.5
Linum usitatissimum	0	-.7	Verbena officinalis	.01	-.5
Lycopus europaeus	.01	-.5	Vinca minor	.03	0
Malva sylvestris	.01	-.5	Viola tricolor	.02	-.2
Mandragora officinalis	.02	-.2	Zizia aurea	.03	0

Holmium

Name	Value	Deviation	Name	Value	Deviation
Osmundo regalis	.31	6.8	Teucrium chamaedrys	.02	-.2
Chelidonium majus	.15	2.9	Ulmus campestris	.02	-.2
Rumex acetosa	.15	2.9	Viola tricolor	.02	-.2
Bellis perennis	.14	2.7	Agrimonia eupatoria	.01	-.5
Taraxacum officinale	.11	2	Anethum graveolens	.01	-.5
Cicuta virosa	.1	1.7	Cardamine pratensis	.01	-.5
Gnaphalium leontopodiu	.07	1	Cardamine pratensis R	.01	-.5
Tormentilla erecta	.07	1	Cinnamodendron cortiso	.01	-.5
Centaurium erythraea	.06	.8	Clematis erecta	.01	-.5
Urtica dioica	.06	.8	Echinacea angustifolia	.01	-.5
Collinsonia canadensis	.05	.5	Escholtzia californica	.01	-.5
Lapsana communis	.05	.5	Faba vulgaris	.01	-.5
Petroselinum crispum	.05	.5	Fagopyrum esculentum	.01	-.5
Aconitum napellus	.04	.3	Glechoma hederacea	.01	-.5
Adonis vernalis	.04	.3	Hedera helix	.01	-.5
Alchemilla vulgaris	.04	.3	Hydrophyllum virginicu	.01	-.5
Angelica archangelica	.04	.3	Hyssopus officinalis	.01	-.5
Bryonia dioica	.04	.3	Laurocerasus	.01	-.5
Cimicifuga racemosa	.04	.3	Leonurus cardiaca	.01	-.5
Mercurialis perennis	.04	.3	Lespedeza sieboldii	.01	-.5
Oenanthe crocata	.04	.3	Lycopus europaeus	.01	-.5
Cardiospermum halicaca	.03	0	Malva sylvestris	.01	-.5
Cupressus lawsoniana	.03	0	Melilotus officinalis	.01	-.5
Lappa arctium	.03	0	Milium solis	.01	-.5
Melissa officinalis	.03	0	Prunus padus	.01	-.5
Ocimum canum	.03	0	Rhus toxicodendron	.01	-.5
Ocimum canum R	.03	0	Ruta graveolens	.01	-.5
Ononis spinosa	.03	0	Salicaria purpurea	.01	-.5
Paeonia officinalis	.03	0	Salix purpurea	.01	-.5
Primula veris	.03	0	Salvia sclarea	.01	-.5
Scutellaria lateriflor	.03	0	Solidago virgaurea	.01	-.5
Vinca minor	.03	0	Spilanthes oleracea	.01	-.5
Zizia aurea	.03	0	Syringa vulgaris	.01	-.5
Anthemis nobilis	.02	-.2	Thuja occidentalis	.01	-.5
Belladonna	.02	-.2	Tussilago farfara	.01	-.5
Castanea vesca	.02	-.2	Tussilago petasites	.01	-.5
Chamomilla	.02	-.2	Urtica urens	.01	-.5
Fraxinus excelsior	.02	-.2	Urtica urens 2	.01	-.5
Galium aparine	.02	-.2	Verbena officinalis	.01	-.5
Hyoscyamus niger	.02	-.2	Aquilegia vulgaris	0	-.7
Mandragora officinalis	.02	-.2	Chenopodium anthelmint	0	-.7
Marrubium vulgare	.02	-.2	Conium maculatum	0	-.7
Mentha arvensis	.02	-.2	Echinacea purpurea	0	-.7
Plantago major	.02	-.2	Linum usitatissimum	0	-.7
Polygonum aviculare	.02	-.2	Osmundo regalis 2	0	-.7
Psoralea bituminosa	.02	-.2	Sanguisorba officinali	0	-.7
Scrophularia nodosa	.02	-.2	Stachys officinalis	0	-.7
Stachys sylvatica	.02	-.2	Stachys officinalis R	0	-.7

Indium

Name	Value	Deviation	Name	Value	Deviation
Aconitum napellus	5.66	-.1	Marrubium vulgare	3.81	-.3
Adonis vernalis	4.46	-.2	Melilotus officinalis	2.97	-.4
Agrimonia eupatoria	4.19	-.3	Melissa officinalis	6.59	0
Alchemilla vulgaris	7.12	.1	Mentha arvensis	8.56	.2
Anethum graveolens	1.45	-.5	Mercurialis perennis	5.5	-.1
Angelica archangelica	2.94	-.4	Milium solis	3.95	-.3
Anthemis nobilis	3.32	-.3	Ocimum canum	3.86	-.3
Aquilegia vulgaris	4.6	-.2	Ocimum canum R	5.64	-.1
Belladonna	6.05	-.1	Oenanthe crocata	3.7	-.3
Bellis perennis	13.48	.7	Ononis spinosa	6.53	0
Bryonia dioica	4.44	-.2	Osmundo regalis	25.07	2
Cardamine pratensis	6.03	-.1	Osmundo regalis 2	1.59	-.5
Cardamine pratensis R	6.65	0	Paeonia officinalis	4.05	-.3
Cardiospermum halicaca	2.5	-.4	Petroselinum crispum	5.61	-.1
Castanea vesca	4.78	-.2	Plantago major	1.86	-.5
Centaurium erythraea	10.15	.4	Polygonum aviculare	3.29	-.4
Chamomilla	2.98	-.4	Primula veris	6.63	0
Chelidonium majus	18.46	1.3	Prunus padus	4.42	-.2
Chenopodium anthelmint	2.65	-.4	Psoralea bituminosa	3.01	-.4
Cicuta virosa	9.04	.3	Rhus toxicodendron	3.48	-.3
Cimicifuga racemosa	2.21	-.5	Rumex acetosa	13.12	.7
Cinnamodendron cortiso	5.85	-.1	Ruta graveolens	4.82	-.2
Clematis erecta	6.73	0	Salicaria purpurea	4.6	-.2
Collinsonia canadensis	3.22	-.4	Salix purpurea	9.06	.3
Conium maculatum	6.47	0	Salvia sclarea	4.23	-.3
Cupressus lawsoniana	43.53	3.9	Sanguisorba officinali	2.53	-.4
Echinacea angustifolia	3.01	-.4	Scrophularia nodosa	13.26	.7
Echinacea purpurea	2.62	-.4	Scutellaria lateriflor	5.27	-.1
Escholtzia californica	1.7	-.5	Solidago virgaurea	1.7	-.5
Faba vulgaris	3.52	-.3	Spilanthes oleracea	1.23	-.6
Fagopyrum esculentum	0	-.7	Stachys officinalis	1.97	-.5
Fraxinus excelsior	8.96	.3	Stachys officinalis R	2.49	-.4
Galium aparine	4.46	-.2	Stachys sylvatica	6.97	0
Glechoma hederacea	7.18	.1	Syringa vulgaris	79.71	7.8
Gnaphalium leontopodiu	11.3	.5	Taraxacum officinale	9.54	.3
Hedera helix	12.58	.6	Teucrium chamaedrys	3.89	-.3
Hydrophyllum virginicu	6.73	0	Thuja occidentalis	7.92	.1
Hyoscyamus niger	4.02	-.3	Tormentilla erecta	6.11	-.1
Hyssopus officinalis	3.08	-.4	Tussilago farfara	3.13	-.4
Lappa arctium	6.38	0	Tussilago petasites	3.66	-.3
Lapsana communis	5.56	-.1	Ulmus campestris	8.54	.2
Laurocerasus	3.5	-.3	Urtica dioica	9.42	.3
Leonurus cardiaca	4.15	-.3	Urtica urens	1.06	-.6
Lespedeza sieboldii	4.39	-.2	Urtica urens 2	2.08	-.5
Linum usitatissimum	1.65	-.5	Verbena officinalis	0	-.7
Lycopus europaeus	3.12	-.4	Vinca minor	17.52	1.2
Malva sylvestris	2.85	-.4	Viola tricolor	1.97	-.5
Mandragora officinalis	1.73	-.5	Zizia aurea	5.26	-.1

Indium

Name	Value	Deviation	Name	Value	Deviation
Syringa vulgaris	79.71	7.8	Ruta graveolens	4.82	-.2
Cupressus lawsoniana	43.53	3.9	Salicaria purpurea	4.6	-.2
Osmundo regalis	25.07	2	Agrimonia eupatoria	4.19	-.3
Chelidonium majus	18.46	1.3	Anthemis nobilis	3.32	-.3
Vinca minor	17.52	1.2	Faba vulgaris	3.52	-.3
Bellis perennis	13.48	.7	Hyoscyamus niger	4.02	-.3
Rumex acetosa	13.12	.7	Laurocerasus	3.5	-.3
Scrophularia nodosa	13.26	.7	Leonurus cardiaca	4.15	-.3
Hedera helix	12.58	.6	Marrubium vulgare	3.81	-.3
Gnaphalium leontopodiu	11.3	.5	Milium solis	3.95	-.3
Centaurium erythraea	10.15	.4	Ocimum canum	3.86	-.3
Cicuta virosa	9.04	.3	Oenanthe crocata	3.7	-.3
Fraxinus excelsior	8.96	.3	Paeonia officinalis	4.05	-.3
Salix purpurea	9.06	.3	Rhus toxicodendron	3.48	-.3
Taraxacum officinale	9.54	.3	Salvia sclarea	4.23	-.3
Urtica dioica	9.42	.3	Teucrium chamaedrys	3.89	-.3
Mentha arvensis	8.56	.2	Tussilago petasites	3.66	-.3
Ulmus campestris	8.54	.2	Angelica archangelica	2.94	-.4
Alchemilla vulgaris	7.12	.1	Cardiospermum halicaca	2.5	-.4
Glechoma hederacea	7.18	.1	Chamomilla	2.98	-.4
Thuja occidentalis	7.92	.1	Chenopodium anthelmint	2.65	-.4
Cardamine pratensis R	6.65	0	Collinsonia canadensis	3.22	-.4
Clematis erecta	6.73	0	Echinacea angustifolia	3.01	-.4
Conium maculatum	6.47	0	Echinacea purpurea	2.62	-.4
Hydrophyllum virginicu	6.73	0	Hyssopus officinalis	3.08	-.4
Lappa arctium	6.38	0	Lycopus europaeus	3.12	-.4
Melissa officinalis	6.59	0	Malva sylvestris	2.85	-.4
Ononis spinosa	6.53	0	Melilotus officinalis	2.97	-.4
Primula veris	6.63	0	Polygonum aviculare	3.29	-.4
Stachys sylvatica	6.97	0	Psoralea bituminosa	3.01	-.4
Aconitum napellus	5.66	-.1	Sanguisorba officinali	2.53	-.4
Belladonna	6.05	-.1	Stachys officinalis R	2.49	-.4
Cardamine pratensis	6.03	-.1	Tussilago farfara	3.13	-.4
Cinnamodendron cortiso	5.85	-.1	Anethum graveolens	1.45	-.5
Lapsana communis	5.56	-.1	Cimicifuga racemosa	2.21	-.5
Mercurialis perennis	5.5	-.1	Escholtzia californica	1.7	-.5
Ocimum canum R	5.64	-.1	Linum usitatissimum	1.65	-.5
Petroselinum crispum	5.61	-.1	Mandragora officinalis	1.73	-.5
Scutellaria lateriflor	5.27	-.1	Osmundo regalis 2	1.59	-.5
Tormentilla erecta	6.11	-.1	Plantago major	1.86	-.5
Zizia aurea	5.26	-.1	Solidago virgaurea	1.7	-.5
Adonis vernalis	4.46	-.2	Stachys officinalis	1.97	-.5
Aquilegia vulgaris	4.6	-.2	Urtica urens 2	2.08	-.5
Bryonia dioica	4.44	-.2	Viola tricolor	1.97	-.5
Castanea vesca	4.78	-.2	Spilanthes oleracea	1.23	-.6
Galium aparine	4.46	-.2	Urtica urens	1.06	-.6
Lespedeza sieboldii	4.39	-.2	Fagopyrum esculentum	0	-.7
Prunus padus	4.42	-.2	Verbena officinalis	0	-.7

Lanthanum

Name	Value	Deviation	Name	Value	Deviation
Aconitum napellus	2.13	.4	Marrubium vulgare	.87	-.3
Adonis vernalis	2	.3	Melilotus officinalis	.58	-.5
Agrimonia eupatoria	.32	-.6	Melissa officinalis	1.36	-.1
Alchemilla vulgaris	2.01	.3	Mentha arvensis	.94	-.3
Anethum graveolens	.34	-.6	Mercurialis perennis	2.18	.4
Angelica archangelica	1.76	.2	Milium solis	.69	-.4
Anthemis nobilis	1.28	-.1	Ocimum canum	1.26	-.1
Aquilegia vulgaris	.28	-.7	Ocimum canum R	1.24	-.1
Belladonna	1.13	-.2	Oenanthe crocata	1.94	.3
Bellis perennis	7.37	3.3	Ononis spinosa	1.25	-.1
Bryonia dioica	2.3	.5	Osmundo regalis	12.59	6.2
Cardamine pratensis	1.88	.2	Osmundo regalis 2	.09	-.8
Cardamine pratensis R	1.66	.1	Paeonia officinalis	1.52	0
Cardiospermum halicaca	1.26	-.1	Petroselinum crispum	2.48	.6
Castanea vesca	.75	-.4	Plantago major	2.14	.4
Centaurium erythraea	3.27	1	Polygonum aviculare	.98	-.3
Chamomilla	1.1	-.2	Primula veris	1.58	.1
Chelidonium majus	6.93	3	Prunus padus	.8	-.4
Chenopodium anthelmint	.34	-.6	Psoralea bituminosa	.99	-.3
Cicuta virosa	4.82	1.9	Rhus toxicodendron	.6	-.5
Cimicifuga racemosa	2.04	.3	Rumex acetosa	6.52	2.8
Cinnamodendron cortiso	.69	-.4	Ruta graveolens	.53	-.5
Clematis erecta	.52	-.5	Salicaria purpurea	.54	-.5
Collinsonia canadensis	2.44	.5	Salix purpurea	.9	-.3
Conium maculatum	.34	-.6	Salvia sclarea	.53	-.5
Cupressus lawsoniana	2.03	.3	Sanguisorba officinali	.3	-.6
Echinacea angustifolia	.72	-.4	Scrophularia nodosa	1.08	-.2
Echinacea purpurea	.37	-.6	Scutellaria lateriflor	.96	-.3
Escholtzia californica	.42	-.6	Solidago virgaurea	.25	-.7
Faba vulgaris	.51	-.5	Spilanthes oleracea	.52	-.5
Fagopyrum esculentum	.31	-.6	Stachys officinalis	.25	-.7
Fraxinus excelsior	1.2	-.1	Stachys officinalis R	.18	-.7
Galium aparine	1.2	-.1	Stachys sylvatica	1.22	-.1
Glechoma hederacea	1.03	-.2	Syringa vulgaris	1.02	-.2
Gnaphalium leontopodiu	3.61	1.2	Taraxacum officinale	5.48	2.2
Hedera helix	.9	-.3	Teucrium chamaedrys	.7	-.4
Hydrophyllum virginicu	.46	-.6	Thuja occidentalis	.68	-.4
Hyoscyamus niger	1.08	-.2	Tormentilla erecta	3.2	1
Hyssopus officinalis	.45	-.6	Tussilago farfara	.3	-.6
Lappa arctium	1.92	.2	Tussilago petasites	.7	-.4
Lapsana communis	2.3	.5	Ulmus campestris	1.08	-.2
Laurocerasus	.36	-.6	Urtica dioica	2.93	.8
Leonurus cardiaca	.46	-.6	Urtica urens	.49	-.5
Lespedeza sieboldii	.42	-.6	Urtica urens 2	.56	-.5
Linum usitatissimum	.23	-.7	Verbena officinalis	.28	-.7
Lycopus europaeus	.38	-.6	Vinca minor	1.35	-.1
Malva sylvestris	.51	-.5	Viola tricolor	.72	-.4
Mandragora officinalis	.61	-.5	Zizia aurea	1.34	-.1

Lanthanum

Name	Value	Deviation	Name	Value	Deviation
Osmundo regalis	12.59	6.2	Mentha arvensis	.94	-.3
Bellis perennis	7.37	3.3	Polygonum aviculare	.98	-.3
Chelidonium majus	6.93	3	Psoralea bituminosa	.99	-.3
Rumex acetosa	6.52	2.8	Salix purpurea	.9	-.3
Taraxacum officinale	5.48	2.2	Scutellaria lateriflor	.96	-.3
Cicuta virosa	4.82	1.9	Castanea vesca	.75	-.4
Gnaphalium leontopodiu	3.61	1.2	Cinnamodendron cortiso	.69	-.4
Centaurium erythraea	3.27	1	Echinacea angustifolia	.72	-.4
Tormentilla erecta	3.2	1	Milium solis	.69	-.4
Urtica dioica	2.93	.8	Prunus padus	.8	-.4
Petroselinum crispum	2.48	.6	Teucrium chamaedrys	.7	-.4
Bryonia dioica	2.3	.5	Thuja occidentalis	.68	-.4
Collinsonia canadensis	2.44	.5	Tussilago petasites	.7	-.4
Lapsana communis	2.3	.5	Viola tricolor	.72	-.4
Aconitum napellus	2.13	.4	Clematis erecta	.52	-.5
Mercurialis perennis	2.18	.4	Faba vulgaris	.51	-.5
Plantago major	2.14	.4	Malva sylvestris	.51	-.5
Adonis vernalis	2	.3	Mandragora officinalis	.61	-.5
Alchemilla vulgaris	2.01	.3	Melilotus officinalis	.58	-.5
Cimicifuga racemosa	2.04	.3	Rhus toxicodendron	.6	-.5
Cupressus lawsoniana	2.03	.3	Ruta graveolens	.53	-.5
Oenanthe crocata	1.94	.3	Salicaria purpurea	.54	-.5
Angelica archangelica	1.76	.2	Salvia sclarea	.53	-.5
Cardamine pratensis	1.88	.2	Spilanthes oleracea	.52	-.5
Lappa arctium	1.92	.2	Urtica urens	.49	-.5
Cardamine pratensis R	1.66	.1	Urtica urens 2	.56	-.5
Primula veris	1.58	.1	Agrimonia eupatoria	.32	-.6
Paeonia officinalis	1.52	0	Anethum graveolens	.34	-.6
Anthemis nobilis	1.28	-.1	Chenopodium anthelmint	.34	-.6
Cardiospermum halicaca	1.26	-.1	Conium maculatum	.34	-.6
Fraxinus excelsior	1.2	-.1	Echinacea purpurea	.37	-.6
Galium aparine	1.2	-.1	Escholtzia californica	.42	-.6
Melissa officinalis	1.36	-.1	Fagopyrum esculentum	.31	-.6
Ocimum canum	1.26	-.1	Hydrophyllum virginicu	.46	-.6
Ocimum canum R	1.24	-.1	Hyssopus officinalis	.45	-.6
Ononis spinosa	1.25	-.1	Laurocerasus	.36	-.6
Stachys sylvatica	1.22	-.1	Leonurus cardiaca	.46	-.6
Vinca minor	1.35	-.1	Lespedeza sieboldii	.42	-.6
Zizia aurea	1.34	-.1	Lycopus europaeus	.38	-.6
Belladonna	1.13	-.2	Sanguisorba officinali	.3	-.6
Chamomilla	1.1	-.2	Tussilago farfara	.3	-.6
Glechoma hederacea	1.03	-.2	Aquilegia vulgaris	.28	-.7
Hyoscyamus niger	1.08	-.2	Linum usitatissimum	.23	-.7
Scrophularia nodosa	1.08	-.2	Solidago virgaurea	.25	-.7
Syringa vulgaris	1.02	-.2	Stachys officinalis	.25	-.7
Ulmus campestris	1.08	-.2	Stachys officinalis R	.18	-.7
Hedera helix	.9	-.3	Verbena officinalis	.28	-.7
Marrubium vulgare	.87	-.3	Osmundo regalis 2	.09	-.8

Lithium

Name	Value	Deviation	Name	Value	Deviation
Aconitum napellus	5.47	.3	Marrubium vulgare	1.89	-.5
Adonis vernalis	3.85	-.1	Melilotus officinalis	1.59	-.6
Agrimonia eupatoria	.71	-.8	Melissa officinalis	2.76	-.3
Alchemilla vulgaris	4.05	0	Mentha arvensis	1.95	-.5
Anethum graveolens	1.05	-.7	Mercurialis perennis	9.31	1.2
Angelica archangelica	3.53	-.2	Milium solis	1.86	-.5
Anthemis nobilis	3.71	-.1	Ocimum canum	2.85	-.3
Aquilegia vulgaris	1.63	-.6	Ocimum canum R	2.84	-.3
Belladonna	10.71	1.5	Oenanthe crocata	4.03	0
Bellis perennis	9.29	1.2	Ononis spinosa	2.32	-.4
Bryonia dioica	5.34	.3	Osmundo regalis 1	15.92	2.7
Cardamine pratensis	3.72	-.1	Osmundo regalis 2	.81	-.8
Cardamine pratensis R	3.49	-.2	Paeonia officinalis	13.73	2.2
Cardiospermum halicaca	5.23	.2	Petroselinum crispum	5.42	.3
Castanea vesca	5.06	.2	Plantago major	2.46	-.4
Centaurium erythraea	6.66	.6	Polygonum aviculare	2.72	-.3
Chamomilla	3.08	-.3	Primula veris	2.55	-.4
Chelidonium majus	14	2.2	Prunus padus	1.93	-.5
Chenopodium anthelmint	.52	-.8	Psoralea bituminosa	6.83	.6
Cicuta virosa	10.49	1.4	Rhus toxicodendron	3.2	-.2
Cimicifuga racemosa	1.77	-.6	Rumex acetosa	12.52	1.9
Cinnamodendron cortiso	5.98	.4	Ruta graveolens	15.9	2.7
Clematis erecta	1.72	-.6	Salicaria purpurea	14.21	2.3
Collinsonia canadensis	4.36	0	Salix purpurea	2.71	-.3
Conium maculatum	1.13	-.7	Salvia sclarea	20.48	3.7
Cupressus lawsoniana	21.35	3.9	Sanguisorba officinali	1.28	-.7
Echinacea angustifolia	3.66	-.1	Scrophularia nodosa	5.29	.2
Echinacea purpurea	.7	-.8	Scutellaria lateriflor	2.63	-.4
Escholtzia californica	.75	-.8	Solidago virgaurea	.87	-.8
Faba vulgaris	1.83	-.5	Spilanthes oleracea	1.26	-.7
Fagopyrum esculentum	0	-1	Stachys officinalis	0	-1
Fraxinus excelsior	1.52	-.6	Stachys officinalis R	.78	-.8
Galium aparine	1.96	-.5	Stachys sylvatica	2.64	-.4
Glechoma hederacea	2.03	-.5	Syringa vulgaris	1.9	-.5
Gnaphalium leontopodiu	5.5	.3	Taraxacum officinale	7.47	.7
Hedera helix	5.42	.3	Teucrium chamaedrys	2.03	-.5
Hydrophyllum virginicu	.56	-.8	Thuja occidentalis	2.16	-.5
Hyoscyamus niger	10.69	1.5	Tormentilla erecta	4.75	.1
Hyssopus officinalis	.85	-.8	Tussilago farfara	1.61	-.6
Lappa arctium	4.12	0	Tussilago petasites	2.43	-.4
Lapsana communis	3.49	-.2	Ulmus campestris	4.51	.1
Laurocerasus	2.32	-.4	Urtica dioica	6.54	.5
Leonurus cardiaca	.95	-.8	Urtica urens 1	1.74	-.6
Lespedeza sieboldii	.79	-.8	Urtica urens 2	2.15	-.5
Linum usitatissimum	.69	-.8	Verbena officinalis	.97	-.7
Lycopus europaeus	2.37	-.4	Vinca minor	2.99	-.3
Malva sylvestris	1.1	-.7	Viola tricolor	1.86	-.5
Mandragora officinalis	1.36	-.7	Zizia aurea	3.51	-.2

Lithium

Name	Value	Deviation	Name	Value	Deviation
Cupressus lawsoniana	21.35	3.9	Laurocerasus	2.32	-.4
Salvia sclarea	20.48	3.7	Lycopus europaeus	2.37	-.4
Osmundo regalis 1	15.92	2.7	Ononis spinosa	2.32	-.4
Ruta graveolens	15.9	2.7	Plantago major	2.46	-.4
Salicaria purpurea	14.21	2.3	Primula veris	2.55	-.4
Chelidonium majus	14	2.2	Scutellaria lateriflor	2.63	-.4
Paeonia officinalis	13.73	2.2	Stachys sylvatica	2.64	-.4
Rumex acetosa	12.52	1.9	Tussilago petasites	2.43	-.4
Belladonna	10.71	1.5	Faba vulgaris	1.83	-.5
Hyoscyamus niger	10.69	1.5	Galium aparine	1.96	-.5
Cicuta virosa	10.49	1.4	Glechoma hederacea	2.03	-.5
Bellis perennis	9.29	1.2	Marrubium vulgare	1.89	-.5
Mercurialis perennis	9.31	1.2	Mentha arvensis	1.95	-.5
Taraxacum officinale	7.47	.7	Milium solis	1.86	-.5
Centaurium erythraea	6.66	.6	Prunus padus	1.93	-.5
Psoralea bituminosa	6.83	.6	Syringa vulgaris	1.9	-.5
Urtica dioica	6.54	.5	Teucrium chamaedrys	2.03	-.5
Cinnamodendron cortiso	5.98	.4	Thuja occidentalis	2.16	-.5
Aconitum napellus	5.47	.3	Urtica urens 2	2.15	-.5
Bryonia dioica	5.34	.3	Viola tricolor	1.86	-.5
Gnaphalium leontopodiu	5.5	.3	Aquilegia vulgaris	1.63	-.6
Hedera helix	5.42	.3	Cimicifuga racemosa	1.77	-.6
Petroselinum crispum	5.42	.3	Clematis erecta	1.72	-.6
Cardiospermum halicaca	5.23	.2	Fraxinus excelsior	1.52	-.6
Castanea vesca	5.06	.2	Melilotus officinalis	1.59	-.6
Scrophularia nodosa	5.29	.2	Tussilago farfara	1.61	-.6
Tormentilla erecta	4.75	.1	Urtica urens 1	1.74	-.6
Ulmus campestris	4.51	.1	Anethum graveolens	1.05	-.7
Alchemilla vulgaris	4.05	0	Conium maculatum	1.13	-.7
Collinsonia canadensis	4.36	0	Malva sylvestris	1.1	-.7
Lappa arctium	4.12	0	Mandragora officinalis	1.36	-.7
Oenanthe crocata	4.03	0	Sanguisorba officinali	1.28	-.7
Adonis vernalis	3.85	-.1	Spilanthes oleracea	1.26	-.7
Anthemis nobilis	3.71	-.1	Verbena officinalis	.97	-.7
Cardamine pratensis	3.72	-.1	Agrimonia eupatoria	.71	-.8
Echinacea angustifolia	3.66	-.1	Chenopodium anthelmint	.52	-.8
Angelica archangelica	3.53	-.2	Echinacea purpurea	.7	-.8
Cardamine pratensis R	3.49	-.2	Escholtzia californica	.75	-.8
Lapsana communis	3.49	-.2	Hydrophyllum virginicu	.56	-.8
Rhus toxicodendron	3.2	-.2	Hyssopus officinalis	.85	-.8
Zizia aurea	3.51	-.2	Leonurus cardiaca	.95	-.8
Chamomilla	3.08	-.3	Lespedeza sieboldii	.79	-.8
Melissa officinalis	2.76	-.3	Linum usitatissimum	.69	-.8
Ocimum canum	2.85	-.3	Osmundo regalis 2	.81	-.8
Ocimum canum R	2.84	-.3	Solidago virgaurea	.87	-.8
Polygonum aviculare	2.72	-.3	Stachys officinalis R	.78	-.8
Salix purpurea	2.71	-.3	Fagopyrum esculentum	0	-1
Vinca minor	2.99	-.3	Stachys officinalis	0	-1

Lutetium

Name	Value	Deviation	Name	Value	Deviation
Aconitum napellus	.01	.1	Marrubium vulgare	.01	.1
Adonis vernalis	.01	.1	Melilotus officinalis	0	-.6
Agrimonia eupatoria	0	-.6	Melissa officinalis	.01	.1
Alchemilla vulgaris	.01	.1	Mentha arvensis	0	-.6
Anethum graveolens	0	-.6	Mercurialis perennis	.01	.1
Angelica archangelica	.02	.8	Milium solis	0	-.6
Anthemis nobilis	.01	.1	Ocimum canum	.01	.1
Aquilegia vulgaris	0	-.6	Ocimum canum R	.01	.1
Belladonna	.01	.1	Oenanthe crocata	.01	.1
Bellis perennis	.05	2.8	Ononis spinosa	.01	.1
Bryonia dioica	.02	.8	Osmundo regalis	.1	6.3
Cardamine pratensis	0	-.6	Osmundo regalis 2	0	-.6
Cardamine pratensis R	0	-.6	Paeonia officinalis	.01	.1
Cardiospermum halicaca	.01	.1	Petroselinum crispum	.02	.8
Castanea vesca	0	-.6	Plantago major	.01	.1
Centaurium erythraea	.02	.8	Polygonum aviculare	.01	.1
Chamomilla	.01	.1	Primula veris	.01	.1
Chelidonium majus	.05	2.8	Prunus padus	0	-.6
Chenopodium anthelmint	0	-.6	Psoralea bituminosa	0	-.6
Cicuta virosa	.04	2.1	Rhus toxicodendron	0	-.6
Cimicifuga racemosa	.02	.8	Rumex acetosa	.05	2.8
Cinnamodendron cortiso	.01	.1	Ruta graveolens	0	-.6
Clematis erecta	0	-.6	Salicaria purpurea	0	-.6
Collinsonia canadensis	.02	.8	Salix purpurea	0	-.6
Conium maculatum	0	-.6	Salvia sclarea	0	-.6
Cupressus lawsoniana	.01	.1	Sanguisorba officinali	0	-.6
Echinacea angustifolia	0	-.6	Scrophularia nodosa	.01	.1
Echinacea purpurea	0	-.6	Scutellaria lateriflor	.01	.1
Escholtzia californica	0	-.6	Solidago virgaurea	0	-.6
Faba vulgaris	0	-.6	Spilanthes oleracea	0	-.6
Fagopyrum esculentum	0	-.6	Stachys officinalis	0	-.6
Fraxinus excelsior	.01	.1	Stachys officinalis R	0	-.6
Galium aparine	.01	.1	Stachys sylvatica	.01	.1
Glechoma hederacea	.01	.1	Syringa vulgaris	0	-.6
Gnaphalium leontopodiu	.03	1.4	Taraxacum officinale	.03	1.4
Hedera helix	.01	.1	Teucrium chamaedrys	.01	.1
Hydrophyllum virginicu	0	-.6	Thuja occidentalis	0	-.6
Hyoscyamus niger	.01	.1	Tormentilla erecta	.02	.8
Hyssopus officinalis	0	-.6	Tussilago farfara	0	-.6
Lappa arctium	.01	.1	Tussilago petasites	0	-.6
Lapsana communis	.02	.8	Ulmus campestris	0	-.6
Laurocerasus	0	-.6	Urtica dioica	.02	.8
Leonurus cardiaca	0	-.6	Urtica urens	0	-.6
Lespedeza sieboldii	0	-.6	Urtica urens 2	0	-.6
Linum usitatissimum	0	-.6	Verbena officinalis	0	-.6
Lycopus europaeus	0	-.6	Vinca minor	.01	.1
Malva sylvestris	0	-.6	Viola tricolor	.01	.1
Mandragora officinalis	.01	.1	Zizia aurea	.01	.1

Lutetium

Name	Value	Deviation	Name	Value	Deviation
Osmundo regalis	.1	6.3	Viola tricolor	.01	.1
Bellis perennis	.05	2.8	Zizia aurea	.01	.1
Chelidonium majus	.05	2.8	Agrimonia eupatoria	0	-.6
Rumex acetosa	.05	2.8	Anethum graveolens	0	-.6
Cicuta virosa	.04	2.1	Aquilegia vulgaris	0	-.6
Gnaphalium leontopodiu	.03	1.4	Cardamine pratensis	0	-.6
Taraxacum officinale	.03	1.4	Cardamine pratensis R	0	-.6
Angelica archangelica	.02	.8	Castanea vesca	0	-.6
Bryonia dioica	.02	.8	Chenopodium anthelmint	0	-.6
Centaurium erythraea	.02	.8	Clematis erecta	0	-.6
Cimicifuga racemosa	.02	.8	Conium maculatum	0	-.6
Collinsonia canadensis	.02	.8	Echinacea angustifolia	0	-.6
Lapsana communis	.02	.8	Echinacea purpurea	0	-.6
Petroselinum crispum	.02	.8	Escholtzia californica	0	-.6
Tormentilla erecta	.02	.8	Faba vulgaris	0	-.6
Urtica dioica	.02	.8	Fagopyrum esculentum	0	-.6
Aconitum napellus	.01	.1	Hydrophyllum virginicu	0	-.6
Adonis vernalis	.01	.1	Hyssopus officinalis	0	-.6
Alchemilla vulgaris	.01	.1	Laurocerasus	0	-.6
Anthemis nobilis	.01	.1	Leonurus cardiaca	0	-.6
Belladonna	.01	.1	Lespedeza sieboldii	0	-.6
Cardiospermum halicaca	.01	.1	Linum usitatissimum	0	-.6
Chamomilla	.01	.1	Lycopus europaeus	0	-.6
Cinnamodendron cortiso	.01	.1	Malva sylvestris	0	-.6
Cupressus lawsoniana	.01	.1	Melilotus officinalis	0	-.6
Fraxinus excelsior	.01	.1	Mentha arvensis	0	-.6
Galium aparine	.01	.1	Milium solis	0	-.6
Glechoma hederacea	.01	.1	Osmundo regalis 2	0	-.6
Hedera helix	.01	.1	Prunus padus	0	-.6
Hyoscyamus niger	.01	.1	Psoralea bituminosa	0	-.6
Lappa arctium	.01	.1	Rhus toxicodendron	0	-.6
Mandragora officinalis	.01	.1	Ruta graveolens	0	-.6
Marrubium vulgare	.01	.1	Salicaria purpurea	0	-.6
Melissa officinalis	.01	.1	Salix purpurea	0	-.6
Mercurialis perennis	.01	.1	Salvia sclarea	0	-.6
Ocimum canum	.01	.1	Sanguisorba officinali	0	-.6
Ocimum canum R	.01	.1	Solidago virgaurea	0	-.6
Oenanthe crocata	.01	.1	Spilanthes oleracea	0	-.6
Ononis spinosa	.01	.1	Stachys officinalis	0	-.6
Paeonia officinalis	.01	.1	Stachys officinalis R	0	-.6
Plantago major	.01	.1	Syringa vulgaris	0	-.6
Polygonum aviculare	.01	.1	Thuja occidentalis	0	-.6
Primula veris	.01	.1	Tussilago farfara	0	-.6
Scrophularia nodosa	.01	.1	Tussilago petasites	0	-.6
Scutellaria lateriflor	.01	.1	Ulmus campestris	0	-.6
Stachys sylvatica	.01	.1	Urtica urens	0	-.6
Teucrium chamaedrys	.01	.1	Urtica urens 2	0	-.6
Vinca minor	.01	.1	Verbena officinalis	0	-.6

Magnesium

Name	Value	Deviation	Name	Value	Deviation
Aconitum napellus	20124.46	-.7	Marrubium vulgare	19510.13	-.7
Adonis vernalis	14556.98	-.8	Melilotus officinalis	150000	2.1
Agrimonia eupatoria	36159.8	-.3	Melissa officinalis	150000	2.1
Alchemilla vulgaris	150000	2.1	Mentha arvensis	34978.45	-.4
Anethum graveolens	29332.79	-.5	Mercurialis perennis	40372.82	-.2
Angelica archangelica	30031.14	-.5	Milium solis	31279.83	-.4
Anthemis nobilis	19631.53	-.7	Ocimum canum	150000	2.1
Aquilegia vulgaris	24488.34	-.6	Ocimum canum R	150000	2.1
Belladonna	25268.54	-.6	Oenanthe crocata	35550.85	-.3
Bellis perennis	4965.97	-1	Ononis spinosa	26426.59	-.5
Bryonia dioica	31536.29	-.4	Osmundo regalis	150000	2.1
Cardamine pratensis	150000	2.1	Osmundo regalis 2	28954.08	-.5
Cardamine pratensis R	150000	2.1	Paeonia officinalis	29100.22	-.5
Cardiospermum halic	38657.76	-.3	Petroselinum crispum	17536.21	-.7
Castanea vesca	150000	2.1	Plantago major	19147.11	-.7
Centaurium erythraea	21186.01	-.6	Polygonum aviculare	56810.42	.1
Chamomilla	20973.75	-.7	Primula veris	31921.31	-.4
Chelidonium majus	12575.63	-.8	Prunus padus	59863.47	.2
Chenopodium anthelm	21621.05	-.6	Psoralea bituminosa	150000	2.1
Cicuta virosa	150000	2.1	Rhus toxicodendron	37166.07	-.3
Cimicifuga racemosa	23185.79	-.6	Rumex acetosa	33408.55	-.4
Cinnamodendron cort	150000	2.1	Ruta graveolens	41143.67	-.2
Clematis erecta	29308.86	-.5	Salicaria purpurea	45003.1	-.1
Collinsonia canadensis	150000	2.1	Salix purpurea	15658.32	-.8
Conium maculatum	40377.49	-.2	Salvia sclarea	39324.53	-.3
Cupressus lawsoniana	24141.96	-.6	Sanguisorba officin	48677.9	-.1
Echinacea angustif	38835.54	-.3	Scrophularia nodosa	78691.06	.6
Echinacea purpurea	150000	2.1	Scutellaria lateriflor	150000	2.1
Escholtzia californica	25483.15	-.6	Solidago virgaurea	31801.01	-.4
Faba vulgaris	26642.78	-.5	Spilanthes oleracea	32249.76	-.4
Fagopyrum esculentum	37809.67	-.3	Stachys officinalis	28067.73	-.5
Fraxinus excelsior	9778.88	-.9	Stachys officinalis R	29853.32	-.5
Galium aparine	21163.09	-.6	Stachys sylvatica	59248.84	.2
Glechoma hederacea	29086.17	-.5	Syringa vulgaris	38464.79	-.3
Gnaphalium leontop	16972.49	-.7	Taraxacum officinale	9722.49	-.9
Hedera helix	38933.63	-.3	Teucrium chamaedrys	31648.13	-.4
Hydrophyllum virgin	39268.75	-.3	Thuja occidentalis	35989.89	-.3
Hyoscyamus niger	29484.98	-.5	Tormentilla erecta	31359.45	-.4
Hyssopus officinalis	27858.46	-.5	Tussilago farfara	56882.81	.1
Lappa arctium	41057.72	-.2	Tussilago petasites	31420.58	-.4
Lapsana communis	33802.8	-.4	Ulmus campestris	31884.62	-.4
Laurocerasus	150000	2.1	Urtica dioica	28428.42	-.5
Leonurus cardiaca	29649.29	-.5	Urtica urens	23126.44	-.6
Lespedeza sieboldii	24808.94	-.6	Urtica urens 2	21589.59	-.6
Linum usitatissimum	150000	2.1	Verbena officinalis	16164.3	-.8
Lycopus europaeus	33333.4	-.4	Vinca minor	45427.12	-.1
Malva sylvestris	33858.33	-.4	Viola tricolor	22176.67	-.6
Mandragora officinalis	16838.09	-.7	Zizia aurea	21295.63	-.6

Magnesium

Name	Value	Deviation	Name	Value	Deviation
Alchemilla vulgaris	150000	2.1	Solidago virgaurea	31801.01	-.4
Cardamine pratensis	150000	2.1	Spilanthes oleracea	32249.76	-.4
Cardamine pratensis R	150000	2.1	Teucrium chamaedrys	31648.13	-.4
Castanea vesca	150000	2.1	Tormentilla erecta	31359.45	-.4
Cicuta virosa	150000	2.1	Tussilago petasites	31420.58	-.4
Cinnamodendron cortiso	150000	2.1	Ulmus campestris	31884.62	-.4
Collinsonia canadensis	150000	2.1	Anethum graveolens	29332.79	-.5
Echinacea purpurea	150000	2.1	Angelica archangelica	30031.14	-.5
Laurocerasus	150000	2.1	Clematis erecta	29308.86	-.5
Linum usitatissimum	150000	2.1	Faba vulgaris	26642.78	-.5
Melilotus officinalis	150000	2.1	Glechoma hederacea	29086.17	-.5
Melissa officinalis	150000	2.1	Hyoscyamus niger	29484.98	-.5
Ocimum canum	150000	2.1	Hyssopus officinalis	27858.46	-.5
Ocimum canum R	150000	2.1	Leonurus cardiaca	29649.29	-.5
Osmundo regalis	150000	2.1	Ononis spinosa	26426.59	-.5
Psoralea bituminosa	150000	2.1	Osmundo regalis 2	28954.08	-.5
Scutellaria lateriflor	150000	2.1	Paeonia officinalis	29100.22	-.5
Scrophularia nodosa	78691.06	.6	Stachys officinalis	28067.73	-.5
Prunus padus	59863.47	.2	Stachys officinalis R	29853.32	-.5
Stachys sylvatica	59248.84	.2	Urtica dioica	28428.42	-.5
Polygonum aviculare	56810.42	.1	Aquilegia vulgaris	24488.34	-.6
Tussilago farfara	56882.81	.1	Belladonna	25268.54	-.6
Salicaria purpurea	45003.1	-.1	Centaurium erythraea	21186.01	-.6
Sanguisorba officinali	48677.9	-.1	Chenopodium anthel	21621.05	-.6
Vinca minor	45427.12	-.1	Cimicifuga racemosa	23185.79	-.6
Conium maculatum	40377.49	-.2	Cupressus lawsoniana	24141.96	-.6
Lappa arctium	41057.72	-.2	Escholtzia californica	25483.15	-.6
Mercurialis perennis	40372.82	-.2	Galium aparine	21163.09	-.6
Ruta graveolens	41143.67	-.2	Lespedeza sieboldii	24808.94	-.6
Agrimonia eupatoria	36159.8	-.3	Urtica urens	23126.44	-.6
Cardiospermum halic	38657.76	-.3	Urtica urens 2	21589.59	-.6
Echinacea angustifolia	38835.54	-.3	Viola tricolor	22176.67	-.6
Fagopyrum esculent	37809.67	-.3	Zizia aurea	21295.63	-.6
Hedera helix	38933.63	-.3	Aconitum napellus	20124.46	-.7
Hydrophyllum virginicu	39268.75	-.3	Anthemis nobilis	19631.53	-.7
Oenanthe crocata	35550.85	-.3	Chamomilla	20973.75	-.7
Rhus toxicodendron	37166.07	-.3	Gnaphalium leontop	16972.49	-.7
Salvia sclarea	39324.53	-.3	Mandragora officin	16838.09	-.7
Syringa vulgaris	38464.79	-.3	Marrubium vulgare	19510.13	-.7
Thuja occidentalis	35989.89	-.3	Petroselinum crispum	17536.21	-.7
Bryonia dioica	31536.29	-.4	Plantago major	19147.11	-.7
Lapsana communis	33802.8	-.4	Adonis vernalis	14556.98	-.8
Lycopus europaeus	33333.4	-.4	Chelidonium majus	12575.63	-.8
Malva sylvestris	33858.33	-.4	Salix purpurea	15658.32	-.8
Mentha arvensis	34978.45	-.4	Verbena officinalis	16164.3	-.8
Milium solis	31279.83	-.4	Fraxinus excelsior	9778.88	-.9
Primula veris	31921.31	-.4	Taraxacum officinale	9722.49	-.9
Rumex acetosa	33408.55	-.4	Bellis perennis	4965.97	-1

Manganum

Name	Value	Deviation	Name	Value	Deviation
Aconitum napellus	343.52	-.2	Marrubium vulgare	193.06	-.2
Adonis vernalis	191.01	-.2	Melilotus officinalis	767.74	-.1
Agrimonia eupatoria	266.65	-.2	Melissa officinalis	357.79	-.2
Alchemilla vulgaris	893.53	-.1	Mentha arvensis	439.82	-.2
Anethum graveolens	356.94	-.2	Mercurialis perennis	996.08	-.1
Angelica archangelica	304.97	-.2	Milium solis	465.52	-.2
Anthemis nobilis	368.69	-.2	Ocimum canum	314.94	-.2
Aquilegia vulgaris	340.79	-.2	Ocimum canum R	354.4	-.2
Belladonna	285.23	-.2	Oenanthe crocata	317.94	-.2
Bellis perennis	354.21	-.2	Ononis spinosa	272.24	-.2
Bryonia dioica	377.88	-.2	Osmundo regalis	2480	0
Cardamine pratensis	700.77	-.1	Osmundo regalis 2	877.65	-.1
Cardamine pratensis R	696.11	-.1	Paeonia officinalis	346.45	-.2
Cardiospermum halicaca	696.33	-.1	Petroselinum crispum	337.43	-.2
Castanea vesca	150000	9.5	Plantago major	154.82	-.2
Centaurium erythraea	704.16	-.1	Polygonum aviculare	217.98	-.2
Chamomilla	664.07	-.2	Primula veris	418.96	-.2
Chelidonium majus	988.28	-.1	Prunus padus	2040	-.1
Chenopodium anthelmint	133.77	-.2	Psoralea bituminosa	1860	-.1
Cicuta virosa	22600	1.3	Rhus toxicodendron	7500	.3
Cimicifuga racemosa	850.64	-.1	Rumex acetosa	3620	0
Cinnamodendron cortiso	2460	0	Ruta graveolens	4740	.1
Clematis erecta	547.61	-.2	Salicaria purpurea	8460	.4
Collinsonia canadensis	493.39	-.2	Salix purpurea	6030	.2
Conium maculatum	526.85	-.2	Salvia sclarea	1100	-.1
Cupressus lawsoniana	15400	.8	Sanguisorba officinali	1410	-.1
Echinacea angustifolia	301.49	-.2	Scrophularia nodosa	1170	-.1
Echinacea purpurea	565.24	-.2	Scutellaria lateriflor	3000	0
Escholtzia californica	487.42	-.2	Solidago virgaurea	861.33	-.1
Faba vulgaris	424.03	-.2	Spilanthes oleracea	203.54	-.2
Fagopyrum esculentum	273.1	-.2	Stachys officinalis	613.43	-.2
Fraxinus excelsior	350.61	-.2	Stachys officinalis R	647.68	-.2
Galium aparine	1020	-.1	Stachys sylvatica	722.55	-.1
Glechoma hederacea	1180	-.1	Syringa vulgaris	6470	.2
Gnaphalium leontopodiu	945.1	-.1	Taraxacum officinale	389.59	-.2
Hedera helix	4790	.1	Teucrium chamaedrys	261.29	-.2
Hydrophyllum virginicu	464.69	-.2	Thuja occidentalis	2920	0
Hyoscyamus niger	201.53	-.2	Tormentilla erecta	1130	-.1
Hyssopus officinalis	284.63	-.2	Tussilago farfara	324.73	-.2
Lappa arctium	1060	-.1	Tussilago petasites	335.83	-.2
Lapsana communis	626.33	-.2	Ulmus campestris	231.23	-.2
Laurocerasus	3140	0	Urtica dioica	488.27	-.2
Leonurus cardiaca	355.76	-.2	Urtica urens	348.08	-.2
Lespedeza sieboldii	529.03	-.2	Urtica urens 2	68.47	-.2
Linum usitatissimum	256.43	-.2	Verbena officinalis	1150	-.1
Lycopus europaeus	807.07	-.1	Vinca minor	655.33	-.2
Malva sylvestris	384.64	-.2	Viola tricolor	295.58	-.2
Mandragora officinalis	175.72	-.2	Zizia aurea	660.71	-.2

Manganum

Name	Value	Deviation	Name	Value	Deviation
Castanea vesca	150000	9.5	Bryonia dioica	377.88	-.2
Cicuta virosa	22600	1.3	Chamomilla	664.07	-.2
Cupressus lawsoniana	15400	.8	Chenopodium anthelmint	133.77	-.2
Salicaria purpurea	8460	.4	Clematis erecta	547.61	-.2
Rhus toxicodendron	7500	.3	Collinsonia canadensis	493.39	-.2
Salix purpurea	6030	.2	Conium maculatum	526.85	-.2
Syringa vulgaris	6470	.2	Echinacea angustifolia	301.49	-.2
Hedera helix	4790	.1	Echinacea purpurea	565.24	-.2
Ruta graveolens	4740	.1	Escholtzia californica	487.42	-.2
Cinnamodendron cortiso	2460	0	Faba vulgaris	424.03	-.2
Laurocerasus	3140	0	Fagopyrum esculentum	273.1	-.2
Osmundo regalis	2480	0	Fraxinus excelsior	350.61	-.2
Rumex acetosa	3620	0	Hydrophyllum virginicu	464.69	-.2
Scutellaria lateriflor	3000	0	Hyoscyamus niger	201.53	-.2
Thuja occidentalis	2920	0	Hyssopus officinalis	284.63	-.2
Alchemilla vulgaris	893.53	-.1	Lapsana communis	626.33	-.2
Cardamine pratensis	700.77	-.1	Leonurus cardiaca	355.76	-.2
Cardamine pratensis R	696.11	-.1	Lespedeza sieboldii	529.03	-.2
Cardiospermum halicaca	696.33	-.1	Linum usitatissimum	256.43	-.2
Centaurium erythraea	704.16	-.1	Malva sylvestris	384.64	-.2
Chelidonium majus	988.28	-.1	Mandragora officinalis	175.72	-.2
Cimicifuga racemosa	850.64	-.1	Marrubium vulgare	193.06	-.2
Galium aparine	1020	-.1	Melissa officinalis	357.79	-.2
Glechoma hederacea	1180	-.1	Mentha arvensis	439.82	-.2
Gnaphalium leontopodiu	945.1	-.1	Milium solis	465.52	-.2
Lappa arctium	1060	-.1	Ocimum canum	314.94	-.2
Lycopus europaeus	807.07	-.1	Ocimum canum R	354.4	-.2
Melilotus officinalis	767.74	-.1	Oenanthe crocata	317.94	-.2
Mercurialis perennis	996.08	-.1	Ononis spinosa	272.24	-.2
Osmundo regalis 2	877.65	-.1	Paeonia officinalis	346.45	-.2
Prunus padus	2040	-.1	Petroselinum crispum	337.43	-.2
Psoralea bituminosa	1860	-.1	Plantago major	154.82	-.2
Salvia sclarea	1100	-.1	Polygonum aviculare	217.98	-.2
Sanguisorba officinali	1410	-.1	Primula veris	418.96	-.2
Scrophularia nodosa	1170	-.1	Spilanthes oleracea	203.54	-.2
Solidago virgaurea	861.33	-.1	Stachys officinalis	613.43	-.2
Stachys sylvatica	722.55	-.1	Stachys officinalis R	647.68	-.2
Tormentilla erecta	1130	-.1	Taraxacum officinale	389.59	-.2
Verbena officinalis	1150	-.1	Teucrium chamaedrys	261.29	-.2
Aconitum napellus	343.52	-.2	Tussilago farfara	324.73	-.2
Adonis vernalis	191.01	-.2	Tussilago petasites	335.83	-.2
Agrimonia eupatoria	266.65	-.2	Ulmus campestris	231.23	-.2
Anethum graveolens	356.94	-.2	Urtica dioica	488.27	-.2
Angelica archangelica	304.97	-.2	Urtica urens	348.08	-.2
Anthemis nobilis	368.69	-.2	Urtica urens 2	68.47	-.2
Aquilegia vulgaris	340.79	-.2	Vinca minor	655.33	-.2
Belladonna	285.23	-.2	Viola tricolor	295.58	-.2
Bellis perennis	354.21	-.2	Zizia aurea	660.71	-.2

Molybdenum

Name	Value	Deviation	Name	Value	Deviation
Aconitum napellus	8.81	-.2	Marrubium vulgare	9.11	-.2
Adonis vernalis	1.35	-.4	Melilotus officinalis	58.11	.7
Agrimonia eupatoria	13.7	-.1	Melissa officinalis	72	.9
Alchemilla vulgaris	28.43	.1	Mentha arvensis	30.66	.2
Anethum graveolens	9.11	-.2	Mercurialis perennis	10.88	-.2
Angelica archangelica	5.91	-.3	Milium solis	6.95	-.3
Anthemis nobilis	42.9	.4	Ocimum canum	8.89	-.2
Aquilegia vulgaris	42.56	.4	Ocimum canum R	9.55	-.2
Belladonna	11.49	-.2	Oenanthe crocata	3.64	-.3
Bellis perennis	1.3	-.4	Ononis spinosa	48.76	.5
Bryonia dioica	14.5	-.1	Osmundo regalis	21.65	0
Cardamine pratensis	31.19	.2	Osmundo regalis 2	13.11	-.2
Cardamine pratensis R	29.81	.1	Paeonia officinalis	5.65	-.3
Cardiospermum halicaca	14.46	-.1	Petroselinum crispum	13.05	-.2
Castanea vesca	10.74	-.2	Plantago major	13.81	-.1
Centaurium erythraea	12.7	-.2	Polygonum aviculare	25.39	.1
Chamomilla	11.77	-.2	Primula veris	15.24	-.1
Chelidonium majus	4.54	-.3	Prunus padus	2.65	-.3
Chenopodium anthelmint	10.05	-.2	Psoralea bituminosa	94.55	1.3
Cicuta virosa	1.63	-.4	Rhus toxicodendron	3.99	-.3
Cimicifuga racemosa	6.33	-.3	Rumex acetosa	33.27	.2
Cinnamodendron cortiso	15.13	-.1	Ruta graveolens	4.91	-.3
Clematis erecta	6.14	-.3	Salicaria purpurea	7.34	-.3
Collinsonia canadensis	12.19	-.2	Salix purpurea	.93	-.4
Conium maculatum	12.14	-.2	Salvia sclarea	4.35	-.3
Cupressus lawsoniana	18.03	-.1	Sanguisorba officinali	8.01	-.2
Echinacea angustifolia	3.39	-.3	Scrophularia nodosa	11.46	-.2
Echinacea purpurea	6.86	-.3	Scutellaria lateriflor	6.89	-.3
Escholtzia californica	10.34	-.2	Solidago virgaurea	7.62	-.2
Faba vulgaris	28.01	.1	Spilanthes oleracea	2.46	-.3
Fagopyrum esculentum	13.66	-.1	Stachys officinalis	2.49	-.3
Fraxinus excelsior	3.61	-.3	Stachys officinalis R	2.83	-.3
Galium aparine	31.34	.2	Stachys sylvatica	21.92	0
Glechoma hederacea	10.26	-.2	Syringa vulgaris	1.73	-.4
Gnaphalium leontopodiu	24.8	.1	Taraxacum officinale	4.36	-.3
Hedera helix	4.9	-.3	Teucrium chamaedrys	121.82	1.8
Hydrophyllum virginicu	6.53	-.3	Thuja occidentalis	3.73	-.3
Hyoscyamus niger	4.47	-.3	Tormentilla erecta	9.59	-.2
Hyssopus officinalis	17.11	-.1	Tussilago farfara	11.82	-.2
Lappa arctium	14.76	-.1	Tussilago petasites	1.83	-.4
Lapsana communis	17.3	-.1	Ulmus campestris	20.28	0
Laurocerasus	9.92	-.2	Urtica dioica	36.87	.3
Leonurus cardiaca	20.36	0	Urtica urens	12.2	-.2
Lespedeza sieboldii	530.57	9.1	Urtica urens 2	14.1	-.1
Linum usitatissimum	12.94	-.2	Verbena officinalis	10.46	-.2
Lycopus europaeus	14.31	-.1	Vinca minor	9.91	-.2
Malva sylvestris	16.41	-.1	Viola tricolor	37.89	.3
Mandragora officinalis	5.53	-.3	Zizia aurea	3.19	-.3

Molybdenum

Name	Value	Deviation	Name	Value	Deviation
Lespedeza sieboldii	530.57	9.1	Laurocerasus	9.92	-.2
Teucrium chamaedrys	121.82	1.8	Linum usitatissimum	12.94	-.2
Psoralea bituminosa	94.55	1.3	Marrubium vulgare	9.11	-.2
Melissa officinalis	72	.9	Mercurialis perennis	10.88	-.2
Melilotus officinalis	58.11	.7	Ocimum canum	8.89	-.2
Ononis spinosa	48.76	.5	Ocimum canum R	9.55	-.2
Anthemis nobilis	42.9	.4	Osmundo regalis 2	13.11	-.2
Aquilegia vulgaris	42.56	.4	Petroselinum crispum	13.05	-.2
Urtica dioica	36.87	.3	Sanguisorba officinali	8.01	-.2
Viola tricolor	37.89	.3	Scrophularia nodosa	11.46	-.2
Cardamine pratensis	31.19	.2	Solidago virgaurea	7.62	-.2
Galium aparine	31.34	.2	Tormentilla erecta	9.59	-.2
Mentha arvensis	30.66	.2	Tussilago farfara	11.82	-.2
Rumex acetosa	33.27	.2	Urtica urens	12.2	-.2
Alchemilla vulgaris	28.43	.1	Verbena officinalis	10.46	-.2
Cardamine pratensis R	29.81	.1	Vinca minor	9.91	-.2
Faba vulgaris	28.01	.1	Angelica archangelica	5.91	-.3
Gnaphalium leontopodiu	24.8	.1	Chelidonium majus	4.54	-.3
Polygonum aviculare	25.39	.1	Cimicifuga racemosa	6.33	-.3
Leonurus cardiaca	20.36	0	Clematis erecta	6.14	-.3
Osmundo regalis	21.65	0	Echinacea angustifolia	3.39	-.3
Stachys sylvatica	21.92	0	Echinacea purpurea	6.86	-.3
Ulmus campestris	20.28	0	Fraxinus excelsior	3.61	-.3
Agrimonia eupatoria	13.7	-.1	Hedera helix	4.9	-.3
Bryonia dioica	14.5	-.1	Hydrophyllum virginicu	6.53	-.3
Cardiospermum halicaca	14.46	-.1	Hyoscyamus niger	4.47	-.3
Cinnamodendron cortiso	15.13	-.1	Mandragora officinalis	5.53	-.3
Cupressus lawsoniana	18.03	-.1	Milium solis	6.95	-.3
Fagopyrum esculentum	13.66	-.1	Oenanthe crocata	3.64	-.3
Hyssopus officinalis	17.11	-.1	Paeonia officinalis	5.65	-.3
Lappa arctium	14.76	-.1	Prunus padus	2.65	-.3
Lapsana communis	17.3	-.1	Rhus toxicodendron	3.99	-.3
Lycopus europaeus	14.31	-.1	Ruta graveolens	4.91	-.3
Malva sylvestris	16.41	-.1	Salicaria purpurea	7.34	-.3
Plantago major	13.81	-.1	Salvia sclarea	4.35	-.3
Primula veris	15.24	-.1	Scutellaria lateriflor	6.89	-.3
Urtica urens 2	14.1	-.1	Spilanthes oleracea	2.46	-.3
Aconitum napellus	8.81	-.2	Stachys officinalis	2.49	-.3
Anethum graveolens	9.11	-.2	Stachys officinalis R	2.83	-.3
Belladonna	11.49	-.2	Taraxacum officinale	4.36	-.3
Castanea vesca	10.74	-.2	Thuja occidentalis	3.73	-.3
Centaurium erythraea	12.7	-.2	Zizia aurea	3.19	-.3
Chamomilla	11.77	-.2	Adonis vernalis	1.35	-.4
Chenopodium anthelmint	10.05	-.2	Bellis perennis	1.3	-.4
Collinsonia canadensis	12.19	-.2	Cicuta virosa	1.63	-.4
Conium maculatum	12.14	-.2	Salix purpurea	.93	-.4
Escholtzia californica	10.34	-.2	Syringa vulgaris	1.73	-.4
Glechoma hederacea	10.26	-.2	Tussilago petasites	1.83	-.4

Natrium

Name	Value	Deviation	Name	Value	Deviation
Aconitum napellus	3401.9	-.4	Marrubium vulgare	2181.71	-.5
Adonis vernalis	3910.37	-.4	Melilotus officinalis	6365.55	-.2
Agrimonia eupatoria	1809.94	-.5	Melissa officinalis	1926.66	-.5
Alchemilla vulgaris	4655.21	-.3	Mentha arvensis	5974.74	-.2
Anethum graveolens	60000	3.6	Mercurialis perennis	8080.44	-.1
Angelica archangelica	6374.07	-.2	Milium solis	1021.9	-.6
Anthemis nobilis	60000	3.6	Ocimum canum	1828.62	-.5
Aquilegia vulgaris	1519.45	-.6	Ocimum canum R	2082.08	-.5
Belladonna	60000	3.6	Oenanthe crocata	19865.9	.7
Bellis perennis	10313.14	.1	Ononis spinosa	3534.07	-.4
Bryonia dioica	9920.45	0	Osmundo regalis	22729.95	.9
Cardamine pratensis	5902.97	-.2	Osmundo regalis 2	11166.67	.1
Cardamine pratensis R	5698.69	-.3	Paeonia officinalis	60000	3.6
Cardiospermum halic	1204.88	-.6	Petroselinum crispum	14041.81	.3
Castanea vesca	15185.98	.4	Plantago major	1553.84	-.6
Centaurium erythraea	9207.62	0	Polygonum aviculare	1626.58	-.5
Chamomilla	10148.74	.1	Primula veris	2498.95	-.5
Chelidonium majus	2079.82	-.5	Prunus padus	5431.5	-.3
Chenopodium anthelm	2047.22	-.5	Psoralea bituminosa	7969.2	-.1
Cicuta virosa	60000	3.6	Rhus toxicodendron	3151.79	-.4
Cimicifuga racemosa	3694.19	-.4	Rumex acetosa	12337.57	.2
Cinnamodendron cor	11119.78	.1	Ruta graveolens	10369.96	.1
Clematis erecta	4750.16	-.3	Salicaria purpurea	29136.5	1.4
Collinsonia canadensis	3828.36	-.4	Salix purpurea	5277.95	-.3
Conium maculatum	3894.69	-.4	Salvia sclarea	12278.88	.2
Cupressus lawsoniana	6923.31	-.2	Sanguisorba officinali	5107.05	-.3
Echinacea angustifolia	7776.43	-.1	Scrophularia nodosa	2776.81	-.5
Echinacea purpurea	1987.58	-.5	Scutellaria lateriflor	8022.94	-.1
Escholtzia californica	2157.27	-.5	Solidago virgaurea	2057.43	-.5
Faba vulgaris	16240.82	.5	Spilanthes oleracea	716.59	-.6
Fagopyrum esculentum	2084.61	-.5	Stachys officinalis	803.34	-.6
Fraxinus excelsior	2779.29	-.5	Stachys officinalis R	886.6	-.6
Galium aparine	1693.98	-.5	Stachys sylvatica	5270.58	-.3
Glechoma hederacea	5046.87	-.3	Syringa vulgaris	7732.65	-.1
Gnaphalium leontop	6362.67	-.2	Taraxacum officin	25299.36	1.1
Hedera helix	11050.82	.1	Teucrium chamaedrys	6596.94	-.2
Hydrophyllum virgin	363.28	-.6	Thuja occidentalis	4245.57	-.4
Hyoscyamus niger	2082.52	-.5	Tormentilla erecta	8485.86	-.1
Hyssopus officinalis	1535.51	-.6	Tussilago farfara	1641.55	-.5
Lappa arctium	3602.84	-.4	Tussilago petasites	8993.25	0
Lapsana communis	4703.05	-.3	Ulmus campestris	7364.25	-.1
Laurocerasus	1110.86	-.6	Urtica dioica	4639.03	-.3
Leonurus cardiaca	4839.01	-.3	Urtica urens	940.04	-.6
Lespedeza sieboldii	1740.6	-.5	Urtica urens 2	2666.76	-.5
Linum usitatissimum	60000	3.6	Verbena officinalis	2198.2	-.5
Lycopus europaeus	2527.36	-.5	Vinca minor	9357.24	0
Malva sylvestris	6061.11	-.2	Viola tricolor	1032.53	-.6
Mandragora officinalis	2675.51	-.5	Zizia aurea	12695.19	.2

Natrium

Name	Value	Deviation	Name	Value	Deviation
Anethum graveolens	60000	3.6	Salix purpurea	5277.95	-.3
Anthemis nobilis	60000	3.6	Sanguisorba officinali	5107.05	-.3
Belladonna	60000	3.6	Stachys sylvatica	5270.58	-.3
Cicuta virosa	60000	3.6	Urtica dioica	4639.03	-.3
Linum usitatissimum	60000	3.6	Aconitum napellus	3401.9	-.4
Paeonia officinalis	60000	3.6	Adonis vernalis	3910.37	-.4
Salicaria purpurea	29136.5	1.4	Cimicifuga racemosa	3694.19	-.4
Taraxacum officinale	25299.36	1.1	Collinsonia canadensis	3828.36	-.4
Osmundo regalis	22729.95	.9	Conium maculatum	3894.69	-.4
Oenanthe crocata	19865.9	.7	Lappa arctium	3602.84	-.4
Faba vulgaris	16240.82	.5	Ononis spinosa	3534.07	-.4
Castanea vesca	15185.98	.4	Rhus toxicodendron	3151.79	-.4
Petroselinum crispum	14041.81	.3	Thuja occidentalis	4245.57	-.4
Rumex acetosa	12337.57	.2	Agrimonia eupatoria	1809.94	-.5
Salvia sclarea	12278.88	.2	Chelidonium majus	2079.82	-.5
Zizia aurea	12695.19	.2	Chenopodium anthel	2047.22	-.5
Bellis perennis	10313.14	.1	Echinacea purpurea	1987.58	-.5
Chamomilla	10148.74	.1	Escholtzia californica	2157.27	-.5
Cinnamodendron cort	11119.78	.1	Fagopyrum esculentum	2084.61	-.5
Hedera helix	11050.82	.1	Fraxinus excelsior	2779.29	-.5
Osmundo regalis 2	11166.67	.1	Galium aparine	1693.98	-.5
Ruta graveolens	10369.96	.1	Hyoscyamus niger	2082.52	-.5
Bryonia dioica	9920.45	0	Lespedeza sieboldii	1740.6	-.5
Centaurium erythraea	9207.62	0	Lycopus europaeus	2527.36	-.5
Tussilago petasites	8993.25	0	Mandragora officinalis	2675.51	-.5
Vinca minor	9357.24	0	Marrubium vulgare	2181.71	-.5
Echinacea angustifolia	7776.43	-.1	Melissa officinalis	1926.66	-.5
Mercurialis perennis	8080.44	-.1	Ocimum canum	1828.62	-.5
Psoralea bituminosa	7969.2	-.1	Ocimum canum R	2082.08	-.5
Scutellaria lateriflor	8022.94	-.1	Polygonum aviculare	1626.58	-.5
Syringa vulgaris	7732.65	-.1	Primula veris	2498.95	-.5
Tormentilla erecta	8485.86	-.1	Scrophularia nodosa	2776.81	-.5
Ulmus campestris	7364.25	-.1	Solidago virgaurea	2057.43	-.5
Angelica archangelica	6374.07	-.2	Tussilago farfara	1641.55	-.5
Cardamine pratensis	5902.97	-.2	Urtica urens 2	2666.76	-.5
Cupressus lawsoniana	6923.31	-.2	Verbena officinalis	2198.2	-.5
Gnaphalium leontop	6362.67	-.2	Aquilegia vulgaris	1519.45	-.6
Malva sylvestris	6061.11	-.2	Cardiospermum halic	1204.88	-.6
Melilotus officinalis	6365.55	-.2	Hydrophyllum virginicu	363.28	-.6
Mentha arvensis	5974.74	-.2	Hyssopus officinalis	1535.51	-.6
Teucrium chamaedrys	6596.94	-.2	Laurocerasus	1110.86	-.6
Alchemilla vulgaris	4655.21	-.3	Milium solis	1021.9	-.6
Cardamine pratensis R	5698.69	-.3	Plantago major	1553.84	-.6
Clematis erecta	4750.16	-.3	Spilanthes oleracea	716.59	-.6
Glechoma hederacea	5046.87	-.3	Stachys officinalis	803.34	-.6
Lapsana communis	4703.05	-.3	Stachys officinalis R	886.6	-.6
Leonurus cardiaca	4839.01	-.3	Urtica urens	940.04	-.6
Prunus padus	5431.5	-.3	Viola tricolor	1032.53	-.6

Neodymium

Name	Value	Deviation	Name	Value	Deviation
Aconitum napellus	1.69	.3	Marrubium vulgare	.79	-.2
Adonis vernalis	1.52	.2	Melilotus officinalis	.38	-.5
Agrimonia eupatoria	.22	-.6	Melissa officinalis	1.12	0
Alchemilla vulgaris	1.63	.3	Mentha arvensis	.68	-.3
Anethum graveolens	.25	-.6	Mercurialis perennis	1.74	.4
Angelica archangelica	1.51	.2	Milium solis	.54	-.4
Anthemis nobilis	1.09	-.1	Ocimum canum	1.12	0
Aquilegia vulgaris	.21	-.6	Ocimum canum R	1.09	-.1
Belladonna	.9	-.2	Oenanthe crocata	1.54	.2
Bellis perennis	5.54	2.8	Ononis spinosa	1.14	0
Bryonia dioica	1.86	.4	Osmundo regalis	11.54	6.6
Cardamine pratensis	1.03	-.1	Osmundo regalis 2	.06	-.7
Cardamine pratensis R	.87	-.2	Paeonia officinalis	1.25	0
Cardiospermum halicaca	1	-.1	Petroselinum crispum	2.13	.6
Castanea vesca	.63	-.3	Plantago major	1.4	.1
Centaurium erythraea	2.37	.8	Polygonum aviculare	.84	-.2
Chamomilla	.94	-.1	Primula veris	1.18	0
Chelidonium majus	6.2	3.2	Prunus padus	.57	-.4
Chenopodium anthelmint	.21	-.6	Psoralea bituminosa	.87	-.2
Cicuta virosa	3.91	1.7	Rhus toxicodendron	.39	-.5
Cimicifuga racemosa	1.56	.2	Rumex acetosa	5.92	3
Cinnamodendron cortiso	.56	-.4	Ruta graveolens	.35	-.5
Clematis erecta	.41	-.5	Salicaria purpurea	.42	-.5
Collinsonia canadensis	1.94	.5	Salix purpurea	.55	-.4
Conium maculatum	.2	-.6	Salvia sclarea	.34	-.5
Cupressus lawsoniana	1.29	.1	Sanguisorba officinali	.17	-.6
Echinacea angustifolia	.55	-.4	Scrophularia nodosa	.75	-.3
Echinacea purpurea	.18	-.6	Scutellaria lateriflor	.91	-.2
Escholtzia californica	.47	-.4	Solidago virgaurea	.24	-.6
Faba vulgaris	.38	-.5	Spilanthes oleracea	.43	-.5
Fagopyrum esculentum	.2	-.6	Stachys officinalis	.19	-.6
Fraxinus excelsior	.78	-.2	Stachys officinalis R	.13	-.7
Galium aparine	.84	-.2	Stachys sylvatica	.74	-.3
Glechoma hederacea	.71	-.3	Syringa vulgaris	.64	-.3
Gnaphalium leontopodiu	2.71	1	Taraxacum officinale	4.23	1.9
Hedera helix	.57	-.4	Teucrium chamaedrys	.58	-.4
Hydrophyllum virginicu	.32	-.5	Thuja occidentalis	.44	-.5
Hyoscyamus niger	.82	-.2	Tormentilla erecta	2.51	.9
Hyssopus officinalis	.4	-.5	Tussilago farfara	.28	-.6
Lappa arctium	1.54	.2	Tussilago petasites	.57	-.4
Lapsana communis	1.9	.5	Ulmus campestris	.77	-.3
Laurocerasus	.29	-.6	Urtica dioica	2.53	.9
Leonurus cardiaca	.28	-.6	Urtica urens	.4	-.5
Lespedeza sieboldii	.27	-.6	Urtica urens 2	.45	-.5
Linum usitatissimum	.15	-.6	Verbena officinalis	.24	-.6
Lycopus europaeus	.28	-.6	Vinca minor	1.21	0
Malva sylvestris	.48	-.4	Viola tricolor	.69	-.3
Mandragora officinalis	.58	-.4	Zizia aurea	1.19	0

Neodymium

Name	Value	Deviation	Name	Value	Deviation
Osmundo regalis	11.54	6.6	Scrophularia nodosa	.75	-.3
Chelidonium majus	6.2	3.2	Stachys sylvatica	.74	-.3
Rumex acetosa	5.92	3	Syringa vulgaris	.64	-.3
Bellis perennis	5.54	2.8	Ulmus campestris	.77	-.3
Taraxacum officinale	4.23	1.9	Viola tricolor	.69	-.3
Cicuta virosa	3.91	1.7	Cinnamodendron cortiso	.56	-.4
Gnaphalium leontopodiu	2.71	1	Echinacea angustifolia	.55	-.4
Tormentilla erecta	2.51	.9	Escholtzia californica	.47	-.4
Urtica dioica	2.53	.9	Hedera helix	.57	-.4
Centaurium erythraea	2.37	.8	Malva sylvestris	.48	-.4
Petroselinum crispum	2.13	.6	Mandragora officinalis	.58	-.4
Collinsonia canadensis	1.94	.5	Milium solis	.54	-.4
Lapsana communis	1.9	.5	Prunus padus	.57	-.4
Bryonia dioica	1.86	.4	Salix purpurea	.55	-.4
Mercurialis perennis	1.74	.4	Teucrium chamaedrys	.58	-.4
Aconitum napellus	1.69	.3	Tussilago petasites	.57	-.4
Alchemilla vulgaris	1.63	.3	Clematis erecta	.41	-.5
Adonis vernalis	1.52	.2	Faba vulgaris	.38	-.5
Angelica archangelica	1.51	.2	Hydrophyllum virginicu	.32	-.5
Cimicifuga racemosa	1.56	.2	Hyssopus officinalis	.4	-.5
Lappa arctium	1.54	.2	Melilotus officinalis	.38	-.5
Oenanthe crocata	1.54	.2	Rhus toxicodendron	.39	-.5
Cupressus lawsoniana	1.29	.1	Ruta graveolens	.35	-.5
Plantago major	1.4	.1	Salicaria purpurea	.42	-.5
Melissa officinalis	1.12	0	Salvia sclarea	.34	-.5
Ocimum canum	1.12	0	Spilanthes oleracea	.43	-.5
Ononis spinosa	1.14	0	Thuja occidentalis	.44	-.5
Paeonia officinalis	1.25	0	Urtica urens	.4	-.5
Primula veris	1.18	0	Urtica urens 2	.45	-.5
Vinca minor	1.21	0	Agrimonia eupatoria	.22	-.6
Zizia aurea	1.19	0	Anethum graveolens	.25	-.6
Anthemis nobilis	1.09	-.1	Aquilegia vulgaris	.21	-.6
Cardamine pratensis	1.03	-.1	Chenopodium anthelmint	.21	-.6
Cardiospermum halicaca	1	-.1	Conium maculatum	.2	-.6
Chamomilla	.94	-.1	Echinacea purpurea	.18	-.6
Ocimum canum R	1.09	-.1	Fagopyrum esculentum	.2	-.6
Belladonna	.9	-.2	Laurocerasus	.29	-.6
Cardamine pratensis R	.87	-.2	Leonurus cardiaca	.28	-.6
Fraxinus excelsior	.78	-.2	Lespedeza sieboldii	.27	-.6
Galium aparine	.84	-.2	Linum usitatissimum	.15	-.6
Hyoscyamus niger	.82	-.2	Lycopus europaeus	.28	-.6
Marrubium vulgare	.79	-.2	Sanguisorba officinali	.17	-.6
Polygonum aviculare	.84	-.2	Solidago virgaurea	.24	-.6
Psoralea bituminosa	.87	-.2	Stachys officinalis	.19	-.6
Scutellaria lateriflor	.91	-.2	Tussilago farfara	.28	-.6
Castanea vesca	.63	-.3	Verbena officinalis	.24	-.6
Glechoma hederacea	.71	-.3	Osmundo regalis 2	.06	-.7
Mentha arvensis	.68	-.3	Stachys officinalis R	.13	-.7

Niccolum

Name	Value	Deviation	Name	Value	Deviation
Aconitum napellus	6.52	-.2	Marrubium vulgare	0	-.8
Adonis vernalis	11.98	.3	Melilotus officinalis	22.83	1.3
Agrimonia eupatoria	0	-.8	Melissa officinalis	6.1	-.3
Alchemilla vulgaris	7.91	-.1	Mentha arvensis	0	-.8
Anethum graveolens	5.18	-.4	Mercurialis perennis	17.98	.8
Angelica archangelica	7.09	-.2	Milium solis	12.63	.3
Anthemis nobilis	0	-.8	Ocimum canum	0	-.8
Aquilegia vulgaris	7.25	-.2	Ocimum canum R	0	-.8
Belladonna	5.55	-.3	Oenanthe crocata	8.39	-.1
Bellis perennis	9.77	.1	Ononis spinosa	6	-.3
Bryonia dioica	11.61	.2	Osmundo regalis	33.21	2.2
Cardamine pratensis	6.23	-.3	Osmundo regalis 2	0	-.8
Cardamine pratensis R	6.01	-.3	Paeonia officinalis	5.98	-.3
Cardiospermum halicaca	10.45	.1	Petroselinum crispum	10.32	.1
Castanea vesca	25.04	1.5	Plantago major	0	-.8
Centaurium erythraea	12.74	.3	Polygonum aviculare	0	-.8
Chamomilla	6.83	-.2	Primula veris	5.78	-.3
Chelidonium majus	14.29	.5	Prunus padus	0	-.8
Chenopodium anthelmint	0	-.8	Psoralea bituminosa	16.75	.7
Cicuta virosa	18.17	.8	Rhus toxicodendron	8.37	-.1
Cimicifuga racemosa	23.51	1.3	Rumex acetosa	48.7	3.7
Cinnamodendron cortiso	8.33	-.1	Ruta graveolens	26.52	1.6
Clematis erecta	10.17	.1	Salicaria purpurea	8.83	0
Collinsonia canadensis	12.79	.3	Salix purpurea	6.59	-.2
Conium maculatum	0	-.8	Salvia sclarea	5.49	-.3
Cupressus lawsoniana	52.95	4.1	Sanguisorba officinali	12.91	.3
Echinacea angustifolia	5.98	-.3	Scrophularia nodosa	7.29	-.2
Echinacea purpurea	6.39	-.3	Scutellaria lateriflor	10.1	.1
Escholtzia californica	0	-.8	Solidago virgaurea	0	-.8
Faba vulgaris	0	-.8	Spilanthes oleracea	0	-.8
Fagopyrum esculentum	0	-.8	Stachys officinalis	0	-.8
Fraxinus excelsior	0	-.8	Stachys officinalis R	0	-.8
Galium aparine	0	-.8	Stachys sylvatica	6.84	-.2
Glechoma hederacea	7.82	-.1	Syringa vulgaris	46.24	3.4
Gnaphalium leontopodiu	9.09	0	Taraxacum officinale	9.36	0
Hedera helix	8.3	-.1	Teucrium chamaedrys	5.25	-.4
Hydrophyllum virginicu	10.07	.1	Thuja occidentalis	28.81	1.8
Hyoscyamus niger	0	-.8	Tormentilla erecta	21.65	1.2
Hyssopus officinalis	0	-.8	Tussilago farfara	0	-.8
Lappa arctium	15.38	.6	Tussilago petasites	7.52	-.2
Lapsana communis	6.46	-.3	Ulmus campestris	6.31	-.3
Laurocerasus	13.8	.4	Urtica dioica	7.72	-.1
Leonurus cardiaca	5.67	-.3	Urtica urens	0	-.8
Lespedeza sieboldii	10.06	.1	Urtica urens 2	0	-.8
Linum usitatissimum	7.02	-.2	Verbena officinalis	43.64	3.2
Lycopus europaeus	10.02	.1	Vinca minor	6.91	-.2
Malva sylvestris	0	-.8	Viola tricolor	0	-.8
Mandragora officinalis	6.41	-.3	Zizia aurea	6.67	-.2

Niccolum

Name	Value	Deviation	Name	Value	Deviation
Cupressus lawsoniana	52.95	4.1	Stachys sylvatica	6.84	-.2
Rumex acetosa	48.7	3.7	Tussilago petasites	7.52	-.2
Syringa vulgaris	46.24	3.4	Vinca minor	6.91	-.2
Verbena officinalis	43.64	3.2	Zizia aurea	6.67	-.2
Osmundo regalis	33.21	2.2	Belladonna	5.55	-.3
Thuja occidentalis	28.81	1.8	Cardamine pratensis	6.23	-.3
Ruta graveolens	26.52	1.6	Cardamine pratensis R	6.01	-.3
Castanea vesca	25.04	1.5	Echinacea angustifolia	5.98	-.3
Cimicifuga racemosa	23.51	1.3	Echinacea purpurea	6.39	-.3
Melilotus officinalis	22.83	1.3	Lapsana communis	6.46	-.3
Tormentilla erecta	21.65	1.2	Leonurus cardiaca	5.67	-.3
Cicuta virosa	18.17	.8	Mandragora officinalis	6.41	-.3
Mercurialis perennis	17.98	.8	Melissa officinalis	6.1	-.3
Psoralea bituminosa	16.75	.7	Ononis spinosa	6	-.3
Lappa arctium	15.38	.6	Paeonia officinalis	5.98	-.3
Chelidonium majus	14.29	.5	Primula veris	5.78	-.3
Laurocerasus	13.8	.4	Salvia sclarea	5.49	-.3
Adonis vernalis	11.98	.3	Ulmus campestris	6.31	-.3
Centaurium erythraea	12.74	.3	Anethum graveolens	5.18	-.4
Collinsonia canadensis	12.79	.3	Teucrium chamaedrys	5.25	-.4
Milium solis	12.63	.3	Agrimonia eupatoria	0	-.8
Sanguisorba officinali	12.91	.3	Anthemis nobilis	0	-.8
Bryonia dioica	11.61	.2	Chenopodium anthelmint	0	-.8
Bellis perennis	9.77	.1	Conium maculatum	0	-.8
Cardiospermum halicaca	10.45	.1	Escholtzia californica	0	-.8
Clematis erecta	10.17	.1	Faba vulgaris	0	-.8
Hydrophyllum virginicu	10.07	.1	Fagopyrum esculentum	0	-.8
Lespedeza sieboldii	10.06	.1	Fraxinus excelsior	0	-.8
Lycopus europaeus	10.02	.1	Galium aparine	0	-.8
Petroselinum crispum	10.32	.1	Hyoscyamus niger	0	-.8
Scutellaria lateriflor	10.1	.1	Hyssopus officinalis	0	-.8
Gnaphalium leontopodiu	9.09	0	Malva sylvestris	0	-.8
Salicaria purpurea	8.83	0	Marrubium vulgare	0	-.8
Taraxacum officinale	9.36	0	Mentha arvensis	0	-.8
Alchemilla vulgaris	7.91	-.1	Ocimum canum	0	-.8
Cinnamodendron cortiso	8.33	-.1	Ocimum canum R	0	-.8
Glechoma hederacea	7.82	-.1	Osmundo regalis 2	0	-.8
Hedera helix	8.3	-.1	Plantago major	0	-.8
Oenanthe crocata	8.39	-.1	Polygonum aviculare	0	-.8
Rhus toxicodendron	8.37	-.1	Prunus padus	0	-.8
Urtica dioica	7.72	-.1	Solidago virgaurea	0	-.8
Aconitum napellus	6.52	-.2	Spilanthes oleracea	0	-.8
Angelica archangelica	7.09	-.2	Stachys officinalis	0	-.8
Aquilegia vulgaris	7.25	-.2	Stachys officinalis R	0	-.8
Chamomilla	6.83	-.2	Tussilago farfara	0	-.8
Linum usitatissimum	7.02	-.2	Urtica urens	0	-.8
Salix purpurea	6.59	-.2	Urtica urens 2	0	-.8
Scrophularia nodosa	7.29	-.2	Viola tricolor	0	-.8

Niobium

Name	Value	Deviation	Name	Value	Deviation
Aconitum napellus	.11	-.2	Marrubium vulgare	.11	-.2
Adonis vernalis	.26	1.1	Melilotus officinalis	.05	-.7
Agrimonia eupatoria	.06	-.7	Melissa officinalis	.27	1.2
Alchemilla vulgaris	.25	1	Mentha arvensis	.17	.3
Anethum graveolens	.05	-.7	Mercurialis perennis	.24	.9
Angelica archangelica	.3	1.4	Milium solis	.16	.2
Anthemis nobilis	.17	.3	Ocimum canum	.22	.7
Aquilegia vulgaris	.04	-.8	Ocimum canum R	.18	.4
Belladonna	.12	-.1	Oenanthe crocata	.24	.9
Bellis perennis	.13	-.1	Ononis spinosa	.11	-.2
Bryonia dioica	.32	1.6	Osmundo regalis	.68	4.7
Cardamine pratensis	.1	-.3	Osmundo regalis 2	.02	-1
Cardamine pratensis R	.08	-.5	Paeonia officinalis	.16	.2
Cardiospermum halicaca	.11	-.2	Petroselinum crispum	.18	.4
Castanea vesca	.07	-.6	Plantago major	.18	.4
Centaurium erythraea	.5	3.1	Polygonum aviculare	.21	.6
Chamomilla	.13	-.1	Primula veris	.32	1.6
Chelidonium majus	.14	0	Prunus padus	.15	.1
Chenopodium anthelmint	.05	-.7	Psoralea bituminosa	.14	0
Cicuta virosa	.44	2.6	Rhus toxicodendron	.06	-.7
Cimicifuga racemosa	.08	-.5	Rumex acetosa	.28	1.2
Cinnamodendron cortiso	.08	-.5	Ruta graveolens	.04	-.8
Clematis erecta	.08	-.5	Salicaria purpurea	.05	-.7
Collinsonia canadensis	.18	.4	Salix purpurea	.09	-.4
Conium maculatum	.05	-.7	Salvia sclarea	.03	-.9
Cupressus lawsoniana	.33	1.7	Sanguisorba officinali	.02	-1
Echinacea angustifolia	.08	-.5	Scrophularia nodosa	.13	-.1
Echinacea purpurea	.04	-.8	Scutellaria lateriflor	.13	-.1
Escholtzia californica	.08	-.5	Solidago virgaurea	.03	-.9
Faba vulgaris	.07	-.6	Spilanthes oleracea	.05	-.7
Fagopyrum esculentum	.04	-.8	Stachys officinalis	.02	-1
Fraxinus excelsior	.1	-.3	Stachys officinalis R	.02	-1
Galium aparine	.09	-.4	Stachys sylvatica	.12	-.1
Glechoma hederacea	.12	-.1	Syringa vulgaris	.2	.5
Gnaphalium leontopodiu	.42	2.5	Taraxacum officinale	.24	.9
Hedera helix	.11	-.2	Teucrium chamaedrys	.08	-.5
Hydrophyllum virginicu	.07	-.6	Thuja occidentalis	.07	-.6
Hyoscyamus niger	.2	.5	Tormentilla erecta	.12	-.1
Hyssopus officinalis	.07	-.6	Tussilago farfara	.05	-.7
Lappa arctium	.15	.1	Tussilago petasites	.07	-.6
Lapsana communis	.43	2.5	Ulmus campestris	.07	-.6
Laurocerasus	.05	-.7	Urtica dioica	.17	.3
Leonurus cardiaca	.08	-.5	Urtica urens	.04	-.8
Lespedeza sieboldii	.04	-.8	Urtica urens 2	.05	-.7
Linum usitatissimum	.02	-1	Verbena officinalis	.05	-.7
Lycopus europaeus	.07	-.6	Vinca minor	.13	-.1
Malva sylvestris	.11	-.2	Viola tricolor	.09	-.4
Mandragora officinalis	.05	-.7	Zizia aurea	.1	-.3

Niobium

Name	Value	Deviation	Name	Value	Deviation
Osmundo regalis	.68	4.7	Cardamine pratensis	.1	-.3
Centaurium erythraea	.5	3.1	Fraxinus excelsior	.1	-.3
Cicuta virosa	.44	2.6	Zizia aurea	.1	-.3
Gnaphalium leontopodiu	.42	2.5	Galium aparine	.09	-.4
Lapsana communis	.43	2.5	Salix purpurea	.09	-.4
Cupressus lawsoniana	.33	1.7	Viola tricolor	.09	-.4
Bryonia dioica	.32	1.6	Cardamine pratensis R	.08	-.5
Primula veris	.32	1.6	Cimicifuga racemosa	.08	-.5
Angelica archangelica	.3	1.4	Cinnamodendron cortiso	.08	-.5
Melissa officinalis	.27	1.2	Clematis erecta	.08	-.5
Rumex acetosa	.28	1.2	Echinacea angustifolia	.08	-.5
Adonis vernalis	.26	1.1	Escholtzia californica	.08	-.5
Alchemilla vulgaris	.25	1	Leonurus cardiaca	.08	-.5
Mercurialis perennis	.24	.9	Teucrium chamaedrys	.08	-.5
Oenanthe crocata	.24	.9	Castanea vesca	.07	-.6
Taraxacum officinale	.24	.9	Faba vulgaris	.07	-.6
Ocimum canum	.22	.7	Hydrophyllum virginicu	.07	-.6
Polygonum aviculare	.21	.6	Hyssopus officinalis	.07	-.6
Hyoscyamus niger	.2	.5	Lycopus europaeus	.07	-.6
Syringa vulgaris	.2	.5	Thuja occidentalis	.07	-.6
Collinsonia canadensis	.18	.4	Tussilago petasites	.07	-.6
Ocimum canum R	.18	.4	Ulmus campestris	.07	-.6
Petroselinum crispum	.18	.4	Agrimonia eupatoria	.06	-.7
Plantago major	.18	.4	Anethum graveolens	.05	-.7
Anthemis nobilis	.17	.3	Chenopodium anthelmint	.05	-.7
Mentha arvensis	.17	.3	Conium maculatum	.05	-.7
Urtica dioica	.17	.3	Laurocerasus	.05	-.7
Milium solis	.16	.2	Mandragora officinalis	.05	-.7
Paeonia officinalis	.16	.2	Melilotus officinalis	.05	-.7
Lappa arctium	.15	.1	Rhus toxicodendron	.06	-.7
Prunus padus	.15	.1	Salicaria purpurea	.05	-.7
Chelidonium majus	.14	0	Spilanthes oleracea	.05	-.7
Psoralea bituminosa	.14	0	Tussilago farfara	.05	-.7
Belladonna	.12	-.1	Urtica urens 2	.05	-.7
Bellis perennis	.13	-.1	Verbena officinalis	.05	-.7
Chamomilla	.13	-.1	Aquilegia vulgaris	.04	-.8
Glechoma hederacea	.12	-.1	Echinacea purpurea	.04	-.8
Scrophularia nodosa	.13	-.1	Fagopyrum esculentum	.04	-.8
Scutellaria lateriflor	.13	-.1	Lespedeza sieboldii	.04	-.8
Stachys sylvatica	.12	-.1	Ruta graveolens	.04	-.8
Tormentilla erecta	.12	-.1	Urtica urens	.04	-.8
Vinca minor	.13	-.1	Salvia sclarea	.03	-.9
Aconitum napellus	.11	-.2	Solidago virgaurea	.03	-.9
Cardiospermum halicaca	.11	-.2	Linum usitatissimum	.02	-1
Hedera helix	.11	-.2	Osmundo regalis 2	.02	-1
Malva sylvestris	.11	-.2	Sanguisorba officinali	.02	-1
Marrubium vulgare	.11	-.2	Stachys officinalis	.02	-1
Ononis spinosa	.11	-.2	Stachys officinalis R	.02	-1

Plumbum

Name	Value	Deviation	Name	Value	Deviation
Aconitum napellus	9.27	-.1	Marrubium vulgare	6.13	-.4
Adonis vernalis	7.55	-.3	Melilotus officinalis	6.11	-.4
Agrimonia eupatoria	6.4	-.4	Melissa officinalis	13.39	.3
Alchemilla vulgaris	11.56	.1	Mentha arvensis	10.67	0
Anethum graveolens	4.62	-.5	Mercurialis perennis	11.37	.1
Angelica archangelica	7.38	-.3	Milium solis	6.11	-.4
Anthemis nobilis	9.55	-.1	Ocimum canum	4.62	-.5
Aquilegia vulgaris	2.82	-.7	Ocimum canum R	5.09	-.5
Belladonna	7.2	-.3	Oenanthe crocata	5.49	-.4
Bellis perennis	35.43	2.2	Ononis spinosa	15.08	.4
Bryonia dioica	20.68	.9	Osmundo regalis	87.09	6.8
Cardamine pratensis	11.16	.1	Osmundo regalis 2	2.19	-.7
Cardamine pratensis R	13.29	.2	Paeonia officinalis	8.37	-.2
Cardiospermum halicaca	6.93	-.3	Petroselinum crispum	18.32	.7
Castanea vesca	4.73	-.5	Plantago major	5.01	-.5
Centaurium erythraea	15.65	.5	Polygonum aviculare	7.77	-.2
Chamomilla	4.86	-.5	Primula veris	9.16	-.1
Chelidonium majus	25.67	1.4	Prunus padus	8.26	-.2
Chenopodium anthelmint	2.54	-.7	Psoralea bituminosa	5.54	-.4
Cicuta virosa	28.39	1.6	Rhus toxicodendron	7.36	-.3
Cimicifuga racemosa	11.6	.1	Rumex acetosa	26.5	1.4
Cinnamodendron cortiso	17.04	.6	Ruta graveolens	5.07	-.5
Clematis erecta	7.73	-.2	Salicaria purpurea	9.21	-.1
Collinsonia canadensis	9.43	-.1	Salix purpurea	13.09	.2
Conium maculatum	6.04	-.4	Salvia sclarea	7.12	-.3
Cupressus lawsoniana	26.56	1.4	Sanguisorba officinali	3.72	-.6
Echinacea angustifolia	5.17	-.5	Scrophularia nodosa	19.06	.8
Echinacea purpurea	4.76	-.5	Scutellaria lateriflor	9.14	-.1
Escholtzia californica	2.07	-.8	Solidago virgaurea	3.78	-.6
Faba vulgaris	7.45	-.3	Spilanthes oleracea	4.2	-.6
Fagopyrum esculentum	3.33	-.6	Stachys officinalis	2.75	-.7
Fraxinus excelsior	20.58	.9	Stachys officinalis R	2.56	-.7
Galium aparine	9.04	-.1	Stachys sylvatica	9.53	-.1
Glechoma hederacea	8.97	-.1	Syringa vulgaris	11.03	0
Gnaphalium leontopodiu	12.53	.2	Taraxacum officinale	19.95	.8
Hedera helix	9.13	-.1	Teucrium chamaedrys	4.86	-.5
Hydrophyllum virginicu	6.17	-.4	Thuja occidentalis	9.18	-.1
Hyoscyamus niger	5.7	-.4	Tormentilla erecta	54.99	4
Hyssopus officinalis	6.25	-.4	Tussilago farfara	3.66	-.6
Lappa arctium	11.43	.1	Tussilago petasites	7.53	-.3
Lapsana communis	12.52	.2	Ulmus campestris	21.64	1
Laurocerasus	4.57	-.5	Urtica dioica	12.19	.1
Leonurus cardiaca	6.43	-.4	Urtica urens	2.71	-.7
Lespedeza sieboldii	6.78	-.3	Urtica urens 2	3.37	-.6
Linum usitatissimum	2.54	-.7	Verbena officinalis	4.74	-.5
Lycopus europaeus	4.38	-.5	Vinca minor	12.68	.2
Malva sylvestris	3.47	-.6	Viola tricolor	3.95	-.6
Mandragora officinalis	2.49	-.7	Zizia aurea	7.69	-.3

Plumbum

Name	Value	Deviation	Name	Value	Deviation
Osmundo regalis	87.09	6.8	Cardiospermum halicaca	6.93	-.3
Tormentilla erecta	54.99	4	Faba vulgaris	7.45	-.3
Bellis perennis	35.43	2.2	Lespedeza sieboldii	6.78	-.3
Cicuta virosa	28.39	1.6	Rhus toxicodendron	7.36	-.3
Chelidonium majus	25.67	1.4	Salvia sclarea	7.12	-.3
Cupressus lawsoniana	26.56	1.4	Tussilago petasites	7.53	-.3
Rumex acetosa	26.5	1.4	Zizia aurea	7.69	-.3
Ulmus campestris	21.64	1	Agrimonia eupatoria	6.4	-.4
Bryonia dioica	20.68	.9	Conium maculatum	6.04	-.4
Fraxinus excelsior	20.58	.9	Hydrophyllum virginicu	6.17	-.4
Scrophularia nodosa	19.06	.8	Hyoscyamus niger	5.7	-.4
Taraxacum officinale	19.95	.8	Hyssopus officinalis	6.25	-.4
Petroselinum crispum	18.32	.7	Leonurus cardiaca	6.43	-.4
Cinnamodendron cortiso	17.04	.6	Marrubium vulgare	6.13	-.4
Centaurium erythraea	15.65	.5	Melilotus officinalis	6.11	-.4
Ononis spinosa	15.08	.4	Milium solis	6.11	-.4
Melissa officinalis	13.39	.3	Oenanthe crocata	5.49	-.4
Cardamine pratensis R	13.29	.2	Psoralea bituminosa	5.54	-.4
Gnaphalium leontopodiu	12.53	.2	Anethum graveolens	4.62	-.5
Lapsana communis	12.52	.2	Castanea vesca	4.73	-.5
Salix purpurea	13.09	.2	Chamomilla	4.86	-.5
Vinca minor	12.68	.2	Echinacea angustifolia	5.17	-.5
Alchemilla vulgaris	11.56	.1	Echinacea purpurea	4.76	-.5
Cardamine pratensis	11.16	.1	Laurocerasus	4.57	-.5
Cimicifuga racemosa	11.6	.1	Lycopus europaeus	4.38	-.5
Lappa arctium	11.43	.1	Ocimum canum	4.62	-.5
Mercurialis perennis	11.37	.1	Ocimum canum R	5.09	-.5
Urtica dioica	12.19	.1	Plantago major	5.01	-.5
Mentha arvensis	10.67	0	Ruta graveolens	5.07	-.5
Syringa vulgaris	11.03	0	Teucrium chamaedrys	4.86	-.5
Aconitum napellus	9.27	-.1	Verbena officinalis	4.74	-.5
Anthemis nobilis	9.55	-.1	Fagopyrum esculentum	3.33	-.6
Collinsonia canadensis	9.43	-.1	Malva sylvestris	3.47	-.6
Galium aparine	9.04	-.1	Sanguisorba officinali	3.72	-.6
Glechoma hederacea	8.97	-.1	Solidago virgaurea	3.78	-.6
Hedera helix	9.13	-.1	Spilanthes oleracea	4.2	-.6
Primula veris	9.16	-.1	Tussilago farfara	3.66	-.6
Salicaria purpurea	9.21	-.1	Urtica urens 2	3.37	-.6
Scutellaria lateriflor	9.14	-.1	Viola tricolor	3.95	-.6
Stachys sylvatica	9.53	-.1	Aquilegia vulgaris	2.82	-.7
Thuja occidentalis	9.18	-.1	Chenopodium anthelmint	2.54	-.7
Clematis erecta	7.73	-.2	Linum usitatissimum	2.54	-.7
Paeonia officinalis	8.37	-.2	Mandragora officinalis	2.49	-.7
Polygonum aviculare	7.77	-.2	Osmundo regalis 2	2.19	-.7
Prunus padus	8.26	-.2	Stachys officinalis	2.75	-.7
Adonis vernalis	7.55	-.3	Stachys officinalis R	2.56	-.7
Angelica archangelica	7.38	-.3	Urtica urens	2.71	-.7
Belladonna	7.2	-.3	Escholtzia californica	2.07	-.8

Praseodymium

Name	Value	Deviation	Name	Value	Deviation
Aconitum napellus	.48	.4	Marrubium vulgare	.2	-.3
Adonis vernalis	.49	.4	Melilotus officinalis	.11	-.5
Agrimonia eupatoria	.07	-.6	Melissa officinalis	.3	-.1
Alchemilla vulgaris	.47	.3	Mentha arvensis	.19	-.3
Anethum graveolens	.08	-.6	Mercurialis perennis	.46	.3
Angelica archangelica	.44	.3	Milium solis	.14	-.5
Anthemis nobilis	.33	0	Ocimum canum	.3	-.1
Aquilegia vulgaris	.07	-.6	Ocimum canum R	.3	-.1
Belladonna	.27	-.1	Oenanthe crocata	.43	.2
Bellis perennis	1.76	3.4	Ononis spinosa	.3	-.1
Bryonia dioica	.56	.5	Osmundo regalis	2.89	6.1
Cardamine pratensis	.41	.2	Osmundo regalis 2	.02	-.7
Cardamine pratensis R	.36	.1	Paeonia officinalis	.32	0
Cardiospermum halicaca	.3	-.1	Petroselinum crispum	.55	.5
Castanea vesca	.18	-.4	Plantago major	.38	.1
Centaurium erythraea	.75	1	Polygonum aviculare	.22	-.3
Chamomilla	.27	-.1	Primula veris	.33	0
Chelidonium majus	1.71	3.3	Prunus padus	.16	-.4
Chenopodium anthelmint	.07	-.6	Psoralea bituminosa	.22	-.3
Cicuta virosa	1.09	1.8	Rhus toxicodendron	.11	-.5
Cimicifuga racemosa	.45	.3	Rumex acetosa	1.44	2.7
Cinnamodendron cortiso	.17	-.4	Ruta graveolens	.09	-.6
Clematis erecta	.12	-.5	Salicaria purpurea	.11	-.5
Collinsonia canadensis	.6	.6	Salix purpurea	.17	-.4
Conium maculatum	.07	-.6	Salvia sclarea	.09	-.6
Cupressus lawsoniana	.42	.2	Sanguisorba officinali	.05	-.7
Echinacea angustifolia	.17	-.4	Scrophularia nodosa	.21	-.3
Echinacea purpurea	.07	-.6	Scutellaria lateriflor	.23	-.2
Escholtzia californica	.13	-.5	Solidago virgaurea	.06	-.6
Faba vulgaris	.11	-.5	Spilanthes oleracea	.12	-.5
Fagopyrum esculentum	.06	-.6	Stachys officinalis	.05	-.7
Fraxinus excelsior	.28	-.1	Stachys officinalis R	.03	-.7
Galium aparine	.27	-.1	Stachys sylvatica	.22	-.3
Glechoma hederacea	.22	-.3	Syringa vulgaris	.2	-.3
Gnaphalium leontopodiu	.83	1.2	Taraxacum officinale	1.2	2.1
Hedera helix	.19	-.3	Teucrium chamaedrys	.16	-.4
Hydrophyllum virginicu	.09	-.6	Thuja occidentalis	.13	-.5
Hyoscyamus niger	.25	-.2	Tormentilla erecta	.68	.8
Hyssopus officinalis	.12	-.5	Tussilago farfara	.08	-.6
Lappa arctium	.46	.3	Tussilago petasites	.15	-.4
Lapsana communis	.56	.5	Ulmus campestris	.22	-.3
Laurocerasus	.08	-.6	Urtica dioica	.65	.8
Leonurus cardiaca	.09	-.6	Urtica urens	.11	-.5
Lespedeza sieboldii	.08	-.6	Urtica urens 2	.12	-.5
Linum usitatissimum	.05	-.7	Verbena officinalis	.06	-.6
Lycopus europaeus	.08	-.6	Vinca minor	.31	-.1
Malva sylvestris	.13	-.5	Viola tricolor	.17	-.4
Mandragora officinalis	.15	-.4	Zizia aurea	.3	-.1

Praseodymium

Name	Value	Deviation	Name	Value	Deviation
Osmundo regalis	2.89	6.1	Scrophularia nodosa	.21	-.3
Bellis perennis	1.76	3.4	Stachys sylvatica	.22	-.3
Chelidonium majus	1.71	3.3	Syringa vulgaris	.2	-.3
Rumex acetosa	1.44	2.7	Ulmus campestris	.22	-.3
Taraxacum officinale	1.2	2.1	Castanea vesca	.18	-.4
Cicuta virosa	1.09	1.8	Cinnamodendron cortiso	.17	-.4
Gnaphalium leontopodiu	.83	1.2	Echinacea angustifolia	.17	-.4
Centaurium erythraea	.75	1	Mandragora officinalis	.15	-.4
Tormentilla erecta	.68	.8	Prunus padus	.16	-.4
Urtica dioica	.65	.8	Salix purpurea	.17	-.4
Collinsonia canadensis	.6	.6	Teucrium chamaedrys	.16	-.4
Bryonia dioica	.56	.5	Tussilago petasites	.15	-.4
Lapsana communis	.56	.5	Viola tricolor	.17	-.4
Petroselinum crispum	.55	.5	Clematis erecta	.12	-.5
Aconitum napellus	.48	.4	Escholtzia californica	.13	-.5
Adonis vernalis	.49	.4	Faba vulgaris	.11	-.5
Alchemilla vulgaris	.47	.3	Hyssopus officinalis	.12	-.5
Angelica archangelica	.44	.3	Malva sylvestris	.13	-.5
Cimicifuga racemosa	.45	.3	Melilotus officinalis	.11	-.5
Lappa arctium	.46	.3	Milium solis	.14	-.5
Mercurialis perennis	.46	.3	Rhus toxicodendron	.11	-.5
Cardamine pratensis	.41	.2	Salicaria purpurea	.11	-.5
Cupressus lawsoniana	.42	.2	Spilanthes oleracea	.12	-.5
Oenanthe crocata	.43	.2	Thuja occidentalis	.13	-.5
Cardamine pratensis R	.36	.1	Urtica urens	.11	-.5
Plantago major	.38	.1	Urtica urens 2	.12	-.5
Anthemis nobilis	.33	0	Agrimonia eupatoria	.07	-.6
Paeonia officinalis	.32	0	Anethum graveolens	.08	-.6
Primula veris	.33	0	Aquilegia vulgaris	.07	-.6
Belladonna	.27	-.1	Chenopodium anthelmint	.07	-.6
Cardiospermum halicaca	.3	-.1	Conium maculatum	.07	-.6
Chamomilla	.27	-.1	Echinacea purpurea	.07	-.6
Fraxinus excelsior	.28	-.1	Fagopyrum esculentum	.06	-.6
Galium aparine	.27	-.1	Hydrophyllum virginicu	.09	-.6
Melissa officinalis	.3	-.1	Laurocerasus	.08	-.6
Ocimum canum	.3	-.1	Leonurus cardiaca	.09	-.6
Ocimum canum R	.3	-.1	Lespedeza sieboldii	.08	-.6
Ononis spinosa	.3	-.1	Lycopus europaeus	.08	-.6
Vinca minor	.31	-.1	Ruta graveolens	.09	-.6
Zizia aurea	.3	-.1	Salvia sclarea	.09	-.6
Hyoscyamus niger	.25	-.2	Solidago virgaurea	.06	-.6
Scutellaria lateriflor	.23	-.2	Tussilago farfara	.08	-.6
Glechoma hederacea	.22	-.3	Verbena officinalis	.06	-.6
Hedera helix	.19	-.3	Linum usitatissimum	.05	-.7
Marrubium vulgare	.2	-.3	Osmundo regalis 2	.02	-.7
Mentha arvensis	.19	-.3	Sanguisorba officinali	.05	-.7
Polygonum aviculare	.22	-.3	Stachys officinalis	.05	-.7
Psoralea bituminosa	.22	-.3	Stachys officinalis R	.03	-.7

Rhenium

Name	Value	Deviation	Name	Value	Deviation
Aconitum napellus	19.34	1.5	Marrubium vulgare	8.75	.2
Adonis vernalis	.25	-.8	Melilotus officinalis	6.8	0
Agrimonia eupatoria	4.79	-.2	Melissa officinalis	13.74	.8
Alchemilla vulgaris	2.92	-.5	Mentha arvensis	3.42	-.4
Anethum graveolens	13.68	.8	Mercurialis perennis	8.52	.2
Angelica archangelica	.54	-.7	Milium solis	1.85	-.6
Anthemis nobilis	5.72	-.1	Ocimum canum	11.87	.6
Aquilegia vulgaris	1.41	-.6	Ocimum canum R	13.27	.8
Belladonna	5.58	-.1	Oenanthe crocata	.7	-.7
Bellis perennis	.66	-.7	Ononis spinosa	3.88	-.3
Bryonia dioica	.53	-.7	Osmundo regalis	2.32	-.5
Cardamine pratensis	1.62	-.6	Osmundo regalis 2	.62	-.7
Cardamine pratensis R	1.73	-.6	Paeonia officinalis	.25	-.8
Cardiospermum halicaca	18.05	1.3	Petroselinum crispum	2.18	-.5
Castanea vesca	3.31	-.4	Plantago major	4.28	-.3
Centaurium erythraea	3.16	-.4	Polygonum aviculare	7.07	0
Chamomilla	9.39	.3	Primula veris	2.07	-.6
Chelidonium majus	.55	-.7	Prunus padus	6.26	-.1
Chenopodium anthelmint	6.45	0	Psoralea bituminosa	25.82	2.2
Cicuta virosa	.6	-.7	Rhus toxicodendron	1.94	-.6
Cimicifuga racemosa	.32	-.8	Rumex acetosa	3.61	-.4
Cinnamodendron cortiso	18.03	1.3	Ruta graveolens	8.73	.2
Clematis erecta	65.39	6.9	Salicaria purpurea	1.79	-.6
Collinsonia canadensis	1.4	-.6	Salix purpurea	.58	-.7
Conium maculatum	20.88	1.7	Salvia sclarea	20.54	1.6
Cupressus lawsoniana	6.2	-.1	Sanguisorba officinali	1.09	-.7
Echinacea angustifolia	13.01	.7	Scrophularia nodosa	14.56	.9
Echinacea purpurea	14.71	.9	Scutellaria lateriflor	2.67	-.5
Escholtzia californica	10.87	.5	Solidago virgaurea	4.74	-.2
Faba vulgaris	4.8	-.2	Spilanthes oleracea	8.33	.2
Fagopyrum esculentum	10.31	.4	Stachys officinalis	5.85	-.1
Fraxinus excelsior	.54	-.7	Stachys officinalis R	4.51	-.3
Galium aparine	2.35	-.5	Stachys sylvatica	6.31	-.1
Glechoma hederacea	7.44	.1	Syringa vulgaris	4.15	-.3
Gnaphalium leontopodiu	2.45	-.5	Taraxacum officinale	.28	-.8
Hedera helix	2.36	-.5	Teucrium chamaedrys	16.4	1.1
Hydrophyllum virginicu	7.58	.1	Thuja occidentalis	3.85	-.3
Hyoscyamus niger	5.86	-.1	Tormentilla erecta	.54	-.7
Hyssopus officinalis	5.83	-.1	Tussilago farfara	4.45	-.3
Lappa arctium	4.16	-.3	Tussilago petasites	1.87	-.6
Lapsana communis	.98	-.7	Ulmus campestris	.62	-.7
Laurocerasus	2.43	-.5	Urtica dioica	3.57	-.4
Leonurus cardiaca	22.72	1.9	Urtica urens	22.47	1.8
Lespedeza sieboldii	7.59	.1	Urtica urens 2	9.48	.3
Linum usitatissimum	6.28	-.1	Verbena officinalis	1.8	-.6
Lycopus europaeus	3.91	-.3	Vinca minor	4.58	-.3
Malva sylvestris	3	-.4	Viola tricolor	13.47	.8
Mandragora officinalis	.56	-.7	Zizia aurea	3.34	-.4

Rhenium

Name	Value	Deviation	Name	Value	Deviation
Clematis erecta	65.39	6.9	Stachys officinalis R	4.51	-.3
Psoralea bituminosa	25.82	2.2	Syringa vulgaris	4.15	-.3
Leonurus cardiaca	22.72	1.9	Thuja occidentalis	3.85	-.3
Urtica urens	22.47	1.8	Tussilago farfara	4.45	-.3
Conium maculatum	20.88	1.7	Vinca minor	4.58	-.3
Salvia sclarea	20.54	1.6	Castanea vesca	3.31	-.4
Aconitum napellus	19.34	1.5	Centaurium erythraea	3.16	-.4
Cardiospermum halicaca	18.05	1.3	Malva sylvestris	3	-.4
Cinnamodendron cortiso	18.03	1.3	Mentha arvensis	3.42	-.4
Teucrium chamaedrys	16.4	1.1	Rumex acetosa	3.61	-.4
Echinacea purpurea	14.71	.9	Urtica dioica	3.57	-.4
Scrophularia nodosa	14.56	.9	Zizia aurea	3.34	-.4
Anethum graveolens	13.68	.8	Alchemilla vulgaris	2.92	-.5
Melissa officinalis	13.74	.8	Galium aparine	2.35	-.5
Ocimum canum R	13.27	.8	Gnaphalium leontopodiu	2.45	-.5
Viola tricolor	13.47	.8	Hedera helix	2.36	-.5
Echinacea angustifolia	13.01	.7	Laurocerasus	2.43	-.5
Ocimum canum	11.87	.6	Osmundo regalis	2.32	-.5
Escholtzia californica	10.87	.5	Petroselinum crispum	2.18	-.5
Fagopyrum esculentum	10.31	.4	Scutellaria lateriflor	2.67	-.5
Chamomilla	9.39	.3	Aquilegia vulgaris	1.41	-.6
Urtica urens 2	9.48	.3	Cardamine pratensis	1.62	-.6
Marrubium vulgare	8.75	.2	Cardamine pratensis R	1.73	-.6
Mercurialis perennis	8.52	.2	Collinsonia canadensis	1.4	-.6
Ruta graveolens	8.73	.2	Milium solis	1.85	-.6
Spilanthes oleracea	8.33	.2	Primula veris	2.07	-.6
Glechoma hederacea	7.44	.1	Rhus toxicodendron	1.94	-.6
Hydrophyllum virginicu	7.58	.1	Salicaria purpurea	1.79	-.6
Lespedeza sieboldii	7.59	.1	Tussilago petasites	1.87	-.6
Chenopodium anthelmint	6.45	0	Verbena officinalis	1.8	-.6
Melilotus officinalis	6.8	0	Angelica archangelica	.54	-.7
Polygonum aviculare	7.07	0	Bellis perennis	.66	-.7
Anthemis nobilis	5.72	-.1	Bryonia dioica	.53	-.7
Belladonna	5.58	-.1	Chelidonium majus	.55	-.7
Cupressus lawsoniana	6.2	-.1	Cicuta virosa	.6	-.7
Hyoscyamus niger	5.86	-.1	Fraxinus excelsior	.54	-.7
Hyssopus officinalis	5.83	-.1	Lapsana communis	.98	-.7
Linum usitatissimum	6.28	-.1	Mandragora officinalis	.56	-.7
Prunus padus	6.26	-.1	Oenanthe crocata	.7	-.7
Stachys officinalis	5.85	-.1	Osmundo regalis 2	.62	-.7
Stachys sylvatica	6.31	-.1	Salix purpurea	.58	-.7
Agrimonia eupatoria	4.79	-.2	Sanguisorba officinali	1.09	-.7
Faba vulgaris	4.8	-.2	Tormentilla erecta	.54	-.7
Solidago virgaurea	4.74	-.2	Ulmus campestris	.62	-.7
Lappa arctium	4.16	-.3	Adonis vernalis	.25	-.8
Lycopus europaeus	3.91	-.3	Cimicifuga racemosa	.32	-.8
Ononis spinosa	3.88	-.3	Paeonia officinalis	.25	-.8
Plantago major	4.28	-.3	Taraxacum officinale	.28	-.8

Rubidium

Name	Value	Deviation	Name	Value	Deviation
Aconitum napellus	27.42	-.4	Marrubium vulgare	31.46	-.4
Adonis vernalis	46.44	-.3	Melilotus officinalis	512.29	.9
Agrimonia eupatoria	91.37	-.2	Melissa officinalis	34.19	-.4
Alchemilla vulgaris	38.03	-.4	Mentha arvensis	28.79	-.4
Anethum graveolens	22.05	-.4	Mercurialis perennis	254.82	.2
Angelica archangelica	92.08	-.2	Milium solis	113.92	-.2
Anthemis nobilis	110.04	-.2	Ocimum canum	14.56	-.4
Aquilegia vulgaris	273.19	.3	Ocimum canum R	15.19	-.4
Belladonna	42.77	-.4	Oenanthe crocata	56.39	-.3
Bellis perennis	26.31	-.4	Ononis spinosa	49.76	-.3
Bryonia dioica	152.24	-.1	Osmundo regalis	265.49	.2
Cardamine pratensis	2590	6.4	Osmundo regalis 2	777.2	1.6
Cardamine pratensis R	2340	5.8	Paeonia officinalis	52.99	-.3
Cardiospermum halicaca	43.03	-.4	Petroselinum crispum	42.67	-.4
Castanea vesca	194.57	.1	Plantago major	20.87	-.4
Centaurium erythraea	38.92	-.4	Polygonum aviculare	31.9	-.4
Chamomilla	117.61	-.2	Primula veris	77.46	-.3
Chelidonium majus	22.74	-.4	Prunus padus	130.3	-.1
Chenopodium anthelmint	43.26	-.3	Psoralea bituminosa	45.37	-.3
Cicuta virosa	263.4	.2	Rhus toxicodendron	32.51	-.4
Cimicifuga racemosa	118.86	-.1	Rumex acetosa	297.03	.3
Cinnamodendron cortiso	206.17	.1	Ruta graveolens	45.25	-.3
Clematis erecta	126.74	-.1	Salicaria purpurea	205.49	.1
Collinsonia canadensis	43.67	-.3	Salix purpurea	26.99	-.4
Conium maculatum	83.81	-.2	Salvia sclarea	24.09	-.4
Cupressus lawsoniana	109.6	-.2	Sanguisorba officinali	174.39	0
Echinacea angustifolia	61.78	-.3	Scrophularia nodosa	444.89	.7
Echinacea purpurea	23.21	-.4	Scutellaria lateriflor	236.23	.2
Escholtzia californica	286.28	.3	Solidago virgaurea	51.55	-.3
Faba vulgaris	156.59	0	Spilanthes oleracea	35.06	-.4
Fagopyrum esculentum	96	-.2	Stachys officinalis	230.84	.1
Fraxinus excelsior	13.53	-.4	Stachys officinalis R	255.48	.2
Galium aparine	81.98	-.2	Stachys sylvatica	42.48	-.4
Glechoma hederacea	102.01	-.2	Syringa vulgaris	862.05	1.8
Gnaphalium leontopodiu	185.34	0	Taraxacum officinale	24.39	-.4
Hedera helix	198.02	.1	Teucrium chamaedrys	35.67	-.4
Hydrophyllum virginicu	53.23	-.3	Thuja occidentalis	14.76	-.4
Hyoscyamus niger	125.74	-.1	Tormentilla erecta	28.37	-.4
Hyssopus officinalis	25.64	-.4	Tussilago farfara	405.02	.6
Lappa arctium	107.71	-.2	Tussilago petasites	928.3	2
Lapsana communis	132.17	-.1	Ulmus campestris	31.35	-.4
Laurocerasus	437.07	.7	Urtica dioica	34.98	-.4
Leonurus cardiaca	46.63	-.3	Urtica urens	10.54	-.4
Lespedeza sieboldii	28.89	-.4	Urtica urens 2	16.99	-.4
Linum usitatissimum	39.73	-.4	Verbena officinalis	22.99	-.4
Lycopus europaeus	23.55	-.4	Vinca minor	26.65	-.4
Malva sylvestris	21.45	-.4	Viola tricolor	23.83	-.4
Mandragora officinalis	18.41	-.4	Zizia aurea	139.31	-.1

Rubidium

Name	Value	Deviation	Name	Value	Deviation
Cardamine pratensis	2590	6.4	Leonurus cardiaca	46.63	-.3
Cardamine pratensis R	2340	5.8	Oenanthe crocata	56.39	-.3
Tussilago petasites	928.3	2	Ononis spinosa	49.76	-.3
Syringa vulgaris	862.05	1.8	Paeonia officinalis	52.99	-.3
Osmundo regalis 2	777.2	1.6	Primula veris	77.46	-.3
Melilotus officinalis	512.29	.9	Psoralea bituminosa	45.37	-.3
Laurocerasus	437.07	.7	Ruta graveolens	45.25	-.3
Scrophularia nodosa	444.89	.7	Solidago virgaurea	51.55	-.3
Tussilago farfara	405.02	.6	Aconitum napellus	27.42	-.4
Aquilegia vulgaris	273.19	.3	Alchemilla vulgaris	38.03	-.4
Escholtzia californica	286.28	.3	Anethum graveolens	22.05	-.4
Rumex acetosa	297.03	.3	Belladonna	42.77	-.4
Cicuta virosa	263.4	.2	Bellis perennis	26.31	-.4
Mercurialis perennis	254.82	.2	Cardiospermum halicaca	43.03	-.4
Osmundo regalis	265.49	.2	Centaurium erythraea	38.92	-.4
Scutellaria lateriflor	236.23	.2	Chelidonium majus	22.74	-.4
Stachys officinalis R	255.48	.2	Echinacea purpurea	23.21	-.4
Castanea vesca	194.57	.1	Fraxinus excelsior	13.53	-.4
Cinnamodendron cortiso	206.17	.1	Hyssopus officinalis	25.64	-.4
Hedera helix	198.02	.1	Lespedeza sieboldii	28.89	-.4
Salicaria purpurea	205.49	.1	Linum usitatissimum	39.73	-.4
Stachys officinalis	230.84	.1	Lycopus europaeus	23.55	-.4
Faba vulgaris	156.59	0	Malva sylvestris	21.45	-.4
Gnaphalium leontopodiu	185.34	0	Mandragora officinalis	18.41	-.4
Sanguisorba officinali	174.39	0	Marrubium vulgare	31.46	-.4
Bryonia dioica	152.24	-.1	Melissa officinalis	34.19	-.4
Cimicifuga racemosa	118.86	-.1	Mentha arvensis	28.79	-.4
Clematis erecta	126.74	-.1	Ocimum canum	14.56	-.4
Hyoscyamus niger	125.74	-.1	Ocimum canum R	15.19	-.4
Lapsana communis	132.17	-.1	Petroselinum crispum	42.67	-.4
Prunus padus	130.3	-.1	Plantago major	20.87	-.4
Zizia aurea	139.31	-.1	Polygonum aviculare	31.9	-.4
Agrimonia eupatoria	91.37	-.2	Rhus toxicodendron	32.51	-.4
Angelica archangelica	92.08	-.2	Salix purpurea	26.99	-.4
Anthemis nobilis	110.04	-.2	Salvia sclarea	24.09	-.4
Chamomilla	117.61	-.2	Spilanthes oleracea	35.06	-.4
Conium maculatum	83.81	-.2	Stachys sylvatica	42.48	-.4
Cupressus lawsoniana	109.6	-.2	Taraxacum officinale	24.39	-.4
Fagopyrum esculentum	96	-.2	Teucrium chamaedrys	35.67	-.4
Galium aparine	81.98	-.2	Thuja occidentalis	14.76	-.4
Glechoma hederacea	102.01	-.2	Tormentilla erecta	28.37	-.4
Lappa arctium	107.71	-.2	Ulmus campestris	31.35	-.4
Milium solis	113.92	-.2	Urtica dioica	34.98	-.4
Adonis vernalis	46.44	-.3	Urtica urens	10.54	-.4
Chenopodium anthelmint	43.26	-.3	Urtica urens 2	16.99	-.4
Collinsonia canadensis	43.67	-.3	Verbena officinalis	22.99	-.4
Echinacea angustifolia	61.78	-.3	Vinca minor	26.65	-.4
Hydrophyllum virginicu	53.23	-.3	Viola tricolor	23.83	-.4

Samarium

Name	Value	Deviation	Name	Value	Deviation
Aconitum napellus	.4	.3	Marrubium vulgare	.19	-.2
Adonis vernalis	.39	.3	Melilotus officinalis	.09	-.5
Agrimonia eupatoria	.05	-.6	Melissa officinalis	.26	-.1
Alchemilla vulgaris	.37	.2	Mentha arvensis	.15	-.3
Anethum graveolens	.06	-.6	Mercurialis perennis	.38	.3
Angelica archangelica	.38	.3	Milium solis	.12	-.4
Anthemis nobilis	.27	0	Ocimum canum	.3	.1
Aquilegia vulgaris	.05	-.6	Ocimum canum R	.27	0
Belladonna	.2	-.2	Oenanthe crocata	.35	.2
Bellis perennis	1.49	3.2	Ononis spinosa	.29	0
Bryonia dioica	.45	.4	Osmundo regalis	2.79	6.6
Cardamine pratensis	.12	-.4	Osmundo regalis 2	.02	-.7
Cardamine pratensis R	.1	-.5	Paeonia officinalis	.3	.1
Cardiospermum halicaca	.24	-.1	Petroselinum crispum	.53	.7
Castanea vesca	.16	-.3	Plantago major	.33	.1
Centaurium erythraea	.58	.8	Polygonum aviculare	.19	-.2
Chamomilla	.23	-.1	Primula veris	.29	0
Chelidonium majus	1.44	3	Prunus padus	.14	-.4
Chenopodium anthelmint	.04	-.6	Psoralea bituminosa	.22	-.2
Cicuta virosa	.92	1.7	Rhus toxicodendron	.09	-.5
Cimicifuga racemosa	.38	.3	Rumex acetosa	1.36	2.8
Cinnamodendron cortiso	.12	-.4	Ruta graveolens	.08	-.5
Clematis erecta	.11	-.4	Salicaria purpurea	.1	-.5
Collinsonia canadensis	.5	.6	Salix purpurea	.12	-.4
Conium maculatum	.04	-.6	Salvia sclarea	.08	-.5
Cupressus lawsoniana	.28	0	Sanguisorba officinali	.04	-.6
Echinacea angustifolia	.13	-.4	Scrophularia nodosa	.17	-.3
Echinacea purpurea	.04	-.6	Scutellaria lateriflor	.24	-.1
Escholtzia californica	.13	-.4	Solidago virgaurea	.08	-.5
Faba vulgaris	.09	-.5	Spilanthes oleracea	.12	-.4
Fagopyrum esculentum	.05	-.6	Stachys officinalis	.05	-.6
Fraxinus excelsior	.17	-.3	Stachys officinalis R	.03	-.7
Galium aparine	.2	-.2	Stachys sylvatica	.17	-.3
Glechoma hederacea	.15	-.3	Syringa vulgaris	.13	-.4
Gnaphalium leontopodiu	.66	1	Taraxacum officinale	1.02	1.9
Hedera helix	.14	-.4	Teucrium chamaedrys	.15	-.3
Hydrophyllum virginicu	.08	-.5	Thuja occidentalis	.09	-.5
Hyoscyamus niger	.19	-.2	Tormentilla erecta	.63	.9
Hyssopus officinalis	.11	-.4	Tussilago farfara	.07	-.5
Lappa arctium	.33	.1	Tussilago petasites	.13	-.4
Lapsana communis	.47	.5	Ulmus campestris	.19	-.2
Laurocerasus	.08	-.5	Urtica dioica	.6	.8
Leonurus cardiaca	.07	-.5	Urtica urens	.1	-.5
Lespedeza sieboldii	.06	-.6	Urtica urens 2	.1	-.5
Linum usitatissimum	.02	-.7	Verbena officinalis	.06	-.6
Lycopus europaeus	.07	-.5	Vinca minor	.27	0
Malva sylvestris	.13	-.4	Viola tricolor	.15	-.3
Mandragora officinalis	.16	-.3	Zizia aurea	.27	0

Samarium

Name	Value	Deviation	Name	Value	Deviation
Osmundo regalis	2.79	6.6	Scrophularia nodosa	.17	-.3
Bellis perennis	1.49	3.2	Stachys sylvatica	.17	-.3
Chelidonium majus	1.44	3	Teucrium chamaedrys	.15	-.3
Rumex acetosa	1.36	2.8	Viola tricolor	.15	-.3
Taraxacum officinale	1.02	1.9	Cardamine pratensis	.12	-.4
Cicuta virosa	.92	1.7	Cinnamodendron cortiso	.12	-.4
Gnaphalium leontopodiu	.66	1	Clematis erecta	.11	-.4
Tormentilla erecta	.63	.9	Echinacea angustifolia	.13	-.4
Centaurium erythraea	.58	.8	Escholtzia californica	.13	-.4
Urtica dioica	.6	.8	Hedera helix	.14	-.4
Petroselinum crispum	.53	.7	Hyssopus officinalis	.11	-.4
Collinsonia canadensis	.5	.6	Malva sylvestris	.13	-.4
Lapsana communis	.47	.5	Milium solis	.12	-.4
Bryonia dioica	.45	.4	Prunus padus	.14	-.4
Aconitum napellus	.4	.3	Salix purpurea	.12	-.4
Adonis vernalis	.39	.3	Spilanthes oleracea	.12	-.4
Angelica archangelica	.38	.3	Syringa vulgaris	.13	-.4
Cimicifuga racemosa	.38	.3	Tussilago petasites	.13	-.4
Mercurialis perennis	.38	.3	Cardamine pratensis R	.1	-.5
Alchemilla vulgaris	.37	.2	Faba vulgaris	.09	-.5
Oenanthe crocata	.35	.2	Hydrophyllum virginicu	.08	-.5
Lappa arctium	.33	.1	Laurocerasus	.08	-.5
Ocimum canum	.3	.1	Leonurus cardiaca	.07	-.5
Paeonia officinalis	.3	.1	Lycopus europaeus	.07	-.5
Plantago major	.33	.1	Melilotus officinalis	.09	-.5
Anthemis nobilis	.27	0	Rhus toxicodendron	.09	-.5
Cupressus lawsoniana	.28	0	Ruta graveolens	.08	-.5
Ocimum canum R	.27	0	Salicaria purpurea	.1	-.5
Ononis spinosa	.29	0	Salvia sclarea	.08	-.5
Primula veris	.29	0	Solidago virgaurea	.08	-.5
Vinca minor	.27	0	Thuja occidentalis	.09	-.5
Zizia aurea	.27	0	Tussilago farfara	.07	-.5
Cardiospermum halicaca	.24	-.1	Urtica urens	.1	-.5
Chamomilla	.23	-.1	Urtica urens 2	.1	-.5
Melissa officinalis	.26	-.1	Agrimonia eupatoria	.05	-.6
Scutellaria lateriflor	.24	-.1	Anethum graveolens	.06	-.6
Belladonna	.2	-.2	Aquilegia vulgaris	.05	-.6
Galium aparine	.2	-.2	Chenopodium anthelmint	.04	-.6
Hyoscyamus niger	.19	-.2	Conium maculatum	.04	-.6
Marrubium vulgare	.19	-.2	Echinacea purpurea	.04	-.6
Polygonum aviculare	.19	-.2	Fagopyrum esculentum	.05	-.6
Psoralea bituminosa	.22	-.2	Lespedeza sieboldii	.06	-.6
Ulmus campestris	.19	-.2	Sanguisorba officinali	.04	-.6
Castanea vesca	.16	-.3	Stachys officinalis	.05	-.6
Fraxinus excelsior	.17	-.3	Verbena officinalis	.06	-.6
Glechoma hederacea	.15	-.3	Linum usitatissimum	.02	-.7
Mandragora officinalis	.16	-.3	Osmundo regalis 2	.02	-.7
Mentha arvensis	.15	-.3	Stachys officinalis R	.03	-.7

Scandium

Name	Value	Deviation	Name	Value	Deviation
Aconitum napellus	.81	-.4	Marrubium vulgare	.97	-.2
Adonis vernalis	.98	-.2	Melilotus officinalis	.97	-.2
Agrimonia eupatoria	0	-1.2	Melissa officinalis	1.8	.6
Alchemilla vulgaris	1.52	.3	Mentha arvensis	1.12	-.1
Anethum graveolens	0	-1.2	Mercurialis perennis	1.58	.4
Angelica archangelica	1.09	-.1	Milium solis	.86	-.3
Anthemis nobilis	1.36	.1	Ocimum canum	2.19	1
Aquilegia vulgaris	0	-1.2	Ocimum canum R	2.12	.9
Belladonna	1.02	-.2	Oenanthe crocata	1.78	.6
Bellis perennis	2.1	.9	Ononis spinosa	1.23	0
Bryonia dioica	2	.8	Osmundo regalis	7.42	6.2
Cardamine pratensis	.51	-.7	Osmundo regalis 2	1.63	.4
Cardamine pratensis R	0	-1.2	Paeonia officinalis	1.38	.2
Cardiospermum halicaca	.64	-.6	Petroselinum crispum	1.96	.7
Castanea vesca	.53	-.7	Plantago major	1.07	-.1
Centaurium erythraea	3.38	2.2	Polygonum aviculare	1.46	.2
Chamomilla	1.05	-.2	Primula veris	1.46	.2
Chelidonium majus	2.7	1.5	Prunus padus	.93	-.3
Chenopodium anthelmint	0	-1.2	Psoralea bituminosa	.96	-.2
Cicuta virosa	1.87	.7	Rhus toxicodendron	.73	-.5
Cimicifuga racemosa	.65	-.6	Rumex acetosa	2.39	1.2
Cinnamodendron cortiso	.57	-.6	Ruta graveolens	.72	-.5
Clematis erecta	.81	-.4	Salicaria purpurea	.56	-.6
Collinsonia canadensis	1.39	.2	Salix purpurea	.66	-.5
Conium maculatum	0	-1.2	Salvia sclarea	.59	-.6
Cupressus lawsoniana	1.45	.2	Sanguisorba officinali	.79	-.4
Echinacea angustifolia	1.28	.1	Scrophularia nodosa	1.08	-.1
Echinacea purpurea	1.39	.2	Scutellaria lateriflor	2.26	1
Escholtzia californica	.78	-.4	Solidago virgaurea	.56	-.6
Faba vulgaris	.67	-.5	Spilanthes oleracea	.79	-.4
Fagopyrum esculentum	0	-1.2	Stachys officinalis	0	-1.2
Fraxinus excelsior	.62	-.6	Stachys officinalis R	0	-1.2
Galium aparine	1.32	.1	Stachys sylvatica	1.18	0
Glechoma hederacea	.85	-.4	Syringa vulgaris	.71	-.5
Gnaphalium leontopodiu	3.13	1.9	Taraxacum officinale	3.06	1.8
Hedera helix	.92	-.3	Teucrium chamaedrys	.76	-.4
Hydrophyllum virginicu	1.06	-.2	Thuja occidentalis	.93	-.3
Hyoscyamus niger	1.09	-.1	Tormentilla erecta	1.01	-.2
Hyssopus officinalis	.67	-.5	Tussilago farfara	1.05	-.2
Lappa arctium	1.24	0	Tussilago petasites	1.25	0
Lapsana communis	1.87	.7	Ulmus campestris	3.1	1.9
Laurocerasus	.62	-.6	Urtica dioica	3.07	1.9
Leonurus cardiaca	.77	-.4	Urtica urens	.87	-.3
Lespedeza sieboldii	1.26	0	Urtica urens 2	1.84	.6
Linum usitatissimum	0	-1.2	Verbena officinalis	.64	-.6
Lycopus europaeus	.81	-.4	Vinca minor	1.66	.4
Malva sylvestris	.92	-.3	Viola tricolor	.68	-.5
Mandragora officinalis	0	-1.2	Zizia aurea	2.7	1.5

Scandium

Name	Value	Deviation	Name	Value	Deviation
Osmundo regalis	7.42	6.2	Melilotus officinalis	.97	-.2
Centaurium erythraea	3.38	2.2	Psoralea bituminosa	.96	-.2
Gnaphalium leontopodiu	3.13	1.9	Tormentilla erecta	1.01	-.2
Ulmus campestris	3.1	1.9	Tussilago farfara	1.05	-.2
Urtica dioica	3.07	1.9	Hedera helix	.92	-.3
Taraxacum officinale	3.06	1.8	Malva sylvestris	.92	-.3
Chelidonium majus	2.7	1.5	Milium solis	.86	-.3
Zizia aurea	2.7	1.5	Prunus padus	.93	-.3
Rumex acetosa	2.39	1.2	Thuja occidentalis	.93	-.3
Ocimum canum	2.19	1	Urtica urens	.87	-.3
Scutellaria lateriflor	2.26	1	Aconitum napellus	.81	-.4
Bellis perennis	2.1	.9	Clematis erecta	.81	-.4
Ocimum canum R	2.12	.9	Escholtzia californica	.78	-.4
Bryonia dioica	2	.8	Glechoma hederacea	.85	-.4
Cicuta virosa	1.87	.7	Leonurus cardiaca	.77	-.4
Lapsana communis	1.87	.7	Lycopus europaeus	.81	-.4
Petroselinum crispum	1.96	.7	Sanguisorba officinali	.79	-.4
Melissa officinalis	1.8	.6	Spilanthes oleracea	.79	-.4
Oenanthe crocata	1.78	.6	Teucrium chamaedrys	.76	-.4
Urtica urens 2	1.84	.6	Faba vulgaris	.67	-.5
Mercurialis perennis	1.58	.4	Hyssopus officinalis	.67	-.5
Osmundo regalis 2	1.63	.4	Rhus toxicodendron	.73	-.5
Vinca minor	1.66	.4	Ruta graveolens	.72	-.5
Alchemilla vulgaris	1.52	.3	Salix purpurea	.66	-.5
Collinsonia canadensis	1.39	.2	Syringa vulgaris	.71	-.5
Cupressus lawsoniana	1.45	.2	Viola tricolor	.68	-.5
Echinacea purpurea	1.39	.2	Cardiospermum halicaca	.64	-.6
Paeonia officinalis	1.38	.2	Cimicifuga racemosa	.65	-.6
Polygonum aviculare	1.46	.2	Cinnamodendron cortiso	.57	-.6
Primula veris	1.46	.2	Fraxinus excelsior	.62	-.6
Anthemis nobilis	1.36	.1	Laurocerasus	.62	-.6
Echinacea angustifolia	1.28	.1	Salicaria purpurea	.56	-.6
Galium aparine	1.32	.1	Salvia sclarea	.59	-.6
Lappa arctium	1.24	0	Solidago virgaurea	.56	-.6
Lespedeza sieboldii	1.26	0	Verbena officinalis	.64	-.6
Ononis spinosa	1.23	0	Cardamine pratensis	.51	-.7
Stachys sylvatica	1.18	0	Castanea vesca	.53	-.7
Tussilago petasites	1.25	0	Agrimonia eupatoria	0	-1.2
Angelica archangelica	1.09	-.1	Anethum graveolens	0	-1.2
Hyoscyamus niger	1.09	-.1	Aquilegia vulgaris	0	-1.2
Mentha arvensis	1.12	-.1	Cardamine pratensis R	0	-1.2
Plantago major	1.07	-.1	Chenopodium anthelmint	0	-1.2
Scrophularia nodosa	1.08	-.1	Conium maculatum	0	-1.2
Adonis vernalis	.98	-.2	Fagopyrum esculentum	0	-1.2
Belladonna	1.02	-.2	Linum usitatissimum	0	-1.2
Chamomilla	1.05	-.2	Mandragora officinalis	0	-1.2
Hydrophyllum virginicu	1.06	-.2	Stachys officinalis	0	-1.2
Marrubium vulgare	.97	-.2	Stachys officinalis R	0	-1.2

Selenium

Name	Value	Deviation
Aconitum napellus	0	-1.3
Adonis vernalis	0	-1.3
Agrimonia eupatoria	1.18	-.6
Alchemilla vulgaris	1.5	-.4
Anethum graveolens	1.21	-.6
Angelica archangelica	0	-1.3
Anthemis nobilis	2.33	.1
Aquilegia vulgaris	2.58	.3
Belladonna	3.37	.8
Bellis perennis	0	-1.3
Bryonia dioica	0	-1.3
Cardamine pratensis	2.75	.4
Cardamine pratensis R	2.6	.3
Cardiospermum halicaca	1.2	-.6
Castanea vesca	0	-1.3
Centaurium erythraea	1.52	-.4
Chamomilla	2.99	.5
Chelidonium majus	1.17	-.6
Chenopodium anthelmint	1.65	-.3
Cicuta virosa	2.28	.1
Cimicifuga racemosa	0	-1.3
Cinnamodendron cortiso	4.19	1.3
Clematis erecta	2.14	0
Collinsonia canadensis	1.18	-.6
Conium maculatum	4.14	1.2
Cupressus lawsoniana	2.18	0
Echinacea angustifolia	1.86	-.2
Echinacea purpurea	1.93	-.1
Escholtzia californica	1.72	-.2
Faba vulgaris	1.88	-.1
Fagopyrum esculentum	2.22	.1
Fraxinus excelsior	0	-1.3
Galium aparine	1.48	-.4
Glechoma hederacea	4.89	1.7
Gnaphalium leontopodiu	3.28	.7
Hedera helix	1.52	-.4
Hydrophyllum virginicu	2.17	0
Hyoscyamus niger	1.95	-.1
Hyssopus officinalis	2.15	0
Lappa arctium	3.35	.8
Lapsana communis	2.42	.2
Laurocerasus	0	-1.3
Leonurus cardiaca	3.19	.7
Lespedeza sieboldii	0	-1.3
Linum usitatissimum	1.3	-.5
Lycopus europaeus	1.52	-.4
Malva sylvestris	2.07	0
Mandragora officinalis	0	-1.3
Marrubium vulgare	1.34	-.5
Melilotus officinalis	2.29	.1
Melissa officinalis	1.83	-.2
Mentha arvensis	4.91	1.7
Mercurialis perennis	3.91	1.1
Milium solis	1.8	-.2
Ocimum canum	4.09	1.2
Ocimum canum R	4.47	1.4
Oenanthe crocata	2.21	.1
Ononis spinosa	3.07	.6
Osmundo regalis	2.7	.4
Osmundo regalis 2	1.3	-.5
Paeonia officinalis	1.47	-.4
Petroselinum crispum	2.13	0
Plantago major	1.73	-.2
Polygonum aviculare	2.57	.3
Primula veris	2.96	.5
Prunus padus	2.74	.4
Psoralea bituminosa	2.38	.2
Rhus toxicodendron	1.68	-.3
Rumex acetosa	1.36	-.5
Ruta graveolens	2.43	.2
Salicaria purpurea	3.23	.7
Salix purpurea	1.04	-.7
Salvia sclarea	0	-1.3
Sanguisorba officinali	1.52	-.4
Scrophularia nodosa	5.99	2.4
Scutellaria lateriflor	2.73	.4
Solidago virgaurea	2.39	.2
Spilanthes oleracea	1.84	-.2
Stachys officinalis	1.96	-.1
Stachys officinalis R	2.18	0
Stachys sylvatica	4.33	1.4
Syringa vulgaris	0	-1.3
Taraxacum officinale	1.75	-.2
Teucrium chamaedrys	0	-1.3
Thuja occidentalis	1.41	-.4
Tormentilla erecta	0	-1.3
Tussilago farfara	8.59	4
Tussilago petasites	9.71	4.7
Ulmus campestris	2.75	.4
Urtica dioica	3.42	.8
Urtica urens	1.65	-.3
Urtica urens 2	2.01	-.1
Verbena officinalis	1.21	-.6
Vinca minor	2.47	.2
Viola tricolor	1.24	-.5
Zizia aurea	1.49	-.4

Selenium

Name	Value	Deviation	Name	Value	Deviation
Tussilago petasites	9.71	4.7	Stachys officinalis	1.96	-.1
Tussilago farfara	8.59	4	Urtica urens 2	2.01	-.1
Scrophularia nodosa	5.99	2.4	Echinacea angustifolia	1.86	-.2
Glechoma hederacea	4.89	1.7	Escholtzia californica	1.72	-.2
Mentha arvensis	4.91	1.7	Melissa officinalis	1.83	-.2
Ocimum canum R	4.47	1.4	Milium solis	1.8	-.2
Stachys sylvatica	4.33	1.4	Plantago major	1.73	-.2
Cinnamodendron cortiso	4.19	1.3	Spilanthes oleracea	1.84	-.2
Conium maculatum	4.14	1.2	Taraxacum officinale	1.75	-.2
Ocimum canum	4.09	1.2	Chenopodium anthelmint	1.65	-.3
Mercurialis perennis	3.91	1.1	Rhus toxicodendron	1.68	-.3
Belladonna	3.37	.8	Urtica urens	1.65	-.3
Lappa arctium	3.35	.8	Alchemilla vulgaris	1.5	-.4
Urtica dioica	3.42	.8	Centaurium erythraea	1.52	-.4
Gnaphalium leontopodiu	3.28	.7	Galium aparine	1.48	-.4
Leonurus cardiaca	3.19	.7	Hedera helix	1.52	-.4
Salicaria purpurea	3.23	.7	Lycopus europaeus	1.52	-.4
Ononis spinosa	3.07	.6	Paeonia officinalis	1.47	-.4
Chamomilla	2.99	.5	Sanguisorba officinali	1.52	-.4
Primula veris	2.96	.5	Thuja occidentalis	1.41	-.4
Cardamine pratensis	2.75	.4	Zizia aurea	1.49	-.4
Osmundo regalis	2.7	.4	Linum usitatissimum	1.3	-.5
Prunus padus	2.74	.4	Marrubium vulgare	1.34	-.5
Scutellaria lateriflor	2.73	.4	Osmundo regalis 2	1.3	-.5
Ulmus campestris	2.75	.4	Rumex acetosa	1.36	-.5
Aquilegia vulgaris	2.58	.3	Viola tricolor	1.24	-.5
Cardamine pratensis R	2.6	.3	Agrimonia eupatoria	1.18	-.6
Polygonum aviculare	2.57	.3	Anethum graveolens	1.21	-.6
Lapsana communis	2.42	.2	Cardiospermum halicaca	1.2	-.6
Psoralea bituminosa	2.38	.2	Chelidonium majus	1.17	-.6
Ruta graveolens	2.43	.2	Collinsonia canadensis	1.18	-.6
Solidago virgaurea	2.39	.2	Verbena officinalis	1.21	-.6
Vinca minor	2.47	.2	Salix purpurea	1.04	-.7
Anthemis nobilis	2.33	.1	Aconitum napellus	0	-1.3
Cicuta virosa	2.28	.1	Adonis vernalis	0	-1.3
Fagopyrum esculentum	2.22	.1	Angelica archangelica	0	-1.3
Melilotus officinalis	2.29	.1	Bellis perennis	0	-1.3
Oenanthe crocata	2.21	.1	Bryonia dioica	0	-1.3
Clematis erecta	2.14	0	Castanea vesca	0	-1.3
Cupressus lawsoniana	2.18	0	Cimicifuga racemosa	0	-1.3
Hydrophyllum virginicu	2.17	0	Fraxinus excelsior	0	-1.3
Hyssopus officinalis	2.15	0	Laurocerasus	0	-1.3
Malva sylvestris	2.07	0	Lespedeza sieboldii	0	-1.3
Petroselinum crispum	2.13	0	Mandragora officinalis	0	-1.3
Stachys officinalis R	2.18	0	Salvia sclarea	0	-1.3
Echinacea purpurea	1.93	-.1	Syringa vulgaris	0	-1.3
Faba vulgaris	1.88	-.1	Teucrium chamaedrys	0	-1.3
Hyoscyamus niger	1.95	-.1	Tormentilla erecta	0	-1.3

Silicium

Name	Value	Deviation	Name	Value	Deviation
Aconitum napellus	0	-.4	Marrubium vulgare	0	-.4
Adonis vernalis	0	-.4	Melilotus officinalis	0	-.4
Agrimonia eupatoria	0	-.4	Melissa officinalis	2478.49	.6
Alchemilla vulgaris	2365.36	.6	Mentha arvensis	0	-.4
Anethum graveolens	0	-.4	Mercurialis perennis	0	-.4
Angelica archangelica	0	-.4	Milium solis	0	-.4
Anthemis nobilis	0	-.4	Ocimum canum	3283.17	1
Aquilegia vulgaris	0	-.4	Ocimum canum R	3135.55	.9
Belladonna	0	-.4	Oenanthe crocata	2404.79	.6
Bellis perennis	0	-.4	Ononis spinosa	0	-.4
Bryonia dioica	5504.86	1.9	Osmundo regalis	17158.57	6.7
Cardamine pratensis	0	-.4	Osmundo regalis 2	2552.63	.7
Cardamine pratensis R	0	-.4	Paeonia officinalis	0	-.4
Cardiospermum halicaca	0	-.4	Petroselinum crispum	2548.97	.7
Castanea vesca	0	-.4	Plantago major	0	-.4
Centaurium erythraea	8694.45	3.2	Polygonum aviculare	0	-.4
Chamomilla	0	-.4	Primula veris	2020.66	.5
Chelidonium majus	0	-.4	Prunus padus	0	-.4
Chenopodium anthelmint	0	-.4	Psoralea bituminosa	0	-.4
Cicuta virosa	0	-.4	Rhus toxicodendron	0	-.4
Cimicifuga racemosa	0	-.4	Rumex acetosa	0	-.4
Cinnamodendron cortiso	0	-.4	Ruta graveolens	0	-.4
Clematis erecta	0	-.4	Salicaria purpurea	0	-.4
Collinsonia canadensis	0	-.4	Salix purpurea	0	-.4
Conium maculatum	0	-.4	Salvia sclarea	0	-.4
Cupressus lawsoniana	0	-.4	Sanguisorba officinali	0	-.4
Echinacea angustifolia	0	-.4	Scrophularia nodosa	0	-.4
Echinacea purpurea	0	-.4	Scutellaria lateriflor	3449.87	1
Escholtzia californica	0	-.4	Solidago virgaurea	0	-.4
Faba vulgaris	0	-.4	Spilanthes oleracea	0	-.4
Fagopyrum esculentum	0	-.4	Stachys officinalis	0	-.4
Fraxinus excelsior	0	-.4	Stachys officinalis R	0	-.4
Galium aparine	0	-.4	Stachys sylvatica	0	-.4
Glechoma hederacea	0	-.4	Syringa vulgaris	0	-.4
Gnaphalium leontopodiu	7297.85	2.6	Taraxacum officinale	5334.08	1.8
Hedera helix	0	-.4	Teucrium chamaedrys	0	-.4
Hydrophyllum virginicu	0	-.4	Thuja occidentalis	0	-.4
Hyoscyamus niger	0	-.4	Tormentilla erecta	0	-.4
Hyssopus officinalis	0	-.4	Tussilago farfara	0	-.4
Lappa arctium	0	-.4	Tussilago petasites	0	-.4
Lapsana communis	0	-.4	Ulmus campestris	4194.29	1.3
Laurocerasus	0	-.4	Urtica dioica	3554.03	1.1
Leonurus cardiaca	0	-.4	Urtica urens	0	-.4
Lespedeza sieboldii	0	-.4	Urtica urens 2	2733.05	.7
Linum usitatissimum	0	-.4	Verbena officinalis	0	-.4
Lycopus europaeus	0	-.4	Vinca minor	3629.71	1.1
Malva sylvestris	0	-.4	Viola tricolor	0	-.4
Mandragora officinalis	0	-.4	Zizia aurea	6423.81	2.3

Silicium

Name	Value	Deviation	Name	Value	Deviation
Osmundo regalis	17158.57	6.7	Galium aparine	0	-.4
Centaurium erythraea	8694.45	3.2	Glechoma hederacea	0	-.4
Gnaphalium leontopodiu	7297.85	2.6	Hedera helix	0	-.4
Zizia aurea	6423.81	2.3	Hydrophyllum virginicu	0	-.4
Bryonia dioica	5504.86	1.9	Hyoscyamus niger	0	-.4
Taraxacum officinale	5334.08	1.8	Hyssopus officinalis	0	-.4
Ulmus campestris	4194.29	1.3	Lappa arctium	0	-.4
Urtica dioica	3554.03	1.1	Lapsana communis	0	-.4
Vinca minor	3629.71	1.1	Laurocerasus	0	-.4
Ocimum canum	3283.17	1	Leonurus cardiaca	0	-.4
Scutellaria lateriflor	3449.87	1	Lespedeza sieboldii	0	-.4
Ocimum canum R	3135.55	.9	Linum usitatissimum	0	-.4
Osmundo regalis 2	2552.63	.7	Lycopus europaeus	0	-.4
Petroselinum crispum	2548.97	.7	Malva sylvestris	0	-.4
Urtica urens 2	2733.05	.7	Mandragora officinalis	0	-.4
Alchemilla vulgaris	2365.36	.6	Marrubium vulgare	0	-.4
Melissa officinalis	2478.49	.6	Melilotus officinalis	0	-.4
Oenanthe crocata	2404.79	.6	Mentha arvensis	0	-.4
Primula veris	2020.66	.5	Mercurialis perennis	0	-.4
Aconitum napellus	0	-.4	Milium solis	0	-.4
Adonis vernalis	0	-.4	Ononis spinosa	0	-.4
Agrimonia eupatoria	0	-.4	Paeonia officinalis	0	-.4
Anethum graveolens	0	-.4	Plantago major	0	-.4
Angelica archangelica	0	-.4	Polygonum aviculare	0	-.4
Anthemis nobilis	0	-.4	Prunus padus	0	-.4
Aquilegia vulgaris	0	-.4	Psoralea bituminosa	0	-.4
Belladonna	0	-.4	Rhus toxicodendron	0	-.4
Bellis perennis	0	-.4	Rumex acetosa	0	-.4
Cardamine pratensis	0	-.4	Ruta graveolens	0	-.4
Cardamine pratensis R	0	-.4	Salicaria purpurea	0	-.4
Cardiospermum halicaca	0	-.4	Salix purpurea	0	-.4
Castanea vesca	0	-.4	Salvia sclarea	0	-.4
Chamomilla	0	-.4	Sanguisorba officinali	0	-.4
Chelidonium majus	0	-.4	Scrophularia nodosa	0	-.4
Chenopodium anthelmint	0	-.4	Solidago virgaurea	0	-.4
Cicuta virosa	0	-.4	Spilanthes oleracea	0	-.4
Cimicifuga racemosa	0	-.4	Stachys officinalis	0	-.4
Cinnamodendron cortiso	0	-.4	Stachys officinalis R	0	-.4
Clematis erecta	0	-.4	Stachys sylvatica	0	-.4
Collinsonia canadensis	0	-.4	Syringa vulgaris	0	-.4
Conium maculatum	0	-.4	Teucrium chamaedrys	0	-.4
Cupressus lawsoniana	0	-.4	Thuja occidentalis	0	-.4
Echinacea angustifolia	0	-.4	Tormentilla erecta	0	-.4
Echinacea purpurea	0	-.4	Tussilago farfara	0	-.4
Escholtzia californica	0	-.4	Tussilago petasites	0	-.4
Faba vulgaris	0	-.4	Urtica urens	0	-.4
Fagopyrum esculentum	0	-.4	Verbena officinalis	0	-.4
Fraxinus excelsior	0	-.4	Viola tricolor	0	-.4

Stibium

Name	Value	Deviation	Name	Value	Deviation
Aconitum napellus	.2	-.4	Marrubium vulgare	.21	-.4
Adonis vernalis	.1	-.7	Melilotus officinalis	.45	.4
Agrimonia eupatoria	.35	.1	Melissa officinalis	.41	.2
Alchemilla vulgaris	.29	-.1	Mentha arvensis	.55	.7
Anethum graveolens	.13	-.6	Mercurialis perennis	.23	-.3
Angelica archangelica	.17	-.5	Milium solis	.29	-.1
Anthemis nobilis	.24	-.3	Ocimum canum	.2	-.4
Aquilegia vulgaris	.32	0	Ocimum canum R	.18	-.4
Belladonna	.37	.1	Oenanthe crocata	.1	-.7
Bellis perennis	.16	-.5	Ononis spinosa	.29	-.1
Bryonia dioica	.1	-.7	Osmundo regalis	1.79	4.4
Cardamine pratensis	.55	.7	Osmundo regalis 2	.04	-.9
Cardamine pratensis R	.53	.6	Paeonia officinalis	.16	-.5
Cardiospermum halicaca	.08	-.7	Petroselinum crispum	.14	-.6
Castanea vesca	.31	-.1	Plantago major	.09	-.7
Centaurium erythraea	.37	.1	Polygonum aviculare	.09	-.7
Chamomilla	.18	-.4	Primula veris	.42	.3
Chelidonium majus	.07	-.8	Prunus padus	1	2
Chenopodium anthelmint	.24	-.3	Psoralea bituminosa	.26	-.2
Cicuta virosa	.2	-.4	Rhus toxicodendron	.3	-.1
Cimicifuga racemosa	.12	-.6	Rumex acetosa	.15	-.5
Cinnamodendron cortiso	.57	.7	Ruta graveolens	.35	.1
Clematis erecta	.48	.5	Salicaria purpurea	.27	-.2
Collinsonia canadensis	.4	.2	Salix purpurea	.64	.9
Conium maculatum	.59	.8	Salvia sclarea	.2	-.4
Cupressus lawsoniana	2.39	6.2	Sanguisorba officinali	.09	-.7
Echinacea angustifolia	.12	-.6	Scrophularia nodosa	.71	1.1
Echinacea purpurea	.13	-.6	Scutellaria lateriflor	.27	-.2
Escholtzia californica	.12	-.6	Solidago virgaurea	.11	-.7
Faba vulgaris	.23	-.3	Spilanthes oleracea	.04	-.9
Fagopyrum esculentum	.13	-.6	Stachys officinalis	.1	-.7
Fraxinus excelsior	.52	.6	Stachys officinalis R	.16	-.5
Galium aparine	.07	-.8	Stachys sylvatica	.33	0
Glechoma hederacea	.59	.8	Syringa vulgaris	1.05	2.2
Gnaphalium leontopodiu	.18	-.4	Taraxacum officinale	.22	-.3
Hedera helix	.89	1.7	Teucrium chamaedrys	.26	-.2
Hydrophyllum virginicu	.37	.1	Thuja occidentalis	.7	1.1
Hyoscyamus niger	.21	-.4	Tormentilla erecta	.3	-.1
Hyssopus officinalis	.22	-.3	Tussilago farfara	.26	-.2
Lappa arctium	.55	.7	Tussilago petasites	.29	-.1
Lapsana communis	.16	-.5	Ulmus campestris	.23	-.3
Laurocerasus	.33	0	Urtica dioica	.09	-.7
Leonurus cardiaca	.47	.4	Urtica urens	.03	-.9
Lespedeza sieboldii	.48	.5	Urtica urens 2	.05	-.8
Linum usitatissimum	.16	-.5	Verbena officinalis	.43	.3
Lycopus europaeus	.34	0	Vinca minor	.48	.5
Malva sylvestris	.36	.1	Viola tricolor	.17	-.5
Mandragora officinalis	.05	-.8	Zizia aurea	.12	-.6

Stibium

Name	Value	Deviation	Name	Value	Deviation
Cupressus lawsoniana	2.39	6.2	Chenopodium anthelmint	.24	-.3
Osmundo regalis	1.79	4.4	Faba vulgaris	.23	-.3
Syringa vulgaris	1.05	2.2	Hyssopus officinalis	.22	-.3
Prunus padus	1	2	Mercurialis perennis	.23	-.3
Hedera helix	.89	1.7	Taraxacum officinale	.22	-.3
Scrophularia nodosa	.71	1.1	Ulmus campestris	.23	-.3
Thuja occidentalis	.7	1.1	Aconitum napellus	.2	-.4
Salix purpurea	.64	.9	Chamomilla	.18	-.4
Conium maculatum	.59	.8	Cicuta virosa	.2	-.4
Glechoma hederacea	.59	.8	Gnaphalium leontopodiu	.18	-.4
Cardamine pratensis	.55	.7	Hyoscyamus niger	.21	-.4
Cinnamodendron cortiso	.57	.7	Marrubium vulgare	.21	-.4
Lappa arctium	.55	.7	Ocimum canum	.2	-.4
Mentha arvensis	.55	.7	Ocimum canum R	.18	-.4
Cardamine pratensis R	.53	.6	Salvia sclarea	.2	-.4
Fraxinus excelsior	.52	.6	Angelica archangelica	.17	-.5
Clematis erecta	.48	.5	Bellis perennis	.16	-.5
Lespedeza sieboldii	.48	.5	Lapsana communis	.16	-.5
Vinca minor	.48	.5	Linum usitatissimum	.16	-.5
Leonurus cardiaca	.47	.4	Paeonia officinalis	.16	-.5
Melilotus officinalis	.45	.4	Rumex acetosa	.15	-.5
Primula veris	.42	.3	Stachys officinalis R	.16	-.5
Verbena officinalis	.43	.3	Viola tricolor	.17	-.5
Collinsonia canadensis	.4	.2	Anethum graveolens	.13	-.6
Melissa officinalis	.41	.2	Cimicifuga racemosa	.12	-.6
Agrimonia eupatoria	.35	.1	Echinacea angustifolia	.12	-.6
Belladonna	.37	.1	Echinacea purpurea	.13	-.6
Centaurium erythraea	.37	.1	Escholtzia californica	.12	-.6
Hydrophyllum virginicu	.37	.1	Fagopyrum esculentum	.13	-.6
Malva sylvestris	.36	.1	Petroselinum crispum	.14	-.6
Ruta graveolens	.35	.1	Zizia aurea	.12	-.6
Aquilegia vulgaris	.32	0	Adonis vernalis	.1	-.7
Laurocerasus	.33	0	Bryonia dioica	.1	-.7
Lycopus europaeus	.34	0	Cardiospermum halicaca	.08	-.7
Stachys sylvatica	.33	0	Oenanthe crocata	.1	-.7
Alchemilla vulgaris	.29	-.1	Plantago major	.09	-.7
Castanea vesca	.31	-.1	Polygonum aviculare	.09	-.7
Milium solis	.29	-.1	Sanguisorba officinali	.09	-.7
Ononis spinosa	.29	-.1	Solidago virgaurea	.11	-.7
Rhus toxicodendron	.3	-.1	Stachys officinalis	.1	-.7
Tormentilla erecta	.3	-.1	Urtica dioica	.09	-.7
Tussilago petasites	.29	-.1	Chelidonium majus	.07	-.8
Psoralea bituminosa	.26	-.2	Galium aparine	.07	-.8
Salicaria purpurea	.27	-.2	Mandragora officinalis	.05	-.8
Scutellaria lateriflor	.27	-.2	Urtica urens 2	.05	-.8
Teucrium chamaedrys	.26	-.2	Osmundo regalis 2	.04	-.9
Tussilago farfara	.26	-.2	Spilanthes oleracea	.04	-.9
Anthemis nobilis	.24	-.3	Urtica urens	.03	-.9

Strontium

Name	Value	Deviation	Name	Value	Deviation
Aconitum napellus	1260	.6	Marrubium vulgare	960	0
Adonis vernalis	241	-1.3	Melilotus officinalis	1670	1.3
Agrimonia eupatoria	619	-.6	Melissa officinalis	1140	.3
Alchemilla vulgaris	667	-.5	Mentha arvensis	1210	.5
Anethum graveolens	1260	.6	Mercurialis perennis	1050	.2
Angelica archangelica	623	-.6	Milium solis	511	-.8
Anthemis nobilis	548	-.7	Ocimum canum	1230	.5
Aquilegia vulgaris	366	-1	Ocimum canum R	1280	.6
Belladonna	506	-.8	Oenanthe crocata	414	-1
Bellis perennis	107	-1.5	Ononis spinosa	1020	.1
Bryonia dioica	732	-.4	Osmundo regalis	892	-.1
Cardamine pratensis	833	-.2	Osmundo regalis 2	127	-1.5
Cardamine pratensis R	770	-.3	Paeonia officinalis	2660	3.1
Cardiospermum halicaca	704	-.4	Petroselinum crispum	659	-.5
Castanea vesca	637	-.6	Plantago major	972	0
Centaurium erythraea	364	-1	Polygonum aviculare	873	-.1
Chamomilla	374	-1	Primula veris	349	-1.1
Chelidonium majus	253	-1.2	Prunus padus	1230	.5
Chenopodium anthelmint	838	-.2	Psoralea bituminosa	1420	.8
Cicuta virosa	664	-.5	Rhus toxicodendron	1830	1.6
Cimicifuga racemosa	1040	.2	Rumex acetosa	2110	2.1
Cinnamodendron cortiso	1490	1	Ruta graveolens	1200	.5
Clematis erecta	878	-.1	Salicaria purpurea	1310	.7
Collinsonia canadensis	409	-1	Salix purpurea	2180	2.2
Conium maculatum	671	-.5	Salvia sclarea	1360	.7
Cupressus lawsoniana	1490	1	Sanguisorba officinali	802	-.3
Echinacea angustifolia	1120	.3	Scrophularia nodosa	1140	.3
Echinacea purpurea	1280	.6	Scutellaria lateriflor	1230	.5
Escholtzia californica	357	-1.1	Solidago virgaurea	455	-.9
Faba vulgaris	650	-.5	Spilanthes oleracea	728	-.4
Fagopyrum esculentum	1180	.4	Stachys officinalis	343	-1.1
Fraxinus excelsior	2250	2.3	Stachys officinalis R	369	-1
Galium aparine	951	0	Stachys sylvatica	677	-.5
Glechoma hederacea	641	-.5	Syringa vulgaris	294	-1.2
Gnaphalium leontopodiu	595	-.6	Taraxacum officinale	228	-1.3
Hedera helix	1190	.4	Teucrium chamaedrys	1120	.3
Hydrophyllum virginicu	768	-.3	Thuja occidentalis	2640	3
Hyoscyamus niger	531	-.7	Tormentilla erecta	807	-.2
Hyssopus officinalis	883	-.1	Tussilago farfara	820	-.2
Lappa arctium	639	-.5	Tussilago petasites	479	-.8
Lapsana communis	748	-.4	Ulmus campestris	2960	3.6
Laurocerasus	1090	.3	Urtica dioica	1180	.4
Leonurus cardiaca	934	0	Urtica urens	1660	1.3
Lespedeza sieboldii	1300	.6	Urtica urens 2	1610	1.2
Linum usitatissimum	564	-.7	Verbena officinalis	570	-.7
Lycopus europaeus	670	-.5	Vinca minor	1510	1
Malva sylvestris	1010	.1	Viola tricolor	426	-.9
Mandragora officinalis	162	-1.4	Zizia aurea	1270	.6

Strontium

Name	Value	Deviation	Name	Value	Deviation
Ulmus campestris	2960	3.6	Chenopodium anthelmint	838	-.2
Paeonia officinalis	2660	3.1	Tormentilla erecta	807	-.2
Thuja occidentalis	2640	3	Tussilago farfara	820	-.2
Fraxinus excelsior	2250	2.3	Cardamine pratensis R	770	-.3
Salix purpurea	2180	2.2	Hydrophyllum virginicu	768	-.3
Rumex acetosa	2110	2.1	Sanguisorba officinali	802	-.3
Rhus toxicodendron	1830	1.6	Bryonia dioica	732	-.4
Melilotus officinalis	1670	1.3	Cardiospermum halicaca	704	-.4
Urtica urens	1660	1.3	Lapsana communis	748	-.4
Urtica urens 2	1610	1.2	Spilanthes oleracea	728	-.4
Cinnamodendron cortiso	1490	1	Alchemilla vulgaris	667	-.5
Cupressus lawsoniana	1490	1	Cicuta virosa	664	-.5
Vinca minor	1510	1	Conium maculatum	671	-.5
Psoralea bituminosa	1420	.8	Faba vulgaris	650	-.5
Salicaria purpurea	1310	.7	Glechoma hederacea	641	-.5
Salvia sclarea	1360	.7	Lappa arctium	639	-.5
Aconitum napellus	1260	.6	Lycopus europaeus	670	-.5
Anethum graveolens	1260	.6	Petroselinum crispum	659	-.5
Echinacea purpurea	1280	.6	Stachys sylvatica	677	-.5
Lespedeza sieboldii	1300	.6	Agrimonia eupatoria	619	-.6
Ocimum canum R	1280	.6	Angelica archangelica	623	-.6
Zizia aurea	1270	.6	Castanea vesca	637	-.6
Mentha arvensis	1210	.5	Gnaphalium leontopodiu	595	-.6
Ocimum canum	1230	.5	Anthemis nobilis	548	-.7
Prunus padus	1230	.5	Hyoscyamus niger	531	-.7
Ruta graveolens	1200	.5	Linum usitatissimum	564	-.7
Scutellaria lateriflor	1230	.5	Verbena officinalis	570	-.7
Fagopyrum esculentum	1180	.4	Belladonna	506	-.8
Hedera helix	1190	.4	Milium solis	511	-.8
Urtica dioica	1180	.4	Tussilago petasites	479	-.8
Echinacea angustifolia	1120	.3	Solidago virgaurea	455	-.9
Laurocerasus	1090	.3	Viola tricolor	426	-.9
Melissa officinalis	1140	.3	Aquilegia vulgaris	366	-1
Scrophularia nodosa	1140	.3	Centaurium erythraea	364	-1
Teucrium chamaedrys	1120	.3	Chamomilla	374	-1
Cimicifuga racemosa	1040	.2	Collinsonia canadensis	409	-1
Mercurialis perennis	1050	.2	Oenanthe crocata	414	-1
Malva sylvestris	1010	.1	Stachys officinalis R	369	-1
Ononis spinosa	1020	.1	Escholtzia californica	357	-1.1
Galium aparine	951	0	Primula veris	349	-1.1
Leonurus cardiaca	934	0	Stachys officinalis	343	-1.1
Marrubium vulgare	960	0	Chelidonium majus	253	-1.2
Plantago major	972	0	Syringa vulgaris	294	-1.2
Clematis erecta	878	-.1	Adonis vernalis	241	-1.3
Hyssopus officinalis	883	-.1	Taraxacum officinale	228	-1.3
Osmundo regalis	892	-.1	Mandragora officinalis	162	-1.4
Polygonum aviculare	873	-.1	Bellis perennis	107	-1.5
Cardamine pratensis	833	-.2	Osmundo regalis 2	127	-1.5

Tantalum

Name	Value	Deviation	Name	Value	Deviation
Aconitum napellus	.01	.6	Marrubium vulgare	.01	.6
Adonis vernalis	.01	.6	Melilotus officinalis	0	-.7
Agrimonia eupatoria	0	-.7	Melissa officinalis	.02	2
Alchemilla vulgaris	.01	.6	Mentha arvensis	.01	.6
Anethum graveolens	0	-.7	Mercurialis perennis	.02	2
Angelica archangelica	.01	.6	Milium solis	.01	.6
Anthemis nobilis	.01	.6	Ocimum canum	.01	.6
Aquilegia vulgaris	0	-.7	Ocimum canum R	.01	.6
Belladonna	.01	.6	Oenanthe crocata	.01	.6
Bellis perennis	.01	.6	Ononis spinosa	0	-.7
Bryonia dioica	.01	.6	Osmundo regalis	.04	4.7
Cardamine pratensis	0	-.7	Osmundo regalis 2	0	-.7
Cardamine pratensis R	0	-.7	Paeonia officinalis	.01	.6
Cardiospermum halicaca	0	-.7	Petroselinum crispum	.01	.6
Castanea vesca	0	-.7	Plantago major	.01	.6
Centaurium erythraea	.03	3.4	Polygonum aviculare	.01	.6
Chamomilla	.01	.6	Primula veris	.02	2
Chelidonium majus	.01	.6	Prunus padus	.01	.6
Chenopodium anthelmint	0	-.7	Psoralea bituminosa	0	-.7
Cicuta virosa	.01	.6	Rhus toxicodendron	0	-.7
Cimicifuga racemosa	0	-.7	Rumex acetosa	.01	.6
Cinnamodendron cortiso	0	-.7	Ruta graveolens	0	-.7
Clematis erecta	0	-.7	Salicaria purpurea	0	-.7
Collinsonia canadensis	0	-.7	Salix purpurea	0	-.7
Conium maculatum	0	-.7	Salvia sclarea	0	-.7
Cupressus lawsoniana	.01	.6	Sanguisorba officinali	0	-.7
Echinacea angustifolia	0	-.7	Scrophularia nodosa	0	-.7
Echinacea purpurea	0	-.7	Scutellaria lateriflor	.01	.6
Escholtzia californica	.01	.6	Solidago virgaurea	0	-.7
Faba vulgaris	0	-.7	Spilanthes oleracea	0	-.7
Fagopyrum esculentum	0	-.7	Stachys officinalis	0	-.7
Fraxinus excelsior	0	-.7	Stachys officinalis R	0	-.7
Galium aparine	0	-.7	Stachys sylvatica	0	-.7
Glechoma hederacea	0	-.7	Syringa vulgaris	.01	.6
Gnaphalium leontopodiu	.02	2	Taraxacum officinale	.01	.6
Hedera helix	0	-.7	Teucrium chamaedrys	0	-.7
Hydrophyllum virginicu	0	-.7	Thuja occidentalis	0	-.7
Hyoscyamus niger	.01	.6	Tormentilla erecta	0	-.7
Hyssopus officinalis	0	-.7	Tussilago farfara	0	-.7
Lappa arctium	.01	.6	Tussilago petasites	0	-.7
Lapsana communis	.02	2	Ulmus campestris	0	-.7
Laurocerasus	0	-.7	Urtica dioica	0	-.7
Leonurus cardiaca	.01	.6	Urtica urens	0	-.7
Lespedeza sieboldii	0	-.7	Urtica urens 2	0	-.7
Linum usitatissimum	0	-.7	Verbena officinalis	0	-.7
Lycopus europaeus	0	-.7	Vinca minor	0	-.7
Malva sylvestris	.01	.6	Viola tricolor	0	-.7
Mandragora officinalis	0	-.7	Zizia aurea	.01	.6

Tantalum

Name	Value	Deviation	Name	Value	Deviation
Osmundo regalis	.04	4.7	Cimicifuga racemosa	0	-.7
Centaurium erythraea	.03	3.4	Cinnamodendron cortiso	0	-.7
Gnaphalium leontopodiu	.02	2	Clematis erecta	0	-.7
Lapsana communis	.02	2	Collinsonia canadensis	0	-.7
Melissa officinalis	.02	2	Conium maculatum	0	-.7
Mercurialis perennis	.02	2	Echinacea angustifolia	0	-.7
Primula veris	.02	2	Echinacea purpurea	0	-.7
Aconitum napellus	.01	.6	Faba vulgaris	0	-.7
Adonis vernalis	.01	.6	Fagopyrum esculentum	0	-.7
Alchemilla vulgaris	.01	.6	Fraxinus excelsior	0	-.7
Angelica archangelica	.01	.6	Galium aparine	0	-.7
Anthemis nobilis	.01	.6	Glechoma hederacea	0	-.7
Belladonna	.01	.6	Hedera helix	0	-.7
Bellis perennis	.01	.6	Hydrophyllum virginicu	0	-.7
Bryonia dioica	.01	.6	Hyssopus officinalis	0	-.7
Chamomilla	.01	.6	Laurocerasus	0	-.7
Chelidonium majus	.01	.6	Lespedeza sieboldii	0	-.7
Cicuta virosa	.01	.6	Linum usitatissimum	0	-.7
Cupressus lawsoniana	.01	.6	Lycopus europaeus	0	-.7
Escholtzia californica	.01	.6	Mandragora officinalis	0	-.7
Hyoscyamus niger	.01	.6	Melilotus officinalis	0	-.7
Lappa arctium	.01	.6	Ononis spinosa	0	-.7
Leonurus cardiaca	.01	.6	Osmundo regalis 2	0	-.7
Malva sylvestris	.01	.6	Psoralea bituminosa	0	-.7
Marrubium vulgare	.01	.6	Rhus toxicodendron	0	-.7
Mentha arvensis	.01	.6	Ruta graveolens	0	-.7
Milium solis	.01	.6	Salicaria purpurea	0	-.7
Ocimum canum	.01	.6	Salix purpurea	0	-.7
Ocimum canum R	.01	.6	Salvia sclarea	0	-.7
Oenanthe crocata	.01	.6	Sanguisorba officinali	0	-.7
Paeonia officinalis	.01	.6	Scrophularia nodosa	0	-.7
Petroselinum crispum	.01	.6	Solidago virgaurea	0	-.7
Plantago major	.01	.6	Spilanthes oleracea	0	-.7
Polygonum aviculare	.01	.6	Stachys officinalis	0	-.7
Prunus padus	.01	.6	Stachys officinalis R	0	-.7
Rumex acetosa	.01	.6	Stachys sylvatica	0	-.7
Scutellaria lateriflor	.01	.6	Teucrium chamaedrys	0	-.7
Syringa vulgaris	.01	.6	Thuja occidentalis	0	-.7
Taraxacum officinale	.01	.6	Tormentilla erecta	0	-.7
Zizia aurea	.01	.6	Tussilago farfara	0	-.7
Agrimonia eupatoria	0	-.7	Tussilago petasites	0	-.7
Anethum graveolens	0	-.7	Ulmus campestris	0	-.7
Aquilegia vulgaris	0	-.7	Urtica dioica	0	-.7
Cardamine pratensis	0	-.7	Urtica urens	0	-.7
Cardamine pratensis R	0	-.7	Urtica urens 2	0	-.7
Cardiospermum halicaca	0	-.7	Verbena officinalis	0	-.7
Castanea vesca	0	-.7	Vinca minor	0	-.7
Chenopodium anthelmint	0	-.7	Viola tricolor	0	-.7

Tellurium

Name	Value	Deviation
Aconitum napellus	.04	.1
Adonis vernalis	.02	-.8
Agrimonia eupatoria	.02	-.8
Alchemilla vulgaris	.06	1
Anethum graveolens	.02	-.8
Angelica archangelica	.02	-.8
Anthemis nobilis	.04	.1
Aquilegia vulgaris	.04	.1
Belladonna	.03	-.4
Bellis perennis	.02	-.8
Bryonia dioica	.03	-.4
Cardamine pratensis	.1	2.7
Cardamine pratensis R	.08	1.8
Cardiospermum halicaca	.02	-.8
Castanea vesca	.03	-.4
Centaurium erythraea	.04	.1
Chamomilla	.02	-.8
Chelidonium majus	.04	.1
Chenopodium anthelmint	.04	.1
Cicuta virosa	.13	4.1
Cimicifuga racemosa	.02	-.8
Cinnamodendron cortiso	.05	.5
Clematis erecta	.07	1.4
Collinsonia canadensis	.03	-.4
Conium maculatum	0	-1.7
Cupressus lawsoniana	.11	3.2
Echinacea angustifolia	.03	-.4
Echinacea purpurea	.03	-.4
Escholtzia californica	.01	-1.3
Faba vulgaris	.05	.5
Fagopyrum esculentum	.04	.1
Fraxinus excelsior	.03	-.4
Galium aparine	.03	-.4
Glechoma hederacea	.06	1
Gnaphalium leontopodiu	.03	-.4
Hedera helix	.05	.5
Hydrophyllum virginicu	.06	1
Hyoscyamus niger	.06	1
Hyssopus officinalis	.02	-.8
Lappa arctium	.04	.1
Lapsana communis	.05	.5
Laurocerasus	.03	-.4
Leonurus cardiaca	.05	.5
Lespedeza sieboldii	.03	-.4
Linum usitatissimum	.04	.1
Lycopus europaeus	.02	-.8
Malva sylvestris	.04	.1
Mandragora officinalis	.02	-.8
Marrubium vulgare	.03	-.4
Melilotus officinalis	.04	.1
Melissa officinalis	.04	.1
Mentha arvensis	.05	.5
Mercurialis perennis	.06	1
Milium solis	.05	.5
Ocimum canum	.05	.5
Ocimum canum R	.06	1
Oenanthe crocata	.02	-.8
Ononis spinosa	.04	.1
Osmundo regalis	.08	1.8
Osmundo regalis 2	0	-1.7
Paeonia officinalis	.03	-.4
Petroselinum crispum	.02	-.8
Plantago major	.04	.1
Polygonum aviculare	.02	-.8
Primula veris	.04	.1
Prunus padus	.06	1
Psoralea bituminosa	.05	.5
Rhus toxicodendron	.01	-1.3
Rumex acetosa	.03	-.4
Ruta graveolens	.04	.1
Salicaria purpurea	.05	.5
Salix purpurea	.05	.5
Salvia sclarea	.03	-.4
Sanguisorba officinali	.04	.1
Scrophularia nodosa	.05	.5
Scutellaria lateriflor	.04	.1
Solidago virgaurea	.02	-.8
Spilanthes oleracea	.02	-.8
Stachys officinalis	.04	.1
Stachys officinalis R	.03	-.4
Stachys sylvatica	.02	-.8
Syringa vulgaris	.03	-.4
Taraxacum officinale	.05	.5
Teucrium chamaedrys	.02	-.8
Thuja occidentalis	.08	1.8
Tormentilla erecta	.02	-.8
Tussilago farfara	.07	1.4
Tussilago petasites	.07	1.4
Ulmus campestris	.05	.5
Urtica dioica	.03	-.4
Urtica urens	.01	-1.3
Urtica urens 2	.01	-1.3
Verbena officinalis	.02	-.8
Vinca minor	.05	.5
Viola tricolor	0	-1.7
Zizia aurea	0	-1.7

Tellurium

Name	Value	Deviation	Name	Value	Deviation
Cicuta virosa	.13	4.1	Scutellaria lateriflor	.04	.1
Cupressus lawsoniana	.11	3.2	Stachys officinalis	.04	.1
Cardamine pratensis	.1	2.7	Belladonna	.03	-.4
Cardamine pratensis R	.08	1.8	Bryonia dioica	.03	-.4
Osmundo regalis	.08	1.8	Castanea vesca	.03	-.4
Thuja occidentalis	.08	1.8	Collinsonia canadensis	.03	-.4
Clematis erecta	.07	1.4	Echinacea angustifolia	.03	-.4
Tussilago farfara	.07	1.4	Echinacea purpurea	.03	-.4
Tussilago petasites	.07	1.4	Fraxinus excelsior	.03	-.4
Alchemilla vulgaris	.06	1	Galium aparine	.03	-.4
Glechoma hederacea	.06	1	Gnaphalium leontopodiu	.03	-.4
Hydrophyllum virginicu	.06	1	Laurocerasus	.03	-.4
Hyoscyamus niger	.06	1	Lespedeza sieboldii	.03	-.4
Mercurialis perennis	.06	1	Marrubium vulgare	.03	-.4
Ocimum canum R	.06	1	Paeonia officinalis	.03	-.4
Prunus padus	.06	1	Rumex acetosa	.03	-.4
Cinnamodendron cortiso	.05	.5	Salvia sclarea	.03	-.4
Faba vulgaris	.05	.5	Stachys officinalis R	.03	-.4
Hedera helix	.05	.5	Syringa vulgaris	.03	-.4
Lapsana communis	.05	.5	Urtica dioica	.03	-.4
Leonurus cardiaca	.05	.5	Adonis vernalis	.02	-.8
Mentha arvensis	.05	.5	Agrimonia eupatoria	.02	-.8
Milium solis	.05	.5	Anethum graveolens	.02	-.8
Ocimum canum	.05	.5	Angelica archangelica	.02	-.8
Psoralea bituminosa	.05	.5	Bellis perennis	.02	-.8
Salicaria purpurea	.05	.5	Cardiospermum halicaca	.02	-.8
Salix purpurea	.05	.5	Chamomilla	.02	-.8
Scrophularia nodosa	.05	.5	Cimicifuga racemosa	.02	-.8
Taraxacum officinale	.05	.5	Hyssopus officinalis	.02	-.8
Ulmus campestris	.05	.5	Lycopus europaeus	.02	-.8
Vinca minor	.05	.5	Mandragora officinalis	.02	-.8
Aconitum napellus	.04	.1	Oenanthe crocata	.02	-.8
Anthemis nobilis	.04	.1	Petroselinum crispum	.02	-.8
Aquilegia vulgaris	.04	.1	Polygonum aviculare	.02	-.8
Centaurium erythraea	.04	.1	Solidago virgaurea	.02	-.8
Chelidonium majus	.04	.1	Spilanthes oleracea	.02	-.8
Chenopodium anthelmint	.04	.1	Stachys sylvatica	.02	-.8
Fagopyrum esculentum	.04	.1	Teucrium chamaedrys	.02	-.8
Lappa arctium	.04	.1	Tormentilla erecta	.02	-.8
Linum usitatissimum	.04	.1	Verbena officinalis	.02	-.8
Malva sylvestris	.04	.1	Escholtzia californica	.01	-1.3
Melilotus officinalis	.04	.1	Rhus toxicodendron	.01	-1.3
Melissa officinalis	.04	.1	Urtica urens	.01	-1.3
Ononis spinosa	.04	.1	Urtica urens 2	.01	-1.3
Plantago major	.04	.1	Conium maculatum	0	-1.7
Primula veris	.04	.1	Osmundo regalis 2	0	-1.7
Ruta graveolens	.04	.1	Viola tricolor	0	-1.7
Sanguisorba officinali	.04	.1	Zizia aurea	0	-1.7

Terbium

Name	Value	Deviation	Name	Value	Deviation
Aconitum napellus	.06	.4	Marrubium vulgare	.03	-.2
Adonis vernalis	.05	.2	Melilotus officinalis	.01	-.5
Agrimonia eupatoria	.01	-.5	Melissa officinalis	.04	0
Alchemilla vulgaris	.05	.2	Mentha arvensis	.02	-.4
Anethum graveolens	.01	-.5	Mercurialis perennis	.05	.2
Angelica archangelica	.05	.2	Milium solis	.02	-.4
Anthemis nobilis	.03	-.2	Ocimum canum	.04	0
Aquilegia vulgaris	.01	-.5	Ocimum canum R	.04	0
Belladonna	.03	-.2	Oenanthe crocata	.05	.2
Bellis perennis	.17	2.5	Ononis spinosa	.04	0
Bryonia dioica	.06	.4	Osmundo regalis	.4	7
Cardamine pratensis	.02	-.4	Osmundo regalis 2	0	-.7
Cardamine pratensis R	.02	-.4	Paeonia officinalis	.04	0
Cardiospermum halicaca	.03	-.2	Petroselinum crispum	.07	.6
Castanea vesca	.02	-.4	Plantago major	.03	-.2
Centaurium erythraea	.07	.6	Polygonum aviculare	.03	-.2
Chamomilla	.03	-.2	Primula veris	.04	0
Chelidonium majus	.19	2.9	Prunus padus	.02	-.4
Chenopodium anthelmint	.01	-.5	Psoralea bituminosa	.03	-.2
Cicuta virosa	.13	1.8	Rhus toxicodendron	.01	-.5
Cimicifuga racemosa	.06	.4	Rumex acetosa	.19	2.9
Cinnamodendron cortiso	.02	-.4	Ruta graveolens	.01	-.5
Clematis erecta	.02	-.4	Salicaria purpurea	.02	-.4
Collinsonia canadensis	.06	.4	Salix purpurea	.02	-.4
Conium maculatum	.01	-.5	Salvia sclarea	.01	-.5
Cupressus lawsoniana	.04	0	Sanguisorba officinali	.01	-.5
Echinacea angustifolia	.02	-.4	Scrophularia nodosa	.02	-.4
Echinacea purpurea	.01	-.5	Scutellaria lateriflor	.04	0
Escholtzia californica	.02	-.4	Solidago virgaurea	.01	-.5
Faba vulgaris	.01	-.5	Spilanthes oleracea	.01	-.5
Fagopyrum esculentum	.01	-.5	Stachys officinalis	.01	-.5
Fraxinus excelsior	.02	-.4	Stachys officinalis R	0	-.7
Galium aparine	.03	-.2	Stachys sylvatica	.02	-.4
Glechoma hederacea	.02	-.4	Syringa vulgaris	.02	-.4
Gnaphalium leontopodiu	.08	.8	Taraxacum officinale	.13	1.8
Hedera helix	.02	-.4	Teucrium chamaedrys	.02	-.4
Hydrophyllum virginicu	.01	-.5	Thuja occidentalis	.01	-.5
Hyoscyamus niger	.03	-.2	Tormentilla erecta	.09	1
Hyssopus officinalis	.02	-.4	Tussilago farfara	.01	-.5
Lappa arctium	.05	.2	Tussilago petasites	.02	-.4
Lapsana communis	.07	.6	Ulmus campestris	.03	-.2
Laurocerasus	.01	-.5	Urtica dioica	.08	.8
Leonurus cardiaca	.01	-.5	Urtica urens	.01	-.5
Lespedeza sieboldii	.01	-.5	Urtica urens 2	.01	-.5
Linum usitatissimum	0	-.7	Verbena officinalis	.01	-.5
Lycopus europaeus	.01	-.5	Vinca minor	.04	0
Malva sylvestris	.02	-.4	Viola tricolor	.02	-.4
Mandragora officinalis	.02	-.4	Zizia aurea	.03	-.2

Terbium

Name	Value	Deviation	Name	Value	Deviation
Osmundo regalis	.4	7	Echinacea angustifolia	.02	-.4
Chelidonium majus	.19	2.9	Escholtzia californica	.02	-.4
Rumex acetosa	.19	2.9	Fraxinus excelsior	.02	-.4
Bellis perennis	.17	2.5	Glechoma hederacea	.02	-.4
Cicuta virosa	.13	1.8	Hedera helix	.02	-.4
Taraxacum officinale	.13	1.8	Hyssopus officinalis	.02	-.4
Tormentilla erecta	.09	1	Malva sylvestris	.02	-.4
Gnaphalium leontopodiu	.08	.8	Mandragora officinalis	.02	-.4
Urtica dioica	.08	.8	Mentha arvensis	.02	-.4
Centaurium erythraea	.07	.6	Milium solis	.02	-.4
Lapsana communis	.07	.6	Prunus padus	.02	-.4
Petroselinum crispum	.07	.6	Salicaria purpurea	.02	-.4
Aconitum napellus	.06	.4	Salix purpurea	.02	-.4
Bryonia dioica	.06	.4	Scrophularia nodosa	.02	-.4
Cimicifuga racemosa	.06	.4	Stachys sylvatica	.02	-.4
Collinsonia canadensis	.06	.4	Syringa vulgaris	.02	-.4
Adonis vernalis	.05	.2	Teucrium chamaedrys	.02	-.4
Alchemilla vulgaris	.05	.2	Tussilago petasites	.02	-.4
Angelica archangelica	.05	.2	Viola tricolor	.02	-.4
Lappa arctium	.05	.2	Agrimonia eupatoria	.01	-.5
Mercurialis perennis	.05	.2	Anethum graveolens	.01	-.5
Oenanthe crocata	.05	.2	Aquilegia vulgaris	.01	-.5
Cupressus lawsoniana	.04	0	Chenopodium anthelmint	.01	-.5
Melissa officinalis	.04	0	Conium maculatum	.01	-.5
Ocimum canum	.04	0	Echinacea purpurea	.01	-.5
Ocimum canum R	.04	0	Faba vulgaris	.01	-.5
Ononis spinosa	.04	0	Fagopyrum esculentum	.01	-.5
Paeonia officinalis	.04	0	Hydrophyllum virginicu	.01	-.5
Primula veris	.04	0	Laurocerasus	.01	-.5
Scutellaria lateriflor	.04	0	Leonurus cardiaca	.01	-.5
Vinca minor	.04	0	Lespedeza sieboldii	.01	-.5
Anthemis nobilis	.03	-.2	Lycopus europaeus	.01	-.5
Belladonna	.03	-.2	Melilotus officinalis	.01	-.5
Cardiospermum halicaca	.03	-.2	Rhus toxicodendron	.01	-.5
Chamomilla	.03	-.2	Ruta graveolens	.01	-.5
Galium aparine	.03	-.2	Salvia sclarea	.01	-.5
Hyoscyamus niger	.03	-.2	Sanguisorba officinali	.01	-.5
Marrubium vulgare	.03	-.2	Solidago virgaurea	.01	-.5
Plantago major	.03	-.2	Spilanthes oleracea	.01	-.5
Polygonum aviculare	.03	-.2	Stachys officinalis	.01	-.5
Psoralea bituminosa	.03	-.2	Thuja occidentalis	.01	-.5
Ulmus campestris	.03	-.2	Tussilago farfara	.01	-.5
Zizia aurea	.03	-.2	Urtica urens	.01	-.5
Cardamine pratensis	.02	-.4	Urtica urens 2	.01	-.5
Cardamine pratensis R	.02	-.4	Verbena officinalis	.01	-.5
Castanea vesca	.02	-.4	Linum usitatissimum	0	-.7
Cinnamodendron cortiso	.02	-.4	Osmundo regalis 2	0	-.7
Clematis erecta	.02	-.4	Stachys officinalis R	0	-.7

Thallium

Name	Value	Deviation	Name	Value	Deviation
Aconitum napellus	.09	-.2	Marrubium vulgare	.02	-.5
Adonis vernalis	.08	-.2	Melilotus officinalis	.01	-.5
Agrimonia eupatoria	.05	-.4	Melissa officinalis	.17	.1
Alchemilla vulgaris	.07	-.3	Mentha arvensis	.03	-.4
Anethum graveolens	.02	-.5	Mercurialis perennis	.12	-.1
Angelica archangelica	.16	.1	Milium solis	.07	-.3
Anthemis nobilis	.11	-.1	Ocimum canum	.08	-.2
Aquilegia vulgaris	.01	-.5	Ocimum canum R	.08	-.2
Belladonna	.02	-.5	Oenanthe crocata	.03	-.4
Bellis perennis	.16	.1	Ononis spinosa	.09	-.2
Bryonia dioica	.16	.1	Osmundo regalis	.13	0
Cardamine pratensis	.02	-.5	Osmundo regalis 2	.02	-.5
Cardamine pratensis R	.02	-.5	Paeonia officinalis	.29	.6
Cardiospermum halicaca	.23	.4	Petroselinum crispum	.06	-.3
Castanea vesca	.13	0	Plantago major	.03	-.4
Centaurium erythraea	.04	-.4	Polygonum aviculare	.03	-.4
Chamomilla	.05	-.4	Primula veris	.05	-.4
Chelidonium majus	1.94	7.3	Prunus padus	.11	-.1
Chenopodium anthelmint	.03	-.4	Psoralea bituminosa	.2	.2
Cicuta virosa	.12	-.1	Rhus toxicodendron	.23	.4
Cimicifuga racemosa	.29	.6	Rumex acetosa	.71	2.3
Cinnamodendron cortiso	.74	2.4	Ruta graveolens	.04	-.4
Clematis erecta	.26	.5	Salicaria purpurea	.39	1
Collinsonia canadensis	.18	.2	Salix purpurea	.02	-.5
Conium maculatum	.06	-.3	Salvia sclarea	.12	-.1
Cupressus lawsoniana	.02	-.5	Sanguisorba officinali	.13	0
Echinacea angustifolia	.05	-.4	Scrophularia nodosa	.06	-.3
Echinacea purpurea	.26	.5	Scutellaria lateriflor	.05	-.4
Escholtzia californica	.03	-.4	Solidago virgaurea	.11	-.1
Faba vulgaris	.09	-.2	Spilanthes oleracea	.12	-.1
Fagopyrum esculentum	.14	0	Stachys officinalis	.04	-.4
Fraxinus excelsior	.04	-.4	Stachys officinalis R	.03	-.4
Galium aparine	.17	.1	Stachys sylvatica	.14	0
Glechoma hederacea	.07	-.3	Syringa vulgaris	.01	-.5
Gnaphalium leontopodiu	.05	-.4	Taraxacum officinale	.04	-.4
Hedera helix	.04	-.4	Teucrium chamaedrys	.05	-.4
Hydrophyllum virginicu	.06	-.3	Thuja occidentalis	.07	-.3
Hyoscyamus niger	.1	-.2	Tormentilla erecta	1.14	4.1
Hyssopus officinalis	.16	.1	Tussilago farfara	.02	-.5
Lappa arctium	.08	-.2	Tussilago petasites	.01	-.5
Lapsana communis	.36	.9	Ulmus campestris	.01	-.5
Laurocerasus	.17	.1	Urtica dioica	.07	-.3
Leonurus cardiaca	.06	-.3	Urtica urens	.15	0
Lespedeza sieboldii	.04	-.4	Urtica urens 2	.06	-.3
Linum usitatissimum	.35	.9	Verbena officinalis	.08	-.2
Lycopus europaeus	.08	-.2	Vinca minor	.21	.3
Malva sylvestris	.02	-.5	Viola tricolor	.09	-.2
Mandragora officinalis	.13	0	Zizia aurea	.04	-.4

Thallium

Name	Value	Deviation	Name	Value	Deviation
Chelidonium majus	1.94	7.3	Alchemilla vulgaris	.07	-.3
Tormentilla erecta	1.14	4.1	Conium maculatum	.06	-.3
Cinnamodendron cortiso	.74	2.4	Glechoma hederacea	.07	-.3
Rumex acetosa	.71	2.3	Hydrophyllum virginicu	.06	-.3
Salicaria purpurea	.39	1	Leonurus cardiaca	.06	-.3
Lapsana communis	.36	.9	Milium solis	.07	-.3
Linum usitatissimum	.35	.9	Petroselinum crispum	.06	-.3
Cimicifuga racemosa	.29	.6	Scrophularia nodosa	.06	-.3
Paeonia officinalis	.29	.6	Thuja occidentalis	.07	-.3
Clematis erecta	.26	.5	Urtica dioica	.07	-.3
Echinacea purpurea	.26	.5	Urtica urens 2	.06	-.3
Cardiospermum halicaca	.23	.4	Agrimonia eupatoria	.05	-.4
Rhus toxicodendron	.23	.4	Centaurium erythraea	.04	-.4
Vinca minor	.21	.3	Chamomilla	.05	-.4
Collinsonia canadensis	.18	.2	Chenopodium anthelmint	.03	-.4
Psoralea bituminosa	.2	.2	Echinacea angustifolia	.05	-.4
Angelica archangelica	.16	.1	Escholtzia californica	.03	-.4
Bellis perennis	.16	.1	Fraxinus excelsior	.04	-.4
Bryonia dioica	.16	.1	Gnaphalium leontopodiu	.05	-.4
Galium aparine	.17	.1	Hedera helix	.04	-.4
Hyssopus officinalis	.16	.1	Lespedeza sieboldii	.04	-.4
Laurocerasus	.17	.1	Mentha arvensis	.03	-.4
Melissa officinalis	.17	.1	Oenanthe crocata	.03	-.4
Castanea vesca	.13	0	Plantago major	.03	-.4
Fagopyrum esculentum	.14	0	Polygonum aviculare	.03	-.4
Mandragora officinalis	.13	0	Primula veris	.05	-.4
Osmundo regalis	.13	0	Ruta graveolens	.04	-.4
Sanguisorba officinali	.13	0	Scutellaria lateriflor	.05	-.4
Stachys sylvatica	.14	0	Stachys officinalis	.04	-.4
Urtica urens	.15	0	Stachys officinalis R	.03	-.4
Anthemis nobilis	.11	-.1	Taraxacum officinale	.04	-.4
Cicuta virosa	.12	-.1	Teucrium chamaedrys	.05	-.4
Mercurialis perennis	.12	-.1	Zizia aurea	.04	-.4
Prunus padus	.11	-.1	Anethum graveolens	.02	-.5
Salvia sclarea	.12	-.1	Aquilegia vulgaris	.01	-.5
Solidago virgaurea	.11	-.1	Belladonna	.02	-.5
Spilanthes oleracea	.12	-.1	Cardamine pratensis	.02	-.5
Aconitum napellus	.09	-.2	Cardamine pratensis R	.02	-.5
Adonis vernalis	.08	-.2	Cupressus lawsoniana	.02	-.5
Faba vulgaris	.09	-.2	Malva sylvestris	.02	-.5
Hyoscyamus niger	.1	-.2	Marrubium vulgare	.02	-.5
Lappa arctium	.08	-.2	Melilotus officinalis	.01	-.5
Lycopus europaeus	.08	-.2	Osmundo regalis 2	.02	-.5
Ocimum canum	.08	-.2	Salix purpurea	.02	-.5
Ocimum canum R	.08	-.2	Syringa vulgaris	.01	-.5
Ononis spinosa	.09	-.2	Tussilago farfara	.02	-.5
Verbena officinalis	.08	-.2	Tussilago petasites	.01	-.5
Viola tricolor	.09	-.2	Ulmus campestris	.01	-.5

Thorium

Name	Value	Deviation	Name	Value	Deviation
Aconitum napellus	.76	1.1	Marrubium vulgare	.18	-.3
Adonis vernalis	.59	.7	Melilotus officinalis	.08	-.6
Agrimonia eupatoria	.08	-.6	Melissa officinalis	.31	0
Alchemilla vulgaris	.47	.4	Mentha arvensis	.18	-.3
Anethum graveolens	.09	-.6	Mercurialis perennis	.53	.5
Angelica archangelica	.4	.2	Milium solis	.13	-.5
Anthemis nobilis	.37	.1	Ocimum canum	.33	0
Aquilegia vulgaris	.07	-.6	Ocimum canum R	.34	0
Belladonna	.26	-.2	Oenanthe crocata	.43	.3
Bellis perennis	1.75	3.5	Ononis spinosa	.28	-.1
Bryonia dioica	.65	.8	Osmundo regalis	2.34	4.9
Cardamine pratensis	.18	-.3	Osmundo regalis 2	.05	-.7
Cardamine pratensis R	.13	-.5	Paeonia officinalis	.28	-.1
Cardiospermum halicaca	.28	-.1	Petroselinum crispum	.56	.6
Castanea vesca	.12	-.5	Plantago major	.35	.1
Centaurium erythraea	.81	1.2	Polygonum aviculare	.27	-.1
Chamomilla	.3	-.1	Primula veris	.36	.1
Chelidonium majus	2.12	4.4	Prunus padus	.17	-.4
Chenopodium anthelmint	.1	-.5	Psoralea bituminosa	.23	-.2
Cicuta virosa	.89	1.4	Rhus toxicodendron	.09	-.6
Cimicifuga racemosa	.2	-.3	Rumex acetosa	1.4	2.6
Cinnamodendron cortiso	.09	-.6	Ruta graveolens	.1	-.5
Clematis erecta	.13	-.5	Salicaria purpurea	.09	-.6
Collinsonia canadensis	.51	.5	Salix purpurea	.11	-.5
Conium maculatum	.04	-.7	Salvia sclarea	.07	-.6
Cupressus lawsoniana	.34	0	Sanguisorba officinali	.03	-.7
Echinacea angustifolia	.16	-.4	Scrophularia nodosa	.2	-.3
Echinacea purpurea	.04	-.7	Scutellaria lateriflor	.28	-.1
Escholtzia californica	.12	-.5	Solidago virgaurea	.06	-.6
Faba vulgaris	.08	-.6	Spilanthes oleracea	.11	-.5
Fagopyrum esculentum	.04	-.7	Stachys officinalis	.05	-.7
Fraxinus excelsior	.17	-.4	Stachys officinalis R	.03	-.7
Galium aparine	.21	-.3	Stachys sylvatica	.22	-.2
Glechoma hederacea	.19	-.3	Syringa vulgaris	.2	-.3
Gnaphalium leontopodiu	.83	1.2	Taraxacum officinale	1.22	2.2
Hedera helix	.18	-.3	Teucrium chamaedrys	.22	-.2
Hydrophyllum virginicu	.09	-.6	Thuja occidentalis	.12	-.5
Hyoscyamus niger	.23	-.2	Tormentilla erecta	.75	1
Hyssopus officinalis	.11	-.5	Tussilago farfara	.1	-.5
Lappa arctium	.44	.3	Tussilago petasites	.15	-.4
Lapsana communis	.54	.5	Ulmus campestris	.17	-.4
Laurocerasus	.1	-.5	Urtica dioica	.85	1.3
Leonurus cardiaca	.09	-.6	Urtica urens	.13	-.5
Lespedeza sieboldii	.06	-.6	Urtica urens 2	.14	-.4
Linum usitatissimum	.03	-.7	Verbena officinalis	.07	-.6
Lycopus europaeus	.07	-.6	Vinca minor	.31	0
Malva sylvestris	.14	-.4	Viola tricolor	.2	-.3
Mandragora officinalis	.08	-.6	Zizia aurea	.28	-.1

Thorium

Name	Value	Deviation	Name	Value	Deviation
Osmundo regalis	2.34	4.9	Scrophularia nodosa	.2	-.3
Chelidonium majus	2.12	4.4	Syringa vulgaris	.2	-.3
Bellis perennis	1.75	3.5	Viola tricolor	.2	-.3
Rumex acetosa	1.4	2.6	Echinacea angustifolia	.16	-.4
Taraxacum officinale	1.22	2.2	Fraxinus excelsior	.17	-.4
Cicuta virosa	.89	1.4	Malva sylvestris	.14	-.4
Urtica dioica	.85	1.3	Prunus padus	.17	-.4
Centaurium erythraea	.81	1.2	Tussilago petasites	.15	-.4
Gnaphalium leontopodiu	.83	1.2	Ulmus campestris	.17	-.4
Aconitum napellus	.76	1.1	Urtica urens 2	.14	-.4
Tormentilla erecta	.75	1	Cardamine pratensis R	.13	-.5
Bryonia dioica	.65	.8	Castanea vesca	.12	-.5
Adonis vernalis	.59	.7	Chenopodium anthelmint	.1	-.5
Petroselinum crispum	.56	.6	Clematis erecta	.13	-.5
Collinsonia canadensis	.51	.5	Escholtzia californica	.12	-.5
Lapsana communis	.54	.5	Hyssopus officinalis	.11	-.5
Mercurialis perennis	.53	.5	Laurocerasus	.1	-.5
Alchemilla vulgaris	.47	.4	Milium solis	.13	-.5
Lappa arctium	.44	.3	Ruta graveolens	.1	-.5
Oenanthe crocata	.43	.3	Salix purpurea	.11	-.5
Angelica archangelica	.4	.2	Spilanthes oleracea	.11	-.5
Anthemis nobilis	.37	.1	Thuja occidentalis	.12	-.5
Plantago major	.35	.1	Tussilago farfara	.1	-.5
Primula veris	.36	.1	Urtica urens	.13	-.5
Cupressus lawsoniana	.34	0	Agrimonia eupatoria	.08	-.6
Melissa officinalis	.31	0	Anethum graveolens	.09	-.6
Ocimum canum	.33	0	Aquilegia vulgaris	.07	-.6
Ocimum canum R	.34	0	Cinnamodendron cortiso	.09	-.6
Vinca minor	.31	0	Faba vulgaris	.08	-.6
Cardiospermum halicaca	.28	-.1	Hydrophyllum virginicu	.09	-.6
Chamomilla	.3	-.1	Leonurus cardiaca	.09	-.6
Ononis spinosa	.28	-.1	Lespedeza sieboldii	.06	-.6
Paeonia officinalis	.28	-.1	Lycopus europaeus	.07	-.6
Polygonum aviculare	.27	-.1	Mandragora officinalis	.08	-.6
Scutellaria lateriflor	.28	-.1	Melilotus officinalis	.08	-.6
Zizia aurea	.28	-.1	Rhus toxicodendron	.09	-.6
Belladonna	.26	-.2	Salicaria purpurea	.09	-.6
Hyoscyamus niger	.23	-.2	Salvia sclarea	.07	-.6
Psoralea bituminosa	.23	-.2	Solidago virgaurea	.06	-.6
Stachys sylvatica	.22	-.2	Verbena officinalis	.07	-.6
Teucrium chamaedrys	.22	-.2	Conium maculatum	.04	-.7
Cardamine pratensis	.18	-.3	Echinacea purpurea	.04	-.7
Cimicifuga racemosa	.2	-.3	Fagopyrum esculentum	.04	-.7
Galium aparine	.21	-.3	Linum usitatissimum	.03	-.7
Glechoma hederacea	.19	-.3	Osmundo regalis 2	.05	-.7
Hedera helix	.18	-.3	Sanguisorba officinali	.03	-.7
Marrubium vulgare	.18	-.3	Stachys officinalis	.05	-.7
Mentha arvensis	.18	-.3	Stachys officinalis R	.03	-.7

Thulium

Name	Value	Deviation	Name	Value	Deviation
Aconitum napellus	.01	.1	Marrubium vulgare	.01	.1
Adonis vernalis	.01	.1	Melilotus officinalis	0	-.6
Agrimonia eupatoria	0	-.6	Melissa officinalis	.01	.1
Alchemilla vulgaris	.01	.1	Mentha arvensis	.01	.1
Anethum graveolens	0	-.6	Mercurialis perennis	.01	.1
Angelica archangelica	.01	.1	Milium solis	0	-.6
Anthemis nobilis	.01	.1	Ocimum canum	.01	.1
Aquilegia vulgaris	0	-.6	Ocimum canum R	.01	.1
Belladonna	.01	.1	Oenanthe crocata	.01	.1
Bellis perennis	.05	2.8	Ononis spinosa	.01	.1
Bryonia dioica	.01	.1	Osmundo regalis	.11	6.8
Cardamine pratensis	0	-.6	Osmundo regalis 2	0	-.6
Cardamine pratensis R	0	-.6	Paeonia officinalis	.01	.1
Cardiospermum halicaca	.01	.1	Petroselinum crispum	.02	.7
Castanea vesca	0	-.6	Plantago major	.01	.1
Centaurium erythraea	.02	.7	Polygonum aviculare	.01	.1
Chamomilla	.01	.1	Primula veris	.01	.1
Chelidonium majus	.04	2.1	Prunus padus	.01	.1
Chenopodium anthelmint	0	-.6	Psoralea bituminosa	.01	.1
Cicuta virosa	.03	1.4	Rhus toxicodendron	0	-.6
Cimicifuga racemosa	.01	.1	Rumex acetosa	.05	2.8
Cinnamodendron cortiso	.01	.1	Ruta graveolens	0	-.6
Clematis erecta	0	-.6	Salicaria purpurea	0	-.6
Collinsonia canadensis	.02	.7	Salix purpurea	0	-.6
Conium maculatum	0	-.6	Salvia sclarea	0	-.6
Cupressus lawsoniana	.01	.1	Sanguisorba officinali	0	-.6
Echinacea angustifolia	0	-.6	Scrophularia nodosa	.01	.1
Echinacea purpurea	0	-.6	Scutellaria lateriflor	.01	.1
Escholtzia californica	0	-.6	Solidago virgaurea	0	-.6
Faba vulgaris	0	-.6	Spilanthes oleracea	0	-.6
Fagopyrum esculentum	0	-.6	Stachys officinalis	0	-.6
Fraxinus excelsior	.01	.1	Stachys officinalis R	0	-.6
Galium aparine	.01	.1	Stachys sylvatica	.01	.1
Glechoma hederacea	0	-.6	Syringa vulgaris	0	-.6
Gnaphalium leontopodiu	.03	1.4	Taraxacum officinale	.04	2.1
Hedera helix	0	-.6	Teucrium chamaedrys	.01	.1
Hydrophyllum virginicu	0	-.6	Thuja occidentalis	0	-.6
Hyoscyamus niger	.01	.1	Tormentilla erecta	.02	.7
Hyssopus officinalis	0	-.6	Tussilago farfara	0	-.6
Lappa arctium	.01	.1	Tussilago petasites	0	-.6
Lapsana communis	.02	.7	Ulmus campestris	.01	.1
Laurocerasus	0	-.6	Urtica dioica	.02	.7
Leonurus cardiaca	0	-.6	Urtica urens	0	-.6
Lespedeza sieboldii	0	-.6	Urtica urens 2	0	-.6
Linum usitatissimum	0	-.6	Verbena officinalis	0	-.6
Lycopus europaeus	0	-.6	Vinca minor	.01	.1
Malva sylvestris	0	-.6	Viola tricolor	.01	.1
Mandragora officinalis	.01	.1	Zizia aurea	.01	.1

Thulium

Name	Value	Deviation	Name	Value	Deviation
Osmundo regalis	.11	6.8	Ulmus campestris	.01	.1
Bellis perennis	.05	2.8	Vinca minor	.01	.1
Rumex acetosa	.05	2.8	Viola tricolor	.01	.1
Chelidonium majus	.04	2.1	Zizia aurea	.01	.1
Taraxacum officinale	.04	2.1	Agrimonia eupatoria	0	-.6
Cicuta virosa	.03	1.4	Anethum graveolens	0	-.6
Gnaphalium leontopodiu	.03	1.4	Aquilegia vulgaris	0	-.6
Centaurium erythraea	.02	.7	Cardamine pratensis	0	-.6
Collinsonia canadensis	.02	.7	Cardamine pratensis R	0	-.6
Lapsana communis	.02	.7	Castanea vesca	0	-.6
Petroselinum crispum	.02	.7	Chenopodium anthelmint	0	-.6
Tormentilla erecta	.02	.7	Clematis erecta	0	-.6
Urtica dioica	.02	.7	Conium maculatum	0	-.6
Aconitum napellus	.01	.1	Echinacea angustifolia	0	-.6
Adonis vernalis	.01	.1	Echinacea purpurea	0	-.6
Alchemilla vulgaris	.01	.1	Escholtzia californica	0	-.6
Angelica archangelica	.01	.1	Faba vulgaris	0	-.6
Anthemis nobilis	.01	.1	Fagopyrum esculentum	0	-.6
Belladonna	.01	.1	Glechoma hederacea	0	-.6
Bryonia dioica	.01	.1	Hedera helix	0	-.6
Cardiospermum halicaca	.01	.1	Hydrophyllum virginicu	0	-.6
Chamomilla	.01	.1	Hyssopus officinalis	0	-.6
Cimicifuga racemosa	.01	.1	Laurocerasus	0	-.6
Cinnamodendron cortiso	.01	.1	Leonurus cardiaca	0	-.6
Cupressus lawsoniana	.01	.1	Lespedeza sieboldii	0	-.6
Fraxinus excelsior	.01	.1	Linum usitatissimum	0	-.6
Galium aparine	.01	.1	Lycopus europaeus	0	-.6
Hyoscyamus niger	.01	.1	Malva sylvestris	0	-.6
Lappa arctium	.01	.1	Melilotus officinalis	0	-.6
Mandragora officinalis	.01	.1	Milium solis	0	-.6
Marrubium vulgare	.01	.1	Osmundo regalis 2	0	-.6
Melissa officinalis	.01	.1	Rhus toxicodendron	0	-.6
Mentha arvensis	.01	.1	Ruta graveolens	0	-.6
Mercurialis perennis	.01	.1	Salicaria purpurea	0	-.6
Ocimum canum	.01	.1	Salix purpurea	0	-.6
Ocimum canum R	.01	.1	Salvia sclarea	0	-.6
Oenanthe crocata	.01	.1	Sanguisorba officinali	0	-.6
Ononis spinosa	.01	.1	Solidago virgaurea	0	-.6
Paeonia officinalis	.01	.1	Spilanthes oleracea	0	-.6
Plantago major	.01	.1	Stachys officinalis	0	-.6
Polygonum aviculare	.01	.1	Stachys officinalis R	0	-.6
Primula veris	.01	.1	Syringa vulgaris	0	-.6
Prunus padus	.01	.1	Thuja occidentalis	0	-.6
Psoralea bituminosa	.01	.1	Tussilago farfara	0	-.6
Scrophularia nodosa	.01	.1	Tussilago petasites	0	-.6
Scutellaria lateriflor	.01	.1	Urtica urens	0	-.6
Stachys sylvatica	.01	.1	Urtica urens 2	0	-.6
Teucrium chamaedrys	.01	.1	Verbena officinalis	0	-.6

Titanium

Name	Value	Deviation	Name	Value	Deviation
Aconitum napellus	161.82	-.9	Marrubium vulgare	159.85	-.9
Adonis vernalis	318.29	.9	Melilotus officinalis	196.83	-.5
Agrimonia eupatoria	176.59	-.7	Melissa officinalis	311.65	.8
Alchemilla vulgaris	306.24	.8	Mentha arvensis	226.66	-.2
Anethum graveolens	160.64	-.9	Mercurialis perennis	301.7	.7
Angelica archangelica	441.09	2.3	Milium solis	224.6	-.2
Anthemis nobilis	288.22	.6	Ocimum canum	233.41	-.1
Aquilegia vulgaris	199.58	-.5	Ocimum canum R	239.31	0
Belladonna	156.29	-1	Oenanthe crocata	371.37	1.5
Bellis perennis	212.12	-.3	Ononis spinosa	160.56	-.9
Bryonia dioica	369.47	1.5	Osmundo regalis	484.69	2.8
Cardamine pratensis	343.14	1.2	Osmundo regalis 2	198.23	-.5
Cardamine pratensis R	314.65	.9	Paeonia officinalis	405.86	1.9
Cardiospermum halicaca	176.73	-.7	Petroselinum crispum	228.5	-.1
Castanea vesca	251.35	.1	Plantago major	183.84	-.7
Centaurium erythraea	425.9	2.2	Polygonum aviculare	308.23	.8
Chamomilla	198.17	-.5	Primula veris	268.6	.3
Chelidonium majus	201.69	-.4	Prunus padus	280.7	.5
Chenopodium anthelmint	160.34	-.9	Psoralea bituminosa	226.74	-.2
Cicuta virosa	337.62	1.1	Rhus toxicodendron	236.66	0
Cimicifuga racemosa	234.03	-.1	Rumex acetosa	335.91	1.1
Cinnamodendron cortiso	131.22	-1.3	Ruta graveolens	238.14	0
Clematis erecta	172.93	-.8	Salicaria purpurea	223.67	-.2
Collinsonia canadensis	404.26	1.9	Salix purpurea	155.64	-1
Conium maculatum	229.07	-.1	Salvia sclarea	100.25	-1.6
Cupressus lawsoniana	390.61	1.7	Sanguisorba officinali	232.52	-.1
Echinacea angustifolia	173.39	-.8	Scrophularia nodosa	343.81	1.2
Echinacea purpurea	128.61	-1.3	Scutellaria lateriflor	326.97	1
Escholtzia californica	201.28	-.4	Solidago virgaurea	176.05	-.7
Faba vulgaris	140.96	-1.1	Spilanthes oleracea	129.55	-1.3
Fagopyrum esculentum	200.97	-.5	Stachys officinalis	173.81	-.8
Fraxinus excelsior	98.28	-1.6	Stachys officinalis R	180.72	-.7
Galium aparine	185.08	-.6	Stachys sylvatica	255.85	.2
Glechoma hederacea	219.08	-.2	Syringa vulgaris	547.71	3.6
Gnaphalium leontopodiu	296.82	.7	Taraxacum officinale	250.9	.1
Hedera helix	210.5	-.3	Teucrium chamaedrys	252.8	.1
Hydrophyllum virginicu	258.95	.2	Thuja occidentalis	176.84	-.7
Hyoscyamus niger	158.38	-.9	Tormentilla erecta	338.07	1.1
Hyssopus officinalis	145.6	-1.1	Tussilago farfara	126.15	-1.3
Lappa arctium	293.41	.6	Tussilago petasites	241.29	0
Lapsana communis	322.72	1	Ulmus campestris	216.85	-.3
Laurocerasus	231.81	-.1	Urtica dioica	249.11	.1
Leonurus cardiaca	163.69	-.9	Urtica urens	124.64	-1.3
Lespedeza sieboldii	280.53	.5	Urtica urens 2	244.09	0
Linum usitatissimum	300.8	.7	Verbena officinalis	169.61	-.8
Lycopus europaeus	175.8	-.7	Vinca minor	238.66	0
Malva sylvestris	268.61	.3	Viola tricolor	181.07	-.7
Mandragora officinalis	165.26	-.9	Zizia aurea	177.96	-.7

Titanium

Name	Value	Deviation	Name	Value	Deviation
Syringa vulgaris	547.71	3.6	Glechoma hederacea	219.08	-.2
Osmundo regalis	484.69	2.8	Mentha arvensis	226.66	-.2
Angelica archangelica	441.09	2.3	Milium solis	224.6	-.2
Centaurium erythraea	425.9	2.2	Psoralea bituminosa	226.74	-.2
Collinsonia canadensis	404.26	1.9	Salicaria purpurea	223.67	-.2
Paeonia officinalis	405.86	1.9	Bellis perennis	212.12	-.3
Cupressus lawsoniana	390.61	1.7	Hedera helix	210.5	-.3
Bryonia dioica	369.47	1.5	Ulmus campestris	216.85	-.3
Oenanthe crocata	371.37	1.5	Chelidonium majus	201.69	-.4
Cardamine pratensis	343.14	1.2	Escholtzia californica	201.28	-.4
Scrophularia nodosa	343.81	1.2	Aquilegia vulgaris	199.58	-.5
Cicuta virosa	337.62	1.1	Chamomilla	198.17	-.5
Rumex acetosa	335.91	1.1	Fagopyrum esculentum	200.97	-.5
Tormentilla erecta	338.07	1.1	Melilotus officinalis	196.83	-.5
Lapsana communis	322.72	1	Osmundo regalis 2	198.23	-.5
Scutellaria lateriflor	326.97	1	Galium aparine	185.08	-.6
Adonis vernalis	318.29	.9	Agrimonia eupatoria	176.59	-.7
Cardamine pratensis R	314.65	.9	Cardiospermum halicaca	176.73	-.7
Alchemilla vulgaris	306.24	.8	Lycopus europaeus	175.8	-.7
Melissa officinalis	311.65	.8	Plantago major	183.84	-.7
Polygonum aviculare	308.23	.8	Solidago virgaurea	176.05	-.7
Gnaphalium leontopodiu	296.82	.7	Stachys officinalis R	180.72	-.7
Linum usitatissimum	300.8	.7	Thuja occidentalis	176.84	-.7
Mercurialis perennis	301.7	.7	Viola tricolor	181.07	-.7
Anthemis nobilis	288.22	.6	Zizia aurea	177.96	-.7
Lappa arctium	293.41	.6	Clematis erecta	172.93	-.8
Lespedeza sieboldii	280.53	.5	Echinacea angustifolia	173.39	-.8
Prunus padus	280.7	.5	Stachys officinalis	173.81	-.8
Malva sylvestris	268.61	.3	Verbena officinalis	169.61	-.8
Primula veris	268.6	.3	Aconitum napellus	161.82	-.9
Hydrophyllum virginicu	258.95	.2	Anethum graveolens	160.64	-.9
Stachys sylvatica	255.85	.2	Chenopodium anthelmint	160.34	-.9
Castanea vesca	251.35	.1	Hyoscyamus niger	158.38	-.9
Taraxacum officinale	250.9	.1	Leonurus cardiaca	163.69	-.9
Teucrium chamaedrys	252.8	.1	Mandragora officinalis	165.26	-.9
Urtica dioica	249.11	.1	Marrubium vulgare	159.85	-.9
Ocimum canum R	239.31	0	Ononis spinosa	160.56	-.9
Rhus toxicodendron	236.66	0	Belladonna	156.29	-1
Ruta graveolens	238.14	0	Salix purpurea	155.64	-1
Tussilago petasites	241.29	0	Faba vulgaris	140.96	-1.1
Urtica urens 2	244.09	0	Hyssopus officinalis	145.6	-1.1
Vinca minor	238.66	0	Cinnamodendron cortiso	131.22	-1.3
Cimicifuga racemosa	234.03	-.1	Echinacea purpurea	128.61	-1.3
Conium maculatum	229.07	-.1	Spilanthes oleracea	129.55	-1.3
Laurocerasus	231.81	-.1	Tussilago farfara	126.15	-1.3
Ocimum canum	233.41	-.1	Urtica urens	124.64	-1.3
Petroselinum crispum	228.5	-.1	Fraxinus excelsior	98.28	-1.6
Sanguisorba officinali	232.52	-.1	Salvia sclarea	100.25	-1.6

Tungstenium

Name	Value	Deviation	Name	Value	Deviation
Aconitum napellus	0	-.6	Marrubium vulgare	0	-.6
Adonis vernalis	0	-.6	Melilotus officinalis	.81	2.2
Agrimonia eupatoria	0	-.6	Melissa officinalis	.54	1.2
Alchemilla vulgaris	.57	1.3	Mentha arvensis	.58	1.4
Anethum graveolens	0	-.6	Mercurialis perennis	.57	1.3
Angelica archangelica	0	-.6	Milium solis	.52	1.2
Anthemis nobilis	0	-.6	Ocimum canum	0	-.6
Aquilegia vulgaris	.53	1.2	Ocimum canum R	0	-.6
Belladonna	0	-.6	Oenanthe crocata	.5	1.1
Bellis perennis	0	-.6	Ononis spinosa	0	-.6
Bryonia dioica	0	-.6	Osmundo regalis	.7	1.8
Cardamine pratensis	.89	2.4	Osmundo regalis 2	0	-.6
Cardamine pratensis R	.87	2.4	Paeonia officinalis	0	-.6
Cardiospermum halicaca	.94	2.6	Petroselinum crispum	.55	1.3
Castanea vesca	0	-.6	Plantago major	0	-.6
Centaurium erythraea	.51	1.1	Polygonum aviculare	0	-.6
Chamomilla	0	-.6	Primula veris	.54	1.2
Chelidonium majus	0	-.6	Prunus padus	0	-.6
Chenopodium anthelmint	0	-.6	Psoralea bituminosa	0	-.6
Cicuta virosa	.76	2	Rhus toxicodendron	0	-.6
Cimicifuga racemosa	.53	1.2	Rumex acetosa	0	-.6
Cinnamodendron cortiso	0	-.6	Ruta graveolens	0	-.6
Clematis erecta	0	-.6	Salicaria purpurea	.59	1.4
Collinsonia canadensis	.51	1.1	Salix purpurea	0	-.6
Conium maculatum	.6	1.4	Salvia sclarea	0	-.6
Cupressus lawsoniana	.6	1.4	Sanguisorba officinali	0	-.6
Echinacea angustifolia	.69	1.8	Scrophularia nodosa	0	-.6
Echinacea purpurea	0	-.6	Scutellaria lateriflor	0	-.6
Escholtzia californica	0	-.6	Solidago virgaurea	0	-.6
Faba vulgaris	.59	1.4	Spilanthes oleracea	0	-.6
Fagopyrum esculentum	0	-.6	Stachys officinalis	0	-.6
Fraxinus excelsior	0	-.6	Stachys officinalis R	0	-.6
Galium aparine	0	-.6	Stachys sylvatica	0	-.6
Glechoma hederacea	0	-.6	Syringa vulgaris	0	-.6
Gnaphalium leontopodiu	.56	1.3	Taraxacum officinale	0	-.6
Hedera helix	0	-.6	Teucrium chamaedrys	0	-.6
Hydrophyllum virginicu	0	-.6	Thuja occidentalis	0	-.6
Hyoscyamus niger	0	-.6	Tormentilla erecta	0	-.6
Hyssopus officinalis	0	-.6	Tussilago farfara	0	-.6
Lappa arctium	0	-.6	Tussilago petasites	0	-.6
Lapsana communis	.57	1.3	Ulmus campestris	0	-.6
Laurocerasus	0	-.6	Urtica dioica	0	-.6
Leonurus cardiaca	0	-.6	Urtica urens	0	-.6
Lespedeza sieboldii	.55	1.3	Urtica urens 2	0	-.6
Linum usitatissimum	0	-.6	Verbena officinalis	0	-.6
Lycopus europaeus	.62	1.5	Vinca minor	0	-.6
Malva sylvestris	.54	1.2	Viola tricolor	0	-.6
Mandragora officinalis	0	-.6	Zizia aurea	0	-.6

Tungstenium

Name	Value	Deviation	Name	Value	Deviation
Cardiospermum halicaca	.94	2.6	Glechoma hederacea	0	-.6
Cardamine pratensis	.89	2.4	Hedera helix	0	-.6
Cardamine pratensis R	.87	2.4	Hydrophyllum virginicu	0	-.6
Melilotus officinalis	.81	2.2	Hyoscyamus niger	0	-.6
Cicuta virosa	.76	2	Hyssopus officinalis	0	-.6
Echinacea angustifolia	.69	1.8	Lappa arctium	0	-.6
Osmundo regalis	.7	1.8	Laurocerasus	0	-.6
Lycopus europaeus	.62	1.5	Leonurus cardiaca	0	-.6
Conium maculatum	.6	1.4	Linum usitatissimum	0	-.6
Cupressus lawsoniana	.6	1.4	Mandragora officinalis	0	-.6
Faba vulgaris	.59	1.4	Marrubium vulgare	0	-.6
Mentha arvensis	.58	1.4	Ocimum canum	0	-.6
Salicaria purpurea	.59	1.4	Ocimum canum R	0	-.6
Alchemilla vulgaris	.57	1.3	Ononis spinosa	0	-.6
Gnaphalium leontopodiu	.56	1.3	Osmundo regalis 2	0	-.6
Lapsana communis	.57	1.3	Paeonia officinalis	0	-.6
Lespedeza sieboldii	.55	1.3	Plantago major	0	-.6
Mercurialis perennis	.57	1.3	Polygonum aviculare	0	-.6
Petroselinum crispum	.55	1.3	Prunus padus	0	-.6
Aquilegia vulgaris	.53	1.2	Psoralea bituminosa	0	-.6
Cimicifuga racemosa	.53	1.2	Rhus toxicodendron	0	-.6
Malva sylvestris	.54	1.2	Rumex acetosa	0	-.6
Melissa officinalis	.54	1.2	Ruta graveolens	0	-.6
Milium solis	.52	1.2	Salix purpurea	0	-.6
Primula veris	.54	1.2	Salvia sclarea	0	-.6
Centaurium erythraea	.51	1.1	Sanguisorba officinali	0	-.6
Collinsonia canadensis	.51	1.1	Scrophularia nodosa	0	-.6
Oenanthe crocata	.5	1.1	Scutellaria lateriflor	0	-.6
Aconitum napellus	0	-.6	Solidago virgaurea	0	-.6
Adonis vernalis	0	-.6	Spilanthes oleracea	0	-.6
Agrimonia eupatoria	0	-.6	Stachys officinalis	0	-.6
Anethum graveolens	0	-.6	Stachys officinalis R	0	-.6
Angelica archangelica	0	-.6	Stachys sylvatica	0	-.6
Anthemis nobilis	0	-.6	Syringa vulgaris	0	-.6
Belladonna	0	-.6	Taraxacum officinale	0	-.6
Bellis perennis	0	-.6	Teucrium chamaedrys	0	-.6
Bryonia dioica	0	-.6	Thuja occidentalis	0	-.6
Castanea vesca	0	-.6	Tormentilla erecta	0	-.6
Chamomilla	0	-.6	Tussilago farfara	0	-.6
Chelidonium majus	0	-.6	Tussilago petasites	0	-.6
Chenopodium anthelmint	0	-.6	Ulmus campestris	0	-.6
Cinnamodendron cortiso	0	-.6	Urtica dioica	0	-.6
Clematis erecta	0	-.6	Urtica urens	0	-.6
Echinacea purpurea	0	-.6	Urtica urens 2	0	-.6
Escholtzia californica	0	-.6	Verbena officinalis	0	-.6
Fagopyrum esculentum	0	-.6	Vinca minor	0	-.6
Fraxinus excelsior	0	-.6	Viola tricolor	0	-.6
Galium aparine	0	-.6	Zizia aurea	0	-.6

Uranium

Name	Value	Deviation	Name	Value	Deviation
Aconitum napellus	.22	.4	Marrubium vulgare	.09	-.2
Adonis vernalis	.16	.1	Melilotus officinalis	.04	-.5
Agrimonia eupatoria	.03	-.5	Melissa officinalis	.16	.1
Alchemilla vulgaris	.17	.1	Mentha arvensis	.08	-.3
Anethum graveolens	.04	-.5	Mercurialis perennis	.17	.1
Angelica archangelica	.19	.2	Milium solis	.06	-.4
Anthemis nobilis	.1	-.2	Ocimum canum	.13	-.1
Aquilegia vulgaris	.02	-.6	Ocimum canum R	.13	-.1
Belladonna	.13	-.1	Oenanthe crocata	.22	.4
Bellis perennis	.46	1.5	Ononis spinosa	.18	.2
Bryonia dioica	.2	.3	Osmundo regalis	1.46	6.3
Cardamine pratensis	.06	-.4	Osmundo regalis 2	.02	-.6
Cardamine pratensis R	.04	-.5	Paeonia officinalis	.3	.8
Cardiospermum halicaca	.12	-.1	Petroselinum crispum	.18	.2
Castanea vesca	.09	-.2	Plantago major	.12	-.1
Centaurium erythraea	.3	.8	Polygonum aviculare	.08	-.3
Chamomilla	.12	-.1	Primula veris	.16	.1
Chelidonium majus	.69	2.6	Prunus padus	.06	-.4
Chenopodium anthelmint	.02	-.6	Psoralea bituminosa	.1	-.2
Cicuta virosa	.73	2.8	Rhus toxicodendron	.03	-.5
Cimicifuga racemosa	.22	.4	Rumex acetosa	1.1	4.6
Cinnamodendron cortiso	.05	-.4	Ruta graveolens	.04	-.5
Clematis erecta	.07	-.3	Salicaria purpurea	.04	-.5
Collinsonia canadensis	.25	.5	Salix purpurea	.06	-.4
Conium maculatum	.02	-.6	Salvia sclarea	.03	-.5
Cupressus lawsoniana	.12	-.1	Sanguisorba officinali	.01	-.6
Echinacea angustifolia	.06	-.4	Scrophularia nodosa	.09	-.2
Echinacea purpurea	.02	-.6	Scutellaria lateriflor	.12	-.1
Escholtzia californica	.06	-.4	Solidago virgaurea	.02	-.6
Faba vulgaris	.04	-.5	Spilanthes oleracea	.06	-.4
Fagopyrum esculentum	.03	-.5	Stachys officinalis	.02	-.6
Fraxinus excelsior	.06	-.4	Stachys officinalis R	.01	-.6
Galium aparine	.1	-.2	Stachys sylvatica	.09	-.2
Glechoma hederacea	.07	-.3	Syringa vulgaris	.06	-.4
Gnaphalium leontopodiu	.23	.4	Taraxacum officinale	.34	1
Hedera helix	.08	-.3	Teucrium chamaedrys	.07	-.3
Hydrophyllum virginicu	.02	-.6	Thuja occidentalis	.05	-.4
Hyoscyamus niger	.11	-.1	Tormentilla erecta	.2	.3
Hyssopus officinalis	.04	-.5	Tussilago farfara	.04	-.5
Lappa arctium	.16	.1	Tussilago petasites	.08	-.3
Lapsana communis	.17	.1	Ulmus campestris	.07	-.3
Laurocerasus	.02	-.6	Urtica dioica	.35	1
Leonurus cardiaca	.05	-.4	Urtica urens	.04	-.5
Lespedeza sieboldii	.03	-.5	Urtica urens 2	.06	-.4
Linum usitatissimum	.04	-.5	Verbena officinalis	.03	-.5
Lycopus europaeus	.07	-.3	Vinca minor	.21	.3
Malva sylvestris	.06	-.4	Viola tricolor	.08	-.3
Mandragora officinalis	.09	-.2	Zizia aurea	.13	-.1

Uranium

Name	Value	Deviation	Name	Value	Deviation
Osmundo regalis	1.46	6.3	Lycopus europaeus	.07	-.3
Rumex acetosa	1.1	4.6	Mentha arvensis	.08	-.3
Cicuta virosa	.73	2.8	Polygonum aviculare	.08	-.3
Chelidonium majus	.69	2.6	Teucrium chamaedrys	.07	-.3
Bellis perennis	.46	1.5	Tussilago petasites	.08	-.3
Taraxacum officinale	.34	1	Ulmus campestris	.07	-.3
Urtica dioica	.35	1	Viola tricolor	.08	-.3
Centaurium erythraea	.3	.8	Cardamine pratensis	.06	-.4
Paeonia officinalis	.3	.8	Cinnamodendron cortiso	.05	-.4
Collinsonia canadensis	.25	.5	Echinacea angustifolia	.06	-.4
Aconitum napellus	.22	.4	Escholtzia californica	.06	-.4
Cimicifuga racemosa	.22	.4	Fraxinus excelsior	.06	-.4
Gnaphalium leontopodiu	.23	.4	Leonurus cardiaca	.05	-.4
Oenanthe crocata	.22	.4	Malva sylvestris	.06	-.4
Bryonia dioica	.2	.3	Milium solis	.06	-.4
Tormentilla erecta	.2	.3	Prunus padus	.06	-.4
Vinca minor	.21	.3	Salix purpurea	.06	-.4
Angelica archangelica	.19	.2	Spilanthes oleracea	.06	-.4
Ononis spinosa	.18	.2	Syringa vulgaris	.06	-.4
Petroselinum crispum	.18	.2	Thuja occidentalis	.05	-.4
Adonis vernalis	.16	.1	Urtica urens 2	.06	-.4
Alchemilla vulgaris	.17	.1	Agrimonia eupatoria	.03	-.5
Lappa arctium	.16	.1	Anethum graveolens	.04	-.5
Lapsana communis	.17	.1	Cardamine pratensis R	.04	-.5
Melissa officinalis	.16	.1	Faba vulgaris	.04	-.5
Mercurialis perennis	.17	.1	Fagopyrum esculentum	.03	-.5
Primula veris	.16	.1	Hyssopus officinalis	.04	-.5
Belladonna	.13	-.1	Lespedeza sieboldii	.03	-.5
Cardiospermum halicaca	.12	-.1	Linum usitatissimum	.04	-.5
Chamomilla	.12	-.1	Melilotus officinalis	.04	-.5
Cupressus lawsoniana	.12	-.1	Rhus toxicodendron	.03	-.5
Hyoscyamus niger	.11	-.1	Ruta graveolens	.04	-.5
Ocimum canum	.13	-.1	Salicaria purpurea	.04	-.5
Ocimum canum R	.13	-.1	Salvia sclarea	.03	-.5
Plantago major	.12	-.1	Tussilago farfara	.04	-.5
Scutellaria lateriflor	.12	-.1	Urtica urens	.04	-.5
Zizia aurea	.13	-.1	Verbena officinalis	.03	-.5
Anthemis nobilis	.1	-.2	Aquilegia vulgaris	.02	-.6
Castanea vesca	.09	-.2	Chenopodium anthelmint	.02	-.6
Galium aparine	.1	-.2	Conium maculatum	.02	-.6
Mandragora officinalis	.09	-.2	Echinacea purpurea	.02	-.6
Marrubium vulgare	.09	-.2	Hydrophyllum virginicu	.02	-.6
Psoralea bituminosa	.1	-.2	Laurocerasus	.02	-.6
Scrophularia nodosa	.09	-.2	Osmundo regalis 2	.02	-.6
Stachys sylvatica	.09	-.2	Sanguisorba officinali	.01	-.6
Clematis erecta	.07	-.3	Solidago virgaurea	.02	-.6
Glechoma hederacea	.07	-.3	Stachys officinalis	.02	-.6
Hedera helix	.08	-.3	Stachys officinalis R	.01	-.6

Vanadium

Name	Value	Deviation	Name	Value	Deviation
Aconitum napellus	9.36	.1	Marrubium vulgare	3.08	-.2
Adonis vernalis	7.05	0	Melilotus officinalis	1.45	-.3
Agrimonia eupatoria	0	-.4	Melissa officinalis	5.69	-.1
Alchemilla vulgaris	8.13	0	Mentha arvensis	2.59	-.2
Anethum graveolens	1.54	-.3	Mercurialis perennis	5.76	-.1
Angelica archangelica	22.9	.7	Milium solis	1.7	-.3
Anthemis nobilis	3.56	-.2	Ocimum canum	4.12	-.2
Aquilegia vulgaris	0	-.4	Ocimum canum R	4.27	-.2
Belladonna	4.77	-.1	Oenanthe crocata	9.7	.1
Bellis perennis	25.7	.8	Ononis spinosa	3.8	-.2
Bryonia dioica	8.86	0	Osmundo regalis	212.16	9
Cardamine pratensis	1.84	-.3	Osmundo regalis 2	0	-.4
Cardamine pratensis R	1.5	-.3	Paeonia officinalis	9.41	.1
Cardiospermum halicaca	4.22	-.2	Petroselinum crispum	8.54	0
Castanea vesca	3.47	-.2	Plantago major	3.96	-.2
Centaurium erythraea	12.4	.2	Polygonum aviculare	2.83	-.2
Chamomilla	2.22	-.3	Primula veris	4.76	-.1
Chelidonium majus	39.69	1.4	Prunus padus	2.64	-.2
Chenopodium anthelmint	0	-.4	Psoralea bituminosa	3.88	-.2
Cicuta virosa	35.37	1.2	Rhus toxicodendron	1.67	-.3
Cimicifuga racemosa	28.9	.9	Rumex acetosa	40.52	1.4
Cinnamodendron cortiso	0	-.4	Ruta graveolens	0	-.4
Clematis erecta	1.89	-.3	Salicaria purpurea	1.04	-.3
Collinsonia canadensis	37.01	1.3	Salix purpurea	5.62	-.1
Conium maculatum	3.23	-.2	Salvia sclarea	2.03	-.3
Cupressus lawsoniana	11.07	.1	Sanguisorba officinali	0	-.4
Echinacea angustifolia	1.94	-.3	Scrophularia nodosa	2.51	-.2
Echinacea purpurea	0	-.4	Scutellaria lateriflor	4.42	-.2
Escholtzia californica	0	-.4	Solidago virgaurea	0	-.4
Faba vulgaris	0	-.4	Spilanthes oleracea	2.48	-.2
Fagopyrum esculentum	0	-.4	Stachys officinalis	0	-.4
Fraxinus excelsior	5.18	-.1	Stachys officinalis R	0	-.4
Galium aparine	4.35	-.2	Stachys sylvatica	2.93	-.2
Glechoma hederacea	1.16	-.3	Syringa vulgaris	4.4	-.2
Gnaphalium leontopodiu	8.3	0	Taraxacum officinale	15.52	.3
Hedera helix	2.8	-.2	Teucrium chamaedrys	3.35	-.2
Hydrophyllum virginicu	0	-.4	Thuja occidentalis	3.87	-.2
Hyoscyamus niger	3.01	-.2	Tormentilla erecta	15.5	.3
Hyssopus officinalis	0	-.4	Tussilago farfara	0	-.4
Lappa arctium	6.92	0	Tussilago petasites	0	-.4
Lapsana communis	14.68	.3	Ulmus campestris	5.46	-.1
Laurocerasus	1.11	-.3	Urtica dioica	18.83	.5
Leonurus cardiaca	0	-.4	Urtica urens	1.49	-.3
Lespedeza sieboldii	1.92	-.3	Urtica urens 2	2.15	-.3
Linum usitatissimum	0	-.4	Verbena officinalis	1.55	-.3
Lycopus europaeus	0	-.4	Vinca minor	4.18	-.2
Malva sylvestris	1.01	-.3	Viola tricolor	2.21	-.3
Mandragora officinalis	3.06	-.2	Zizia aurea	7.63	0

Vanadium

Name	Value	Deviation	Name	Value	Deviation
Osmundo regalis	212.16	9	Scrophularia nodosa	2.51	-.2
Chelidonium majus	39.69	1.4	Scutellaria lateriflor	4.42	-.2
Rumex acetosa	40.52	1.4	Spilanthes oleracea	2.48	-.2
Collinsonia canadensis	37.01	1.3	Stachys sylvatica	2.93	-.2
Cicuta virosa	35.37	1.2	Syringa vulgaris	4.4	-.2
Cimicifuga racemosa	28.9	.9	Teucrium chamaedrys	3.35	-.2
Bellis perennis	25.7	.8	Thuja occidentalis	3.87	-.2
Angelica archangelica	22.9	.7	Vinca minor	4.18	-.2
Urtica dioica	18.83	.5	Anethum graveolens	1.54	-.3
Lapsana communis	14.68	.3	Cardamine pratensis	1.84	-.3
Taraxacum officinale	15.52	.3	Cardamine pratensis R	1.5	-.3
Tormentilla erecta	15.5	.3	Chamomilla	2.22	-.3
Centaurium erythraea	12.4	.2	Clematis erecta	1.89	-.3
Aconitum napellus	9.36	.1	Echinacea angustifolia	1.94	-.3
Cupressus lawsoniana	11.07	.1	Glechoma hederacea	1.16	-.3
Oenanthe crocata	9.7	.1	Laurocerasus	1.11	-.3
Paeonia officinalis	9.41	.1	Lespedeza sieboldii	1.92	-.3
Adonis vernalis	7.05	0	Malva sylvestris	1.01	-.3
Alchemilla vulgaris	8.13	0	Melilotus officinalis	1.45	-.3
Bryonia dioica	8.86	0	Milium solis	1.7	-.3
Gnaphalium leontopodiu	8.3	0	Rhus toxicodendron	1.67	-.3
Lappa arctium	6.92	0	Salicaria purpurea	1.04	-.3
Petroselinum crispum	8.54	0	Salvia sclarea	2.03	-.3
Zizia aurea	7.63	0	Urtica urens	1.49	-.3
Belladonna	4.77	-.1	Urtica urens 2	2.15	-.3
Fraxinus excelsior	5.18	-.1	Verbena officinalis	1.55	-.3
Melissa officinalis	5.69	-.1	Viola tricolor	2.21	-.3
Mercurialis perennis	5.76	-.1	Agrimonia eupatoria	0	-.4
Primula veris	4.76	-.1	Aquilegia vulgaris	0	-.4
Salix purpurea	5.62	-.1	Chenopodium anthelmint	0	-.4
Ulmus campestris	5.46	-.1	Cinnamodendron cortiso	0	-.4
Anthemis nobilis	3.56	-.2	Echinacea purpurea	0	-.4
Cardiospermum halicaca	4.22	-.2	Escholtzia californica	0	-.4
Castanea vesca	3.47	-.2	Faba vulgaris	0	-.4
Conium maculatum	3.23	-.2	Fagopyrum esculentum	0	-.4
Galium aparine	4.35	-.2	Hydrophyllum virginicu	0	-.4
Hedera helix	2.8	-.2	Hyssopus officinalis	0	-.4
Hyoscyamus niger	3.01	-.2	Leonurus cardiaca	0	-.4
Mandragora officinalis	3.06	-.2	Linum usitatissimum	0	-.4
Marrubium vulgare	3.08	-.2	Lycopus europaeus	0	-.4
Mentha arvensis	2.59	-.2	Osmundo regalis 2	0	-.4
Ocimum canum	4.12	-.2	Ruta graveolens	0	-.4
Ocimum canum R	4.27	-.2	Sanguisorba officinali	0	-.4
Ononis spinosa	3.8	-.2	Solidago virgaurea	0	-.4
Plantago major	3.96	-.2	Stachys officinalis	0	-.4
Polygonum aviculare	2.83	-.2	Stachys officinalis R	0	-.4
Prunus padus	2.64	-.2	Tussilago farfara	0	-.4
Psoralea bituminosa	3.88	-.2	Tussilago petasites	0	-.4

Ytterbium

Name	Value	Deviation	Name	Value	Deviation
Aconitum napellus	.09	.3	Marrubium vulgare	.04	-.3
Adonis vernalis	.09	.3	Melilotus officinalis	.02	-.5
Agrimonia eupatoria	.01	-.6	Melissa officinalis	.06	0
Alchemilla vulgaris	.08	.2	Mentha arvensis	.04	-.3
Anethum graveolens	.01	-.6	Mercurialis perennis	.09	.3
Angelica archangelica	.1	.4	Milium solis	.03	-.4
Anthemis nobilis	.06	0	Ocimum canum	.06	0
Aquilegia vulgaris	.01	-.6	Ocimum canum R	.07	.1
Belladonna	.05	-.1	Oenanthe crocata	.09	.3
Bellis perennis	.31	2.6	Ononis spinosa	.07	.1
Bryonia dioica	.1	.4	Osmundo regalis	.72	6.9
Cardamine pratensis	.02	-.5	Osmundo regalis 2	0	-.7
Cardamine pratensis R	.02	-.5	Paeonia officinalis	.07	.1
Cardiospermum halicaca	.06	0	Petroselinum crispum	.11	.5
Castanea vesca	.03	-.4	Plantago major	.05	-.1
Centaurium erythraea	.14	.8	Polygonum aviculare	.05	-.1
Chamomilla	.05	-.1	Primula veris	.06	0
Chelidonium majus	.34	2.9	Prunus padus	.03	-.4
Chenopodium anthelmint	.01	-.6	Psoralea bituminosa	.03	-.4
Cicuta virosa	.24	1.9	Rhus toxicodendron	.02	-.5
Cimicifuga racemosa	.1	.4	Rumex acetosa	.33	2.8
Cinnamodendron cortiso	.03	-.4	Ruta graveolens	.02	-.5
Clematis erecta	.02	-.5	Salicaria purpurea	.02	-.5
Collinsonia canadensis	.12	.6	Salix purpurea	.03	-.4
Conium maculatum	.01	-.6	Salvia sclarea	.02	-.5
Cupressus lawsoniana	.06	0	Sanguisorba officinali	.01	-.6
Echinacea angustifolia	.03	-.4	Scrophularia nodosa	.04	-.3
Echinacea purpurea	.01	-.6	Scutellaria lateriflor	.04	-.3
Escholtzia californica	.02	-.5	Solidago virgaurea	.01	-.6
Faba vulgaris	.02	-.5	Spilanthes oleracea	.02	-.5
Fagopyrum esculentum	.01	-.6	Stachys officinalis	.01	-.6
Fraxinus excelsior	.04	-.3	Stachys officinalis R	.01	-.6
Galium aparine	.04	-.3	Stachys sylvatica	.04	-.3
Glechoma hederacea	.03	-.4	Syringa vulgaris	.03	-.4
Gnaphalium leontopodiu	.18	1.2	Taraxacum officinale	.22	1.6
Hedera helix	.03	-.4	Teucrium chamaedrys	.04	-.3
Hydrophyllum virginicu	.02	-.5	Thuja occidentalis	.02	-.5
Hyoscyamus niger	.04	-.3	Tormentilla erecta	.14	.8
Hyssopus officinalis	.02	-.5	Tussilago farfara	.01	-.6
Lappa arctium	.08	.2	Tussilago petasites	.03	-.4
Lapsana communis	.12	.6	Ulmus campestris	.03	-.4
Laurocerasus	.01	-.6	Urtica dioica	.14	.8
Leonurus cardiaca	.02	-.5	Urtica urens	.02	-.5
Lespedeza sieboldii	.01	-.6	Urtica urens 2	.02	-.5
Linum usitatissimum	0	-.7	Verbena officinalis	.02	-.5
Lycopus europaeus	.02	-.5	Vinca minor	.05	-.1
Malva sylvestris	.02	-.5	Viola tricolor	.04	-.3
Mandragora officinalis	.04	-.3	Zizia aurea	.07	.1

Ytterbium

Name	Value	Deviation	Name	Value	Deviation
Osmundo regalis	.72	6.9	Castanea vesca	.03	-.4
Chelidonium majus	.34	2.9	Cinnamodendron cortiso	.03	-.4
Rumex acetosa	.33	2.8	Echinacea angustifolia	.03	-.4
Bellis perennis	.31	2.6	Glechoma hederacea	.03	-.4
Cicuta virosa	.24	1.9	Hedera helix	.03	-.4
Taraxacum officinale	.22	1.6	Milium solis	.03	-.4
Gnaphalium leontopodiu	.18	1.2	Prunus padus	.03	-.4
Centaurium erythraea	.14	.8	Psoralea bituminosa	.03	-.4
Tormentilla erecta	.14	.8	Salix purpurea	.03	-.4
Urtica dioica	.14	.8	Syringa vulgaris	.03	-.4
Collinsonia canadensis	.12	.6	Tussilago petasites	.03	-.4
Lapsana communis	.12	.6	Ulmus campestris	.03	-.4
Petroselinum crispum	.11	.5	Cardamine pratensis	.02	-.5
Angelica archangelica	.1	.4	Cardamine pratensis R	.02	-.5
Bryonia dioica	.1	.4	Clematis erecta	.02	-.5
Cimicifuga racemosa	.1	.4	Escholtzia californica	.02	-.5
Aconitum napellus	.09	.3	Faba vulgaris	.02	-.5
Adonis vernalis	.09	.3	Hydrophyllum virginicu	.02	-.5
Mercurialis perennis	.09	.3	Hyssopus officinalis	.02	-.5
Oenanthe crocata	.09	.3	Leonurus cardiaca	.02	-.5
Alchemilla vulgaris	.08	.2	Lycopus europaeus	.02	-.5
Lappa arctium	.08	.2	Malva sylvestris	.02	-.5
Ocimum canum R	.07	.1	Melilotus officinalis	.02	-.5
Ononis spinosa	.07	.1	Rhus toxicodendron	.02	-.5
Paeonia officinalis	.07	.1	Ruta graveolens	.02	-.5
Zizia aurea	.07	.1	Salicaria purpurea	.02	-.5
Anthemis nobilis	.06	0	Salvia sclarea	.02	-.5
Cardiospermum halicaca	.06	0	Spilanthes oleracea	.02	-.5
Cupressus lawsoniana	.06	0	Thuja occidentalis	.02	-.5
Melissa officinalis	.06	0	Urtica urens	.02	-.5
Ocimum canum	.06	0	Urtica urens 2	.02	-.5
Primula veris	.06	0	Verbena officinalis	.02	-.5
Belladonna	.05	-.1	Agrimonia eupatoria	.01	-.6
Chamomilla	.05	-.1	Anethum graveolens	.01	-.6
Plantago major	.05	-.1	Aquilegia vulgaris	.01	-.6
Polygonum aviculare	.05	-.1	Chenopodium anthelmint	.01	-.6
Vinca minor	.05	-.1	Conium maculatum	.01	-.6
Fraxinus excelsior	.04	-.3	Echinacea purpurea	.01	-.6
Galium aparine	.04	-.3	Fagopyrum esculentum	.01	-.6
Hyoscyamus niger	.04	-.3	Laurocerasus	.01	-.6
Mandragora officinalis	.04	-.3	Lespedeza sieboldii	.01	-.6
Marrubium vulgare	.04	-.3	Sanguisorba officinali	.01	-.6
Mentha arvensis	.04	-.3	Solidago virgaurea	.01	-.6
Scrophularia nodosa	.04	-.3	Stachys officinalis	.01	-.6
Scutellaria lateriflor	.04	-.3	Stachys officinalis R	.01	-.6
Stachys sylvatica	.04	-.3	Tussilago farfara	.01	-.6
Teucrium chamaedrys	.04	-.3	Linum usitatissimum	0	-.7
Viola tricolor	.04	-.3	Osmundo regalis 2	0	-.7

Yttrium

Name	Value	Deviation	Name	Value	Deviation
Aconitum napellus	1.49	.4	Marrubium vulgare	.61	-.3
Adonis vernalis	1.33	.2	Melilotus officinalis	.38	-.4
Agrimonia eupatoria	.19	-.5	Melissa officinalis	.79	-.1
Alchemilla vulgaris	1.17	.1	Mentha arvensis	.46	-.4
Anethum graveolens	.23	-.5	Mercurialis perennis	1.26	.2
Angelica archangelica	1.57	.4	Milium solis	.37	-.4
Anthemis nobilis	.9	-.1	Ocimum canum	.86	-.1
Aquilegia vulgaris	.13	-.6	Ocimum canum R	.81	-.1
Belladonna	.72	-.2	Oenanthe crocata	1.3	.2
Bellis perennis	5.15	2.9	Ononis spinosa	.99	0
Bryonia dioica	1.59	.4	Osmundo regalis	11.06	7.1
Cardamine pratensis	.4	-.4	Osmundo regalis 2	.05	-.6
Cardamine pratensis R	.35	-.4	Paeonia officinalis	1.1	.1
Cardiospermum halicaca	.88	-.1	Petroselinum crispum	1.57	.4
Castanea vesca	.66	-.2	Plantago major	.7	-.2
Centaurium erythraea	1.99	.7	Polygonum aviculare	.63	-.2
Chamomilla	.77	-.1	Primula veris	.84	-.1
Chelidonium majus	4.76	2.6	Prunus padus	.43	-.4
Chenopodium anthelmint	.11	-.6	Psoralea bituminosa	.66	-.2
Cicuta virosa	3.49	1.8	Rhus toxicodendron	.39	-.4
Cimicifuga racemosa	1.84	.6	Rumex acetosa	4.67	2.6
Cinnamodendron cortiso	.55	-.3	Ruta graveolens	.45	-.4
Clematis erecta	.36	-.4	Salicaria purpurea	.31	-.5
Collinsonia canadensis	2.07	.8	Salix purpurea	.36	-.4
Conium maculatum	.16	-.6	Salvia sclarea	.35	-.4
Cupressus lawsoniana	.96	0	Sanguisorba officinali	.15	-.6
Echinacea angustifolia	.44	-.4	Scrophularia nodosa	.56	-.3
Echinacea purpurea	.2	-.5	Scutellaria lateriflor	.69	-.2
Escholtzia californica	.28	-.5	Solidago virgaurea	.18	-.6
Faba vulgaris	.36	-.4	Spilanthes oleracea	.36	-.4
Fagopyrum esculentum	.23	-.5	Stachys officinalis	.14	-.6
Fraxinus excelsior	.59	-.3	Stachys officinalis R	.12	-.6
Galium aparine	.72	-.2	Stachys sylvatica	.53	-.3
Glechoma hederacea	.52	-.3	Syringa vulgaris	.46	-.4
Gnaphalium leontopodiu	2.54	1.1	Taraxacum officinale	3.44	1.7
Hedera helix	.52	-.3	Teucrium chamaedrys	.5	-.3
Hydrophyllum virginicu	.34	-.4	Thuja occidentalis	.32	-.5
Hyoscyamus niger	.61	-.3	Tormentilla erecta	2.37	1
Hyssopus officinalis	.35	-.4	Tussilago farfara	.18	-.6
Lappa arctium	1.21	.2	Tussilago petasites	.41	-.4
Lapsana communis	1.82	.6	Ulmus campestris	.53	-.3
Laurocerasus	.2	-.5	Urtica dioica	1.85	.6
Leonurus cardiaca	.26	-.5	Urtica urens	.33	-.5
Lespedeza sieboldii	.19	-.5	Urtica urens 2	.34	-.4
Linum usitatissimum	.09	-.6	Verbena officinalis	.23	-.5
Lycopus europaeus	.2	-.5	Vinca minor	.71	-.2
Malva sylvestris	.34	-.4	Viola tricolor	.46	-.4
Mandragora officinalis	.54	-.3	Zizia aurea	.9	-.1

Yttrium

Name	Value	Deviation	Name	Value	Deviation
Osmundo regalis	11.06	7.1	Scrophularia nodosa	.56	-.3
Bellis perennis	5.15	2.9	Stachys sylvatica	.53	-.3
Chelidonium majus	4.76	2.6	Teucrium chamaedrys	.5	-.3
Rumex acetosa	4.67	2.6	Ulmus campestris	.53	-.3
Cicuta virosa	3.49	1.8	Cardamine pratensis	.4	-.4
Taraxacum officinale	3.44	1.7	Cardamine pratensis R	.35	-.4
Gnaphalium leontopodiu	2.54	1.1	Clematis erecta	.36	-.4
Tormentilla erecta	2.37	1	Echinacea angustifolia	.44	-.4
Collinsonia canadensis	2.07	.8	Faba vulgaris	.36	-.4
Centaurium erythraea	1.99	.7	Hydrophyllum virginicu	.34	-.4
Cimicifuga racemosa	1.84	.6	Hyssopus officinalis	.35	-.4
Lapsana communis	1.82	.6	Malva sylvestris	.34	-.4
Urtica dioica	1.85	.6	Melilotus officinalis	.38	-.4
Aconitum napellus	1.49	.4	Mentha arvensis	.46	-.4
Angelica archangelica	1.57	.4	Milium solis	.37	-.4
Bryonia dioica	1.59	.4	Prunus padus	.43	-.4
Petroselinum crispum	1.57	.4	Rhus toxicodendron	.39	-.4
Adonis vernalis	1.33	.2	Ruta graveolens	.45	-.4
Lappa arctium	1.21	.2	Salix purpurea	.36	-.4
Mercurialis perennis	1.26	.2	Salvia sclarea	.35	-.4
Oenanthe crocata	1.3	.2	Spilanthes oleracea	.36	-.4
Alchemilla vulgaris	1.17	.1	Syringa vulgaris	.46	-.4
Paeonia officinalis	1.1	.1	Tussilago petasites	.41	-.4
Cupressus lawsoniana	.96	0	Urtica urens 2	.34	-.4
Ononis spinosa	.99	0	Viola tricolor	.46	-.4
Anthemis nobilis	.9	-.1	Agrimonia eupatoria	.19	-.5
Cardiospermum halicaca	.88	-.1	Anethum graveolens	.23	-.5
Chamomilla	.77	-.1	Echinacea purpurea	.2	-.5
Melissa officinalis	.79	-.1	Escholtzia californica	.28	-.5
Ocimum canum	.86	-.1	Fagopyrum esculentum	.23	-.5
Ocimum canum R	.81	-.1	Laurocerasus	.2	-.5
Primula veris	.84	-.1	Leonurus cardiaca	.26	-.5
Zizia aurea	.9	-.1	Lespedeza sieboldii	.19	-.5
Belladonna	.72	-.2	Lycopus europaeus	.2	-.5
Castanea vesca	.66	-.2	Salicaria purpurea	.31	-.5
Galium aparine	.72	-.2	Thuja occidentalis	.32	-.5
Plantago major	.7	-.2	Urtica urens	.33	-.5
Polygonum aviculare	.63	-.2	Verbena officinalis	.23	-.5
Psoralea bituminosa	.66	-.2	Aquilegia vulgaris	.13	-.6
Scutellaria lateriflor	.69	-.2	Chenopodium anthelmint	.11	-.6
Vinca minor	.71	-.2	Conium maculatum	.16	-.6
Cinnamodendron cortiso	.55	-.3	Linum usitatissimum	.09	-.6
Fraxinus excelsior	.59	-.3	Osmundo regalis 2	.05	-.6
Glechoma hederacea	.52	-.3	Sanguisorba officinali	.15	-.6
Hedera helix	.52	-.3	Solidago virgaurea	.18	-.6
Hyoscyamus niger	.61	-.3	Stachys officinalis	.14	-.6
Mandragora officinalis	.54	-.3	Stachys officinalis R	.12	-.6
Marrubium vulgare	.61	-.3	Tussilago farfara	.18	-.6

Zincum

Name	Value	Deviation	Name	Value	Deviation
Aconitum napellus	228.08	-.7	Marrubium vulgare	238.25	-.7
Adonis vernalis	284.06	-.6	Melilotus officinalis	960.74	.4
Agrimonia eupatoria	254.48	-.6	Melissa officinalis	1130	.6
Alchemilla vulgaris	860.19	.3	Mentha arvensis	385.47	-.4
Anethum graveolens	372.14	-.5	Mercurialis perennis	1400	1
Angelica archangelica	807.91	.2	Milium solis	477.01	-.3
Anthemis nobilis	377.62	-.5	Ocimum canum	280.05	-.6
Aquilegia vulgaris	372.14	-.5	Ocimum canum R	287.23	-.6
Belladonna	237.02	-.7	Oenanthe crocata	234.5	-.7
Bellis perennis	127.8	-.8	Ononis spinosa	287.71	-.6
Bryonia dioica	685.89	0	Osmundo regalis	1590	1.3
Cardamine pratensis	3190	3.6	Osmundo regalis 2	501.17	-.3
Cardamine pratensis R	3110	3.5	Paeonia officinalis	1180	.7
Cardiospermum halicaca	690.82	0	Petroselinum crispum	293.85	-.6
Castanea vesca	972.54	.4	Plantago major	167.04	-.8
Centaurium erythraea	895.94	.3	Polygonum aviculare	1200	.7
Chamomilla	325.74	-.5	Primula veris	292.01	-.6
Chelidonium majus	369.09	-.5	Prunus padus	413.62	-.4
Chenopodium anthelmint	902.18	.3	Psoralea bituminosa	1250	.8
Cicuta virosa	293.39	-.6	Rhus toxicodendron	399.98	-.4
Cimicifuga racemosa	430	-.4	Rumex acetosa	441.65	-.4
Cinnamodendron cortiso	2090	2	Ruta graveolens	542.87	-.2
Clematis erecta	407.57	-.4	Salicaria purpurea	1540	1.2
Collinsonia canadensis	870.04	.3	Salix purpurea	3000	3.4
Conium maculatum	346.44	-.5	Salvia sclarea	370.93	-.5
Cupressus lawsoniana	2330	2.4	Sanguisorba officinali	415.76	-.4
Echinacea angustifolia	215.75	-.7	Scrophularia nodosa	540.38	-.2
Echinacea purpurea	115.76	-.8	Scutellaria lateriflor	313.52	-.5
Escholtzia californica	854.24	.2	Solidago virgaurea	289.36	-.6
Faba vulgaris	1130	.6	Spilanthes oleracea	268.66	-.6
Fagopyrum esculentum	175.03	-.7	Stachys officinalis	210.35	-.7
Fraxinus excelsior	1140	.7	Stachys officinalis R	217.5	-.7
Galium aparine	375.68	-.5	Stachys sylvatica	226.39	-.7
Glechoma hederacea	1310	.9	Syringa vulgaris	2790	3.1
Gnaphalium leontopodiu	200.16	-.7	Taraxacum officinale	141.56	-.8
Hedera helix	1120	.6	Teucrium chamaedrys	352.73	-.5
Hydrophyllum virginicu	278.9	-.6	Thuja occidentalis	593.62	-.1
Hyoscyamus niger	266.44	-.6	Tormentilla erecta	2380	2.5
Hyssopus officinalis	283.15	-.6	Tussilago farfara	158.96	-.8
Lappa arctium	382.17	-.4	Tussilago petasites	259.38	-.6
Lapsana communis	874.3	.3	Ulmus campestris	504.45	-.3
Laurocerasus	409.36	-.4	Urtica dioica	172.97	-.7
Leonurus cardiaca	267.8	-.6	Urtica urens	197.64	-.7
Lespedeza sieboldii	1090	.6	Urtica urens 2	137.64	-.8
Linum usitatissimum	1140	.7	Verbena officinalis	235.23	-.7
Lycopus europaeus	1460	1.1	Vinca minor	552.48	-.2
Malva sylvestris	305.22	-.6	Viola tricolor	395.59	-.4
Mandragora officinalis	85.05	-.9	Zizia aurea	367.43	-.5

Zincum

Name	Value	Deviation	Name	Value	Deviation
Cardamine pratensis	3190	3.6	Viola tricolor	395.59	-.4
Cardamine pratensis R	3110	3.5	Anethum graveolens	372.14	-.5
Salix purpurea	3000	3.4	Anthemis nobilis	377.62	-.5
Syringa vulgaris	2790	3.1	Aquilegia vulgaris	372.14	-.5
Tormentilla erecta	2380	2.5	Chamomilla	325.74	-.5
Cupressus lawsoniana	2330	2.4	Chelidonium majus	369.09	-.5
Cinnamodendron cortiso	2090	2	Conium maculatum	346.44	-.5
Osmundo regalis	1590	1.3	Galium aparine	375.68	-.5
Salicaria purpurea	1540	1.2	Salvia sclarea	370.93	-.5
Lycopus europaeus	1460	1.1	Scutellaria lateriflor	313.52	-.5
Mercurialis perennis	1400	1	Teucrium chamaedrys	352.73	-.5
Glechoma hederacea	1310	.9	Zizia aurea	367.43	-.5
Psoralea bituminosa	1250	.8	Adonis vernalis	284.06	-.6
Fraxinus excelsior	1140	.7	Agrimonia eupatoria	254.48	-.6
Linum usitatissimum	1140	.7	Cicuta virosa	293.39	-.6
Paeonia officinalis	1180	.7	Hydrophyllum virginicu	278.9	-.6
Polygonum aviculare	1200	.7	Hyoscyamus niger	266.44	-.6
Faba vulgaris	1130	.6	Hyssopus officinalis	283.15	-.6
Hedera helix	1120	.6	Leonurus cardiaca	267.8	-.6
Lespedeza sieboldii	1090	.6	Malva sylvestris	305.22	-.6
Melissa officinalis	1130	.6	Ocimum canum	280.05	-.6
Castanea vesca	972.54	.4	Ocimum canum R	287.23	-.6
Melilotus officinalis	960.74	.4	Ononis spinosa	287.71	-.6
Alchemilla vulgaris	860.19	.3	Petroselinum crispum	293.85	-.6
Centaurium erythraea	895.94	.3	Primula veris	292.01	-.6
Chenopodium anthelmint	902.18	.3	Solidago virgaurea	289.36	-.6
Collinsonia canadensis	870.04	.3	Spilanthes oleracea	268.66	-.6
Lapsana communis	874.3	.3	Tussilago petasites	259.38	-.6
Angelica archangelica	807.91	.2	Aconitum napellus	228.08	-.7
Escholtzia californica	854.24	.2	Belladonna	237.02	-.7
Bryonia dioica	685.89	0	Echinacea angustifolia	215.75	-.7
Cardiospermum halicaca	690.82	0	Fagopyrum esculentum	175.03	-.7
Thuja occidentalis	593.62	-.1	Gnaphalium leontopodiu	200.16	-.7
Ruta graveolens	542.87	-.2	Marrubium vulgare	238.25	-.7
Scrophularia nodosa	540.38	-.2	Oenanthe crocata	234.5	-.7
Vinca minor	552.48	-.2	Stachys officinalis	210.35	-.7
Milium solis	477.01	-.3	Stachys officinalis R	217.5	-.7
Osmundo regalis 2	501.17	-.3	Stachys sylvatica	226.39	-.7
Ulmus campestris	504.45	-.3	Urtica dioica	172.97	-.7
Cimicifuga racemosa	430	-.4	Urtica urens	197.64	-.7
Clematis erecta	407.57	-.4	Verbena officinalis	235.23	-.7
Lappa arctium	382.17	-.4	Bellis perennis	127.8	-.8
Laurocerasus	409.36	-.4	Echinacea purpurea	115.76	-.8
Mentha arvensis	385.47	-.4	Plantago major	167.04	-.8
Prunus padus	413.62	-.4	Taraxacum officinale	141.56	-.8
Rhus toxicodendron	399.98	-.4	Tussilago farfara	158.96	-.8
Rumex acetosa	441.65	-.4	Urtica urens 2	137.64	-.8
Sanguisorba officinali	415.76	-.4	Mandragora officinalis	85.05	-.9

Zirconium

Name	Value	Deviation	Name	Value	Deviation
Aconitum napellus	0	-.5	Marrubium vulgare	.51	0
Adonis vernalis	0	-.5	Melilotus officinalis	.6	.1
Agrimonia eupatoria	0	-.5	Melissa officinalis	.93	.5
Alchemilla vulgaris	.85	.4	Mentha arvensis	.87	.4
Anethum graveolens	0	-.5	Mercurialis perennis	1.13	.7
Angelica archangelica	.76	.3	Milium solis	0	-.5
Anthemis nobilis	0	-.5	Ocimum canum	1.09	.7
Aquilegia vulgaris	0	-.5	Ocimum canum R	.65	.2
Belladonna	0	-.5	Oenanthe crocata	0	-.5
Bellis perennis	.72	.3	Ononis spinosa	.56	.1
Bryonia dioica	.71	.3	Osmundo regalis	7.35	7.2
Cardamine pratensis	0	-.5	Osmundo regalis 2	0	-.5
Cardamine pratensis R	0	-.5	Paeonia officinalis	.92	.5
Cardiospermum halicaca	.69	.2	Petroselinum crispum	0	-.5
Castanea vesca	0	-.5	Plantago major	.95	.5
Centaurium erythraea	1.17	.7	Polygonum aviculare	.66	.2
Chamomilla	0	-.5	Primula veris	1.11	.7
Chelidonium majus	.78	.3	Prunus padus	1.04	.6
Chenopodium anthelmint	0	-.5	Psoralea bituminosa	.78	.3
Cicuta virosa	3.43	3.1	Rhus toxicodendron	0	-.5
Cimicifuga racemosa	0	-.5	Rumex acetosa	3.69	3.4
Cinnamodendron cortiso	0	-.5	Ruta graveolens	0	-.5
Clematis erecta	0	-.5	Salicaria purpurea	0	-.5
Collinsonia canadensis	1.32	.9	Salix purpurea	0	-.5
Conium maculatum	0	-.5	Salvia sclarea	0	-.5
Cupressus lawsoniana	2.2	1.8	Sanguisorba officinali	0	-.5
Echinacea angustifolia	0	-.5	Scrophularia nodosa	0	-.5
Echinacea purpurea	0	-.5	Scutellaria lateriflor	0	-.5
Escholtzia californica	0	-.5	Solidago virgaurea	0	-.5
Faba vulgaris	0	-.5	Spilanthes oleracea	0	-.5
Fagopyrum esculentum	0	-.5	Stachys officinalis	0	-.5
Fraxinus excelsior	0	-.5	Stachys officinalis R	0	-.5
Galium aparine	.66	.2	Stachys sylvatica	0	-.5
Glechoma hederacea	0	-.5	Syringa vulgaris	1.17	.7
Gnaphalium leontopodiu	0	-.5	Taraxacum officinale	.64	.2
Hedera helix	0	-.5	Teucrium chamaedrys	.61	.2
Hydrophyllum virginicu	0	-.5	Thuja occidentalis	0	-.5
Hyoscyamus niger	.8	.3	Tormentilla erecta	0	-.5
Hyssopus officinalis	0	-.5	Tussilago farfara	.64	.2
Lappa arctium	0	-.5	Tussilago petasites	0	-.5
Lapsana communis	.64	.2	Ulmus campestris	0	-.5
Laurocerasus	.54	.1	Urtica dioica	.92	.5
Leonurus cardiaca	.57	.1	Urtica urens	0	-.5
Lespedeza sieboldii	0	-.5	Urtica urens 2	0	-.5
Linum usitatissimum	0	-.5	Verbena officinalis	.52	.1
Lycopus europaeus	0	-.5	Vinca minor	0	-.5
Malva sylvestris	.52	.1	Viola tricolor	.53	.1
Mandragora officinalis	0	-.5	Zizia aurea	.54	.1

Zirconium

Name	Value	Deviation	Name	Value	Deviation
Osmundo regalis	7.35	7.2	Cardamine pratensis R	0	-.5
Rumex acetosa	3.69	3.4	Castanea vesca	0	-.5
Cicuta virosa	3.43	3.1	Chamomilla	0	-.5
Cupressus lawsoniana	2.2	1.8	Chenopodium anthelmint	0	-.5
Collinsonia canadensis	1.32	.9	Cimicifuga racemosa	0	-.5
Centaurium erythraea	1.17	.7	Cinnamodendron cortiso	0	-.5
Mercurialis perennis	1.13	.7	Clematis erecta	0	-.5
Ocimum canum	1.09	.7	Conium maculatum	0	-.5
Primula veris	1.11	.7	Echinacea angustifolia	0	-.5
Syringa vulgaris	1.17	.7	Echinacea purpurea	0	-.5
Prunus padus	1.04	.6	Escholtzia californica	0	-.5
Melissa officinalis	.93	.5	Faba vulgaris	0	-.5
Paeonia officinalis	.92	.5	Fagopyrum esculentum	0	-.5
Plantago major	.95	.5	Fraxinus excelsior	0	-.5
Urtica dioica	.92	.5	Glechoma hederacea	0	-.5
Alchemilla vulgaris	.85	.4	Gnaphalium leontopodiu	0	-.5
Mentha arvensis	.87	.4	Hedera helix	0	-.5
Angelica archangelica	.76	.3	Hydrophyllum virginicu	0	-.5
Bellis perennis	.72	.3	Hyssopus officinalis	0	-.5
Bryonia dioica	.71	.3	Lappa arctium	0	-.5
Chelidonium majus	.78	.3	Lespedeza sieboldii	0	-.5
Hyoscyamus niger	.8	.3	Linum usitatissimum	0	-.5
Psoralea bituminosa	.78	.3	Lycopus europaeus	0	-.5
Cardiospermum halicaca	.69	.2	Mandragora officinalis	0	-.5
Galium aparine	.66	.2	Milium solis	0	-.5
Lapsana communis	.64	.2	Oenanthe crocata	0	-.5
Ocimum canum R	.65	.2	Osmundo regalis 2	0	-.5
Polygonum aviculare	.66	.2	Petroselinum crispum	0	-.5
Taraxacum officinale	.64	.2	Rhus toxicodendron	0	-.5
Teucrium chamaedrys	.61	.2	Ruta graveolens	0	-.5
Tussilago farfara	.64	.2	Salicaria purpurea	0	-.5
Laurocerasus	.54	.1	Salix purpurea	0	-.5
Leonurus cardiaca	.57	.1	Salvia sclarea	0	-.5
Malva sylvestris	.52	.1	Sanguisorba officinali	0	-.5
Melilotus officinalis	.6	.1	Scrophularia nodosa	0	-.5
Ononis spinosa	.56	.1	Scutellaria lateriflor	0	-.5
Verbena officinalis	.52	.1	Solidago virgaurea	0	-.5
Viola tricolor	.53	.1	Spilanthes oleracea	0	-.5
Zizia aurea	.54	.1	Stachys officinalis	0	-.5
Marrubium vulgare	.51	0	Stachys officinalis R	0	-.5
Aconitum napellus	0	-.5	Stachys sylvatica	0	-.5
Adonis vernalis	0	-.5	Thuja occidentalis	0	-.5
Agrimonia eupatoria	0	-.5	Tormentilla erecta	0	-.5
Anethum graveolens	0	-.5	Tussilago petasites	0	-.5
Anthemis nobilis	0	-.5	Ulmus campestris	0	-.5
Aquilegia vulgaris	0	-.5	Urtica urens	0	-.5
Belladonna	0	-.5	Urtica urens 2	0	-.5
Cardamine pratensis	0	-.5	Vinca minor	0	-.5